Statistical and Machine-Learning Data Mining

**Techniques for Better Predictive Modeling
and Analysis of Big Data**

Second Edition

Statistical and Machine-Learning Data Mining

Techniques for Better Predictive Modeling
and Analysis of Big Data

Second Edition

Bruce Ratner

CRC Press
Taylor & Francis Group
Boca Raton   London   New York

CRC Press is an imprint of the
Taylor & Francis Group, an **informa** business

CRC Press
Taylor & Francis Group
6000 Broken Sound Parkway NW, Suite 300
Boca Raton, FL 33487-2742

International Standard Book Number: 978-1-4398-6091-5 (Hardback)

Library of Congress Cataloging-in-Publication Data

Ratner, Bruce.
  Statistical and machine-learning data mining : techniques for better predictive modeling and analysis of big data / Bruce Ratner. -- 2nd ed.
    p. cm.
  Rev. ed. of: Statistical modeling and analysis for database marketing. c2003.
  Includes bibliographical references and index.
  ISBN 978-1-4398-6091-5
  1. Database marketing--Statistical methods. 2. Data mining--Statistical methods. I. Ratner, Bruce. II. Ratner, Bruce. Statistical modeling and analysis for database marketing. III. Title.

HF5415.126.R38 2012
658.8'72--dc22   2011014298

Visit the Taylor & Francis Web site at
http://www.taylorandfrancis.com

and the CRC Press Web site at
http://www.crcpress.com

This book is dedicated to

My father Isaac—my role model who taught me by doing, not saying.

My mother Leah—my friend who taught me to love love and hate hate.

Contents

Preface

This book is unique. It is the only book, to date, that distinguishes between statistical data mining and machine-learning data mining. I was an orthodox statistician until I resolved my struggles with the weaknesses of statistics within the big data setting of today. Now, as a reform statistician who is free of the statistical rigors of yesterday, with many degrees of freedom to exercise, I have composed by intellectual might the original and practical statistical data mining techniques in the first part of the book. The GenIQ Model, a machine-learning alternative to statistical regression, led to the creative and useful machine-learning data mining techniques in the remaining part of the book.

This book is a compilation of essays that offer detailed background, discussion, and illustration of specific methods for solving the most commonly experienced problems in predictive modeling and analysis of big data. The common theme among these essays is to address each methodology and assign its application to a specific type of problem. To better ground the reader, I spend considerable time discussing the basic methodologies of predictive modeling and analysis. While this type of overview has been attempted before, my approach offers a truly nitty-gritty, step-by-step approach that both tyros and experts in the field can enjoy playing with. The job of the data analyst is overwhelmingly to predict and explain the result of the target variable, such as RESPONSE or PROFIT. Within that task, the target variable is either a binary variable (RESPONSE is one such example) or a continuous variable (of which PROFIT is a good example). The scope of this book is purposely limited, with one exception, to dependency models, for which the target variable is often referred to as the "left-hand" side of an equation, and the variables that predict and/or explain the target variable is the "right-hand" side. This is in contrast to interdependency models that have no left- or right-hand side, and is covered in but one chapter that is tied in the dependency model. Because interdependency models comprise a minimal proportion of the data analyst's workload, I humbly suggest that the focus of this book will prove utilitarian.

Therefore, these essays have been organized in the following fashion. Chapter 1 reveals the two most influential factors in my professional life: John W. Tukey and the personal computer (PC). The PC has changed everything in the world of statistics. The PC can effortlessly produce precise calculations and eliminate the computational burden associated with statistics. One need only provide the right questions. Unfortunately, the confluence of the PC and the world of statistics has turned generalists with minimal statistical backgrounds into quasi statisticians and affords them a false sense of confidence.

In 1962, in his influential article, "The Future of Data Analysis" [1], John Tukey predicted a movement to unlock the rigidities that characterize statistics. It was not until the publication of *Exploratory Data Analysis* [2] in 1977 that Tukey led statistics away from the rigors that defined it into a new area, known as EDA (from the first initials of the title of his seminal work). At its core, EDA, known presently as data mining or formally as statistical data mining, is an unending effort of numerical, counting, and graphical detective work.

To provide a springboard into more esoteric methodologies, Chapter 2 covers the correlation coefficient. While reviewing the correlation coefficient, I bring to light several issues unfamiliar to many, as well as introduce two useful methods for variable assessment. Building on the concept of smooth scatterplot presented in Chapter 2, I introduce in Chapter 3 the smoother scatterplot based on CHAID (chi-squared automatic interaction detection). The new method has the potential of exposing a more reliable depiction of the unmasked relationship for paired-variable assessment than that of the smoothed scatterplot.

In Chapter 4, I show the importance of straight data for the simplicity and desirability it brings for good model building. In Chapter 5, I introduce the method of symmetrizing ranked data and add it to the paradigm of simplicity and desirability presented in Chapter 4.

Principal component analysis, the popular data reduction technique invented in 1901, is repositioned in Chapter 6 as a data mining method for many-variable assessment. In Chapter 7, I readdress the correlation coefficient. I discuss the effects the distributions of the two variables under consideration have on the correlation coefficient interval. Consequently, I provide a procedure for calculating an adjusted correlation coefficient.

In Chapter 8, I deal with logistic regression, a classification technique familiar to everyone, yet in this book, one that serves as the underlying rationale for a case study in building a response model for an investment product. In doing so, I introduce a variety of new data mining techniques. The continuous side of this target variable is covered in Chapter 9. On the heels of discussing the workhorses of statistical regression in Chapters 8 and 9, I resurface the scope of literature on the weaknesses of variable selection methods, and I enliven anew a notable solution for specifying a well-defined regression model in Chapter 10. Chapter 11 focuses on the interpretation of the logistic regression model with the use of CHAID as a data mining tool. Chapter 12 refocuses on the regression coefficient and offers common misinterpretations of the coefficient that point to its weaknesses. Extending the concept of coefficient, I introduce the average correlation coefficient in Chapter 13 to provide a quantitative criterion for assessing competing predictive models and the importance of the predictor variables.

In Chapter 14, I demonstrate how to increase the predictive power of a model beyond that provided by its variable components. This is accomplished by creating an interaction variable, which is the product of two or more component variables. To test the significance of the interaction variable, I make what I feel to be a compelling case for a rather unconventional use of CHAID. Creative use of well-known techniques is further carried out in Chapter 15, where I solve the problem of market segment classification modeling using not only logistic regression but also CHAID. In Chapter 16, CHAID is yet again utilized in a somewhat unconventional manner—as a method for filling in missing values in one's data. To bring an interesting real-life problem into the picture, I wrote Chapter 17 to describe profiling techniques for the marketer who wants a method for identifying his or her best customers. The benefits of the predictive profiling approach is demonstrated and expanded to a discussion of look-alike profiling.

I take a detour in Chapter 18 to discuss how marketers assess the accuracy of a model. Three concepts of model assessment are discussed: the traditional decile analysis, as well as two additional concepts, precision and separability. In Chapter 19, continuing in this mode, I point to the weaknesses in the way the decile analysis is used and offer a new approach known as the bootstrap for measuring the efficiency of marketing models.

The purpose of Chapter 20 is to introduce the principal features of a bootstrap validation method for the ever-popular logistic regression model. Chapter 21 offers a pair of graphics or visual displays that have value beyond the commonly used exploratory phase of analysis. In this chapter, I demonstrate the hitherto untapped potential for visual displays to describe the functionality of the final model once it has been implemented for prediction.

I close the statistical data mining part of the book with Chapter 22, in which I offer a data-mining alternative measure, the predictive contribution coefficient, to the standardized coefficient.

With the discussions just described behind us, we are ready to venture to new ground. In Chapter 1, I elaborated on the concept of machine-learning data mining and defined it as PC learning without the EDA/statistics component. In Chapter 23, I use a metrical modelogue, "To Fit or Not to Fit Data to a Model," to introduce the machine-learning method of GenIQ and its favorable data mining offshoots.

In Chapter 24, I maintain that the machine-learning paradigm, which lets the data define the model, is especially effective with big data. Consequently, I present an exemplar illustration of genetic logistic regression outperforming statistical logistic regression, whose paradigm, in contrast, is to fit the data to a predefined model. In Chapter 25, I introduce and illustrate brightly, perhaps, the quintessential data mining concept: data reuse. Data reuse is appending new variables, which are found when building a GenIQ Model, to the original dataset. The benefit of data reuse is apparent: The original dataset is enhanced with the addition of new, predictive-full GenIQ data-mined variables.

In Chapters 26–28, I address everyday statistics problems with solutions stemming from the data mining features of the GenIQ Model. In statistics, an outlier is an observation whose position falls outside the overall pattern of the data. Outliers are problematic: Statistical regression models are quite sensitive to outliers, which render an estimated regression model with questionable predictions. The common remedy for handling outliers is "determine and discard" them. In Chapter 26, I present an alternative method of moderating outliers instead of discarding them. In Chapter 27, I introduce a new solution to the old problem of overfitting. I illustrate how the GenIQ Model identifies a structural source (complexity) of overfitting, and subsequently instructs for deletion of the individuals who contribute to the complexity, from the dataset under consideration. Chapter 28 revisits the examples (the importance of straight data) discussed in Chapters 4 and 9, in which I posited the solutions without explanation as the material needed to understand the solution was not introduced at that point. At this point, the background required has been covered. Thus, for completeness, I detail the posited solutions in this chapter.

GenIQ is now presented in Chapter 29 as such a nonstatistical machine-learning model. Moreover, in Chapter 30, GenIQ serves as an effective method for finding the best possible subset of variables for a model. Because GenIQ has no coefficients—and coefficients furnish the key to prediction—Chapter 31 presents a method for calculating a quasi-regression coefficient, thereby providing a reliable, assumption-free alternative to the regression coefficient. Such an alternative provides a frame of reference for evaluating and using coefficient-free models, thus allowing the data analyst a comfort level for exploring new ideas, such as GenIQ.

References

1. Tukey, J.W., The future of data analysis, *Annals of Mathematical Statistics*, 33, 1–67, 1962.
2. Tukey, J.W., *Exploratory Data Analysis*, Addison-Wesley, Reading, MA, 1977.

Acknowledgments

This book, like all books—except the Bible—was written with the assistance of others. First and foremost, I acknowledge Hashem who has kept me alive, sustained me, and brought me to this season.

I am grateful to Lara Zoble, my editor, who contacted me about outdoing myself by writing this book. I am indebted to the staff of the Taylor & Francis Group for their excellent work: Jill Jurgensen, senior project coordinator; Jay Margolis, project editor; Ryan Cole, prepress technician; Kate Brown, copy editor; Gerry Jaffe, proofreader; and Elise Weinger, cover designer.

About the Author

Bruce Ratner, PhD, The Significant Statistician™, is president and founder of DM STAT-1 Consulting, the ensample for statistical modeling, analysis and data mining, and machine-learning data mining in the DM Space. DM STAT-1 specializes in all standard statistical techniques and methods using machine-learning/statistics algorithms, such as its patented *GenIQ Model*, to achieve its clients' goals, across industries including direct and database marketing, banking, insurance, finance, retail, telecommunications, health care, pharmaceutical, publication and circulation, mass and direct advertising, catalog marketing, e-commerce, Web mining, B2B (business to business), human capital management, risk management, and nonprofit fund-raising.

Bruce's par excellence consulting expertise is apparent, as he is the author of the best-selling book *Statistical Modeling and Analysis for Database Marketing: Effective Techniques for Mining Big Data*. Bruce ensures his clients' marketing decision problems are solved with the optimal problem solution methodology and rapid startup and timely delivery of project results. Client projects are executed with the highest level of statistical practice. He is an often-invited speaker at public industry events, such as the SAS Data Mining Conference, and private seminars at the request of *Fortune* magazine's top 100 companies.

Bruce has his footprint in the predictive analytics community as a frequent speaker at industry conferences and as the instructor of the advanced statistics course sponsored by the Direct Marketing Association for over a decade. He is the author of over 100 peer-reviewed articles on statistical and machine-learning procedures and software tools. He is a coauthor of the popular textbook the *New Direct Marketing* and is on the editorial board of the *Journal of Database Marketing*.

Bruce is also active in the online data mining industry. He is a frequent contributor to *KDNuggets Publications,* the top resource of the data mining community. His articles on statistical and machine-learning methodologies draw a huge monthly following. Another online venue in which he participates is the professional network LinkedIN. His seminal articles posted on LinkedIN, covering statistical and machine-learning procedures for big data, have sparked countless rich discussions. In addition, he is the author of his own *DM STAT-1 Newsletter* on the Web.

Bruce holds a doctorate in mathematics and statistics, with a concentration in multivariate statistics and response model simulation. His research interests include developing hybrid modeling techniques, which combine traditional statistics and machine-learning methods. He holds a patent for a unique application in solving the two-group classification problem with genetic programming.

1

Introduction

Whatever you are able to do with your might, do it.

—Kohelet 9:10

1.1 The Personal Computer and Statistics

The personal computer (PC) has changed everything—for both better and worse—in the world of statistics. The PC can effortlessly produce precise calculations and eliminate the computational burden associated with statistics. One need only provide the right questions. With the minimal knowledge required to program (instruct) statistical software, which entails telling it where the input data reside, which statistical procedures and calculations are desired, and where the output should go, tasks such as testing and analyzing, the tabulation of raw data into summary measures, as well as many other statistical criteria are fairly rote. The PC has advanced statistical thinking in the decision-making process, as evidenced by visual displays, such as bar charts and line graphs, animated three-dimensional rotating plots, and interactive marketing models found in management presentations. The PC also facilitates support documentation, which includes the calculations for measures such as the current mean profit across market segments from a marketing database; statistical output is copied from the statistical software and then pasted into the presentation application. Interpreting the output and drawing conclusions still requires human intervention.

Unfortunately, the confluence of the PC and the world of statistics has turned generalists with minimal statistical backgrounds into quasi statisticians and affords them a false sense of confidence because they can now produce statistical output. For instance, calculating the mean profit is standard fare in business. However, the mean provides a "typical value"—only when the distribution of the data is symmetric. In marketing databases, the distribution of profit is commonly right-skewed data.* Thus, the mean profit is not a reliable summary measure.† The quasi statistician would doubtlessly

---

* *Right skewed* or *positive skewed* means the distribution has a long tail in the positive direction.
† For moderately skewed distributions, the mode or median should be considered, and assessed for a reliably typical value.

not know to check this supposition, thus rendering the interpretation of the mean profit as floccinaucinihilipilification.*

Another example of how the PC fosters a "quick-and-dirty"† approach to statistical analysis can be found in the ubiquitous correlation coefficient (second in popularity to the mean as a summary measure), which measures association between two variables. There is an assumption (the underlying relationship between the two variables is a linear or a straight line) that must be met for the proper interpretation of the correlation coefficient. Rare is the quasi statistician who is actually aware of the assumption. Meanwhile, well-trained statisticians often do not check this assumption, a habit developed by the uncritical use of statistics with the PC.

The professional statistician has also been empowered by the computational strength of the PC; without it, the natural seven-step cycle of statistical analysis would not be practical [1]. The PC and the analytical cycle comprise the perfect pairing as long as the steps are followed in order and the information obtained from a step is used in the next step. Unfortunately, statisticians are human and succumb to taking shortcuts through the seven-step cycle. They ignore the cycle and focus solely on the sixth step in the following list. To the point, a careful statistical endeavor requires performance of all the steps in the seven-step cycle,‡ which is described as follows:

1. *Definition of the problem:* Determining the best way to tackle the problem is not always obvious. Management objectives are often expressed qualitatively, in which case the selection of the outcome or target (dependent) variable is subjectively biased. When the objectives are clearly stated, the appropriate dependent variable is often not available, in which case a surrogate must be used.

2. *Determining technique:* The technique first selected is often the one with which the data analyst is most comfortable; it is not necessarily the best technique for solving the problem.

3. *Use of competing techniques:* Applying alternative techniques increases the odds that a thorough analysis is conducted.

4. *Rough comparisons of efficacy:* Comparing variability of results across techniques can suggest additional techniques or the deletion of alternative techniques.

5. *Comparison in terms of a precise (and thereby inadequate) criterion:* An explicit criterion is difficult to define; therefore, precise surrogates are often used.

---

* Floccinaucinihilipilification (FLOK-si-NO-si-NY-HIL-i-PIL-i-fi-KAY-shuhn), noun. Its definition is estimating something as worthless.
† The literal translation of this expression clearly supports my claim that the PC is sometimes not a good thing for statistics. I supplant the former with "thorough and clean."
‡ The seven steps are attributed to Tukey. The annotations are my attributions.

6. *Optimization in terms of a precise and inadequate criterion:* An explicit criterion is difficult to define; therefore, precise surrogates are often used.

7. *Comparison in terms of several optimization criteria:* This constitutes the final step in determining the best solution.

The founding fathers of classical statistics—Karl Pearson[*] and Sir Ronald Fisher[†]—would have delighted in the ability of the PC to free them from time-consuming empirical validations of their concepts. Pearson, whose contributions include, to name but a few, regression analysis, the correlation coefficient, the standard deviation (a term he coined), and the chi-square test of statistical significance, would have likely developed even more concepts with the free time afforded by the PC. One can further speculate that the functionality of the PC would have allowed Fisher's methods (e.g., maximum likelihood estimation, hypothesis testing, and analysis of variance) to have immediate and practical applications.

The PC took the classical statistics of Pearson and Fisher from their theoretical blackboards into the practical classrooms and boardrooms. In the 1970s, statisticians were starting to acknowledge that their methodologies had potential for wider applications. However, they knew an accessible computing device was required to perform their on-demand statistical analyses with an acceptable accuracy and within a reasonable turnaround time. Although the statistical techniques had been developed for a small data setting consisting of one or two handfuls of variables and up to hundreds of records, the hand tabulation of data was computationally demanding and almost insurmountable. Accordingly, conducting the statistical techniques on big data was virtually out of the question. With the inception of the microprocessor in the mid-1970s, statisticians now had their computing device, the PC, to perform statistical analysis on big data with excellent accuracy and turnaround time. The desktop PCs replaced the handheld calculators in the classroom and boardrooms. From the 1990s to the present, the PC has offered statisticians advantages that were imponderable decades earlier.

1.2 Statistics and Data Analysis

As early as 1957, Roy believed that the classical statistical analysis was largely likely to be supplanted by assumption-free, nonparametric

---

[*] Karl Person (1900s) contributions include regression analysis, the correlation coefficient, and the chi-square test of statistical significance. He coined the term *standard deviation* in 1893.
[†] Sir Ronald Fisher (1920s) invented the methods of maximum likelihood estimation, hypothesis testing, and analysis of variance.

approaches, which were more realistic and meaningful [2]. It was an onerous task to understand the robustness of the classical (parametric) techniques to violations of the restrictive and unrealistic assumptions underlying their use. In practical applications, the primary assumption of "a random sample from a multivariate normal population" is virtually untenable. The effects of violating this assumption and additional model-specific assumptions (e.g., linearity between predictor and dependent variables, constant variance among errors, and uncorrelated errors) are difficult to determine with any exactitude. It is difficult to encourage the use of the statistical techniques, given that their limitations are not fully understood.

In 1962, in his influential article, "The Future of Data Analysis," John Tukey expressed concern that the field of statistics was not advancing [1]. He felt there was too much focus on the mathematics of statistics and not enough on the analysis of data and predicted a movement to unlock the rigidities that characterize the discipline. In an act of statistical heresy, Tukey took the first step toward revolutionizing statistics by referring to himself not as a statistician but a data analyst. However, it was not until the publication of his seminal masterpiece *Exploratory Data Analysis* in 1977 that Tukey led the discipline away from the rigors of statistical inference into a new area, known as EDA (stemming from the first letter of each word in the title of the unquestionable masterpiece) [3]. For his part, Tukey tried to advance EDA as a separate and distinct discipline from statistics, an idea that is not universally accepted today. EDA offered a fresh, assumption-free, nonparametric approach to problem solving in which the analysis is guided by the data itself and utilizes self-educating techniques, such as iteratively testing and modifying the analysis as the evaluation of feedback, to improve the final analysis for reliable results.

The essence of EDA is best described in Tukey's own words:

> Exploratory data analysis is detective work—numerical detective work—or counting detective work—or graphical detective work. … [It is] about looking at data to see what it seems to say. It concentrates on simple arithmetic and easy-to-draw pictures. It regards whatever appearances we have recognized as partial descriptions, and tries to look beneath them for new insights. [3, p. 1]

EDA includes the following characteristics:

1. *Flexibility*—techniques with greater flexibility to delve into the data
2. *Practicality*—advice for procedures of analyzing data
3. *Innovation*—techniques for interpreting results
4. *Universality*—use all statistics that apply to analyzing data
5. *Simplicity*—above all, the belief that simplicity is the golden rule

On a personal note, when I learned that Tukey preferred to be called a data analyst, I felt both validated and liberated because many of my own analyses fell outside the realm of the classical statistical framework. Furthermore, I had virtually eliminated the mathematical machinery, such as the calculus of maximum likelihood. In homage to Tukey, I more frequently use the terms *data analyst* and *data analysis* rather than statistical analysis and statistician throughout the book.

1.3 EDA

Tukey's book is more than a collection of new and creative rules and operations; it defines EDA as a discipline, which holds that data analysts fail only if they fail to try many things. It further espouses the belief that data analysts are especially successful if their detective work forces them to notice the unexpected. In other words, the philosophy of EDA is a trinity of *attitude* and *flexibility* to do whatever it takes to refine the analysis and *sharp-sightedness* to observe the unexpected when it does appear. EDA is thus a self-propagating theory; each data analyst adds his or her own contribution, thereby contributing to the discipline, as I hope to accomplish with this book.

The sharp-sightedness of EDA warrants more attention, as it is an important feature of the EDA approach. The data analyst should be a keen observer of indicators that are capable of being dealt with successfully and use them to paint an analytical picture of the data. In addition to the ever-ready visual graphical displays as an indicator of what the data reveal, there are numerical indicators, such as counts, percentages, averages, and the other classical descriptive statistics (e.g., standard deviation, minimum, maximum, and missing values). The data analyst's personal judgment and interpretation of indictors are not considered a bad thing, as the goal is to draw informal inferences, rather than those statistically significant inferences that are the hallmark of statistical formality.

In addition to visual and numerical indicators, there are the indirect messages in the data that force the data analyst to take notice, prompting responses such as "the data look like..." or "It appears to be...." Indirect messages may be vague, but their importance is to help the data analyst draw informal inferences. Thus, indicators do not include any of the hard statistical apparatus, such as confidence limits, significance tests, or standard errors.

With EDA, a new trend in statistics was born. Tukey and Mosteller quickly followed up in 1977 with the second EDA book, commonly referred to as EDA II, *Data Analysis and Regression*. EDA II recasts the basics of classical inferential procedures of data analysis and regression into an assumption-free, nonparametric approach guided by "(a) a sequence of philosophical attitudes... for effective data analysis, and (b) a flow of useful and adaptable techniques that make it possible to put these attitudes to work" [4, p. vii].

Hoaglin, Mosteller, and Tukey in 1983 succeeded in advancing EDA with *Understanding Robust and Exploratory Data Analysis,* which provides an understanding of how badly the classical methods behave when their restrictive assumptions do not hold and offers alternative robust and exploratory methods to broaden the effectiveness of statistical analysis [5]. It includes a collection of methods to cope with data in an informal way, guiding the identification of data structures relatively quickly and easily and trading off optimization of objective for stability of results.

Hoaglin et al. in 1991 continued their fruitful EDA efforts with *Fundamentals of Exploratory Analysis of Variance* [6]. They refashioned the basics of the analysis of variance with the classical statistical apparatus (e.g., degrees of freedom, F-ratios, and p values) into a host of numerical and graphical displays, which often give insight into the structure of the data, such as size effects, patterns, and interaction and behavior of residuals.

EDA set off a burst of activity in the visual portrayal of data. Published in 1983, *Graphical Methods for Data Analysis* (Chambers et al.) presented new and old methods—some of which require a computer, while others only paper and pencil—but all are powerful data analytical tools to learn more about data structure [7]. In 1986, du Toit et al. came out with *Graphical Exploratory Data Analysis,* providing a comprehensive, yet simple presentation of the topic [8]. Jacoby, with *Statistical Graphics for Visualizing Univariate and Bivariate Data* (1997), and *Statistical Graphics for Visualizing Multivariate Data* (1998), carried out his objective to obtain pictorial representations of quantitative information by elucidating histograms, one-dimensional and enhanced scatterplots, and nonparametric smoothing [9, 10]. In addition, he successfully transferred graphical displays of multivariate data on a single sheet of paper, a two-dimensional space.

1.4 The EDA Paradigm

EDA presents a major paradigm shift in the ways models are built. With the mantra, "Let your data be your guide," EDA offers a view that is a complete reversal of the classical principles that govern the usual steps of model building. EDA declares the model must always follow the data, not the other way around, as in the classical approach.

In the classical approach, the problem is stated and formulated in terms of an outcome variable Y. It is assumed that the true model explaining all the variation in Y is known. Specifically, it is assumed that all the structures (predictor variables, X_is) affecting Y and their forms are known and present in the model. For example, if Age affects Y, but the log of Age reflects the true relationship with Y, then log of Age must be present in the model. Once the model is specified, the data are taken through the model-specific analysis,

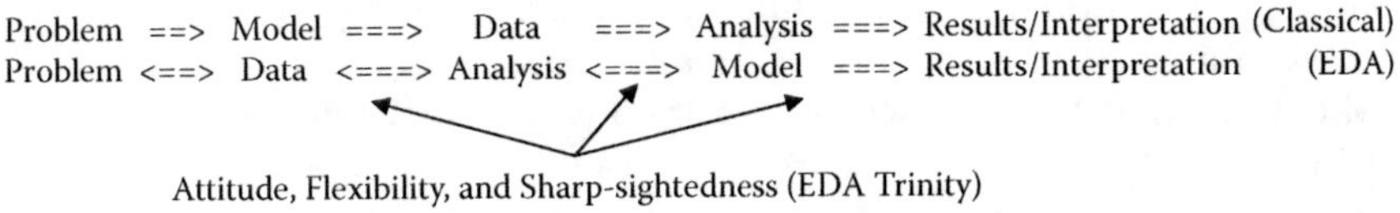

```
Problem ==> Model ===>    Data    ===> Analysis ===> Results/Interpretation (Classical)
Problem <==> Data <===> Analysis <===>  Model   ===> Results/Interpretation     (EDA)
```

Attitude, Flexibility, and Sharp-sightedness (EDA Trinity)

FIGURE 1.1
EDA paradigm.

which provides the results in terms of numerical values associated with the structures or estimates of the coefficients of the true predictor variables. Then, interpretation is made for declaring X_i an important predictor, assessing how X_i affects the prediction of Y, and ranking X_i in order of predictive importance.

Of course, the data analyst never knows the true model. So, familiarity with the content domain of the problem is used to put forth explicitly the true surrogate model, from which good predictions of Y can be made. According to Box, "all models are wrong, but some are useful" [11]. In this case, the model selected provides serviceable predictions of Y. Regardless of the model used, the assumption of knowing the truth about Y sets the statistical logic in motion to cause likely bias in the analysis, results, and interpretation.

In the EDA approach, not much is assumed beyond having some prior experience with content domain of the problem. The right attitude, flexibility, and sharp-sightedness are the forces behind the data analyst, who assesses the problem and lets the data direct the course of the analysis, which then suggests the structures and their forms in the model. If the model passes the validity check, then it is considered final and ready for results and interpretation to be made. If not, with the force still behind the data analyst, revisits of the analysis or data are made until new structures produce a sound and validated model, after which final results and interpretation are made (see Figure 1.1). Without exposure to assumption violations, the EDA paradigm offers a degree of confidence that its prescribed exploratory efforts are not biased, at least in the manner of classical approach. Of course, no analysis is bias free as all analysts admit their own bias into the equation.

1.5 EDA Weaknesses

With all its strengths and determination, EDA as originally developed had two minor weaknesses that could have hindered its wide acceptance and great success. One is of a subjective or psychological nature, and the other is a misconceived notion. Data analysts know that failure to look into a multitude of possibilities can result in a flawed analysis, thus finding themselves in a competitive struggle against the data itself. Thus, EDA can foster data analysts with insecurity that their work is never done. The PC can assist data analysts in being thorough with their analytical due diligence but bears no responsibility for the arrogance EDA engenders.

The belief that EDA, which was originally developed for the small data setting, does not work as well with large samples is a misconception. Indeed, some of the graphical methods, such as the stem-and-leaf plots, and some of the numerical and counting methods, such as folding and binning, do break down with large samples. However, the majority of the EDA methodology is unaffected by data size. Neither the manner by which the methods are carried out nor the reliability of the results is changed. In fact, some of the most powerful EDA techniques scale up nicely, but do require the PC to do the serious number crunching of *big data** [12]. For example, techniques such as ladder of powers, reexpressing,† and smoothing are valuable tools for large-sample or big data applications.

1.6 Small and Big Data

I would like to clarify the general concept of "small" and "big" data, as size, like beauty, is in the mind of the data analyst. In the past, small data fit the conceptual structure of classical statistics. Small always referred to the sample size, not the number of variables, which were always kept to a handful. Depending on the method employed, small was seldom less than 5 individuals; sometimes between 5 and 20; frequently between 30 and 50 or between 50 and 100; and rarely between 100 and 200. In contrast to today's big data, small data are a tabular display of rows (observations or individuals) and columns (variables or features) that fits on a few sheets of paper.

In addition to the compact area they occupy, small data are neat and tidy. They are "clean," in that they contain no improbable or impossible values, except for those due to primal data entry error. They do not include the statistical outliers and influential points or the EDA far-out and outside points. They are in the "ready-to-run" condition required by classical statistical methods.

There are two sides to big data. On one side is classical statistics that considers big as simply not small. Theoretically, big is the sample size after which asymptotic properties of the method "kick in" for valid results. On the other side is contemporary statistics that considers big in terms of lifting

---

* Authors Weiss and Indurkhya and I use the general concept of "big" data. However, we stress different characteristics of the concept.

† Tukey, via his groundbreaking EDA book, put the concept of "reexpression" in the forefront of EDA data mining tools; yet, he never provided any definition. I assume he believed that the term is self-explanatory. Tukey's first mention of reexpression is in a question on page 61 of his work: "What is the single most needed form of re-expression?" I, for one, would like a definition of reexpression, and I provide one further in the book.

observations and learning from the variables. Although it depends on who is analyzing the data, a sample size greater than 50,000 individuals can be considered big. Thus, calculating the average income from a database of 2 million individuals requires heavy-duty lifting (number crunching). In terms of learning or uncovering the structure among the variables, big can be considered 50 variables or more. Regardless of which side the data analyst is working, EDA scales up for both rows and columns of the data table.

1.6.1 Data Size Characteristics

There are three distinguishable characteristics of data size: condition, location, and population. *Condition* refers to the state of readiness of the data for analysis. Data that require minimal time and cost to clean, before reliable analysis can be performed, are said to be well conditioned; data that involve a substantial amount of time and cost are said to be ill conditioned. Small data are typically clean and therefore well conditioned.

Big data are an outgrowth of today's digital environment, which generates data flowing continuously from all directions at unprecedented speed and volume, and these data usually require cleansing. They are considered "dirty" mainly because of the merging of multiple sources. The merging process is inherently a time-intensive process, as multiple passes of the sources must be made to get a sense of how the combined sources fit together. Because of the iterative nature of the process, the logic of matching individual records across sources is at first "fuzzy," then fine-tuned to soundness; until that point, unexplainable, seemingly random, nonsensical values result. Thus, big data are ill conditioned.

Location refers to where the data reside. Unlike the rectangular sheet for small data, big data reside in relational databases consisting of a set of data tables. The link among the data tables can be hierarchical (rank or level dependent) or sequential (time or event dependent). Merging of multiple data sources, each consisting of many rows and columns, produces data of even greater number of rows and columns, clearly suggesting bigness.

Population refers to the group of individuals having qualities or characteristics in common and related to the study under consideration. Small data ideally represent a random sample of a known population that is not expected to encounter changes in its composition in the near future. The data are collected to answer a specific problem, permitting straightforward answers from a given problem-specific method. In contrast, big data often represent multiple, nonrandom samples of unknown populations, shifting in composition within the short term. Big data are "secondary" in nature; that is, they are not collected for an intended purpose. They are available from the hydra of marketing information, for use on any post hoc problem, and may not have a straightforward solution.

It is interesting to note that Tukey never talked specifically about the big data per se. However, he did predict that the cost of computing, in both time and dollars, would be cheap, which arguably suggests that he knew big data were coming. Regarding the cost, clearly today's PC bears this out.

1.6.2 Data Size: Personal Observation of One

The data size discussion raises the following question: "How large should a sample be?" Sample size can be anywhere from folds of 10,000 up to 100,000.

In my experience as a statistical modeler and data mining consultant for over 15 years and a statistics instructor who analyzes deceivingly simple cross tabulations with the basic statistical methods as my data mining tools, I have observed that the less-experienced and -trained data analyst uses sample sizes that are unnecessarily large. I see analyses performed on and models built from samples too large by factors ranging from 20 to 50. Although the PC can perform the heavy calculations, the extra time and cost in getting the larger data out of the data warehouse and then processing them and thinking about it are almost never justified. Of course, the only way a data analyst learns that extra big data are a waste of resources is by performing small versus big data comparisons, a step I recommend.

1.7 Data Mining Paradigm

The term *data mining* emerged from the database marketing community sometime between the late 1970s and early 1980s. Statisticians did not understand the excitement and activity caused by this new technique since the discovery of patterns and relationships (structure) in the data is not new to them. They had known about data mining for a long time, albeit under various names, such as data fishing, snooping, and dredging, and most disparaging, "ransacking" the data. Because any discovery process inherently exploits the data, producing spurious findings, statisticians did not view data mining in a positive light.

To state one of the numerous paraphrases of Maslow's hammer,* "If you have a hammer in hand, you tend eventually to start seeing nails." The statistical version of this maxim is, "Simply looking for something increases the odds that something will be found." Therefore, looking

---

* Abraham Maslow brought to the world of psychology a fresh perspective with his concept of "humanism," which he referred to as the "third force" of psychology after Pavlov's "behaviorism" and Freud's "psychoanalysis." Maslow's hammer is frequently used without anybody seemingly knowing the originator of this unique pithy statement expressing a rule of conduct. Maslow's Jewish parents migrated from Russia to the United States to escape from harsh conditions and sociopolitical turmoil. He was born Brooklyn, New York, in April 1908 and died from a heart attack in June 1970.

for structure typically results in finding structure. All data have spurious structures, which are formed by the "forces" that make things come together, such as chance. The bigger the data, the greater are the odds that spurious structures abound. Thus, an expectation of data mining is that it produces structures, both real and spurious, without distinction between them.

Today, statisticians accept data mining only if it embodies the EDA paradigm. They define *data mining* as any process that finds unexpected structures in data and uses the EDA framework to ensure that the process explores the data, not exploits it (see Figure 1.1). Note the word *unexpected*, which suggests that the process is exploratory rather than a confirmation that an expected structure has been found. By finding what one expects to find, there is no longer uncertainty regarding the existence of the structure.

Statisticians are mindful of the inherent nature of data mining and try to make adjustments to minimize the number of spurious structures identified. In classical statistical analysis, statisticians have explicitly modified most analyses that search for interesting structure, such as adjusting the overall alpha level/type I error rate or inflating the degrees of freedom [13, 14]. In data mining, the statistician has no explicit analytical adjustments available, only the implicit adjustments affected by using the EDA paradigm itself. The steps discussed next outline the data mining/EDA paradigm. As expected from EDA, the steps are defined by *soft* rules.

Suppose the objective is to find structure to help make good predictions of response to a future mail campaign. The following represent the steps that need to be taken:

Obtain the database that has similar mailings to the future mail campaign.

Draw a sample from the database. Size can be several folds of 10,000, up to 100,000.

Perform many exploratory passes of the sample. That is, do all desired calculations to determine interesting or noticeable structures.

Stop the calculations that are used for finding the noticeable structure.

Count the number of noticeable structures that emerge. The structures are not necessarily the results and should not be declared significant findings.

Seek out indicators, visual and numerical, and the indirect messages.

React or respond to all indicators and indirect messages.

Ask questions. Does each structure make sense by itself? Do any of the structures form natural groups? Do the groups make sense; is there consistency among the structures within a group?

Try more techniques. Repeat the many exploratory passes with several fresh samples drawn from the database. Check for consistency across the multiple passes. If results do not behave in a similar way, there may be no structure to predict response to a future mailing, as chance may have infected your data. If results behave similarly, then assess the variability of each structure and each group.

Choose the most stable structures and groups of structures for predicting response to a future mailing.

1.8 Statistics and Machine Learning

Coined by Samuel in 1959, the term *machine learning* (ML) was given to the field of study that assigns computers the ability to learn without being explicitly programmed [15]. In other words, ML investigates ways in which the computer can acquire knowledge directly from data and thus learn to solve problems. It would not be long before ML would influence the statistical community.

In 1963, Morgan and Sonquist led a rebellion against the restrictive assumptions of classical statistics [16]. They developed the automatic interaction detection (AID) regression tree, a methodology without assumptions. AID is a computer-intensive technique that finds or learns multidimensional patterns and relationships in data and serves as an assumption-free, nonparametric alternative to regression prediction and classification analyses. Many statisticians believe that AID marked the beginning of an ML approach to solving statistical problems. There have been many improvements and extensions of AID: THAID, MAID, CHAID (chi-squared automatic interaction detection), and CART, which are now considered viable and quite accessible data mining tools. CHAID and CART have emerged as the most popular today.

I consider AID and its offspring as quasi-ML methods. They are computer-intensive techniques that need the PC machine, a necessary condition for an ML method. However, they are not true ML methods because they use explicitly statistical criteria (e.g., chi squared and the F-tests), for the learning. A genuine ML method has the PC itself learn via mimicking the way humans think. Thus, I must use the term *quasi*. Perhaps a more appropriate and suggestive term for AID-type procedures and other statistical problems using the PC machine is statistical ML.

Independent from the work of Morgan and Sonquist, ML researchers had been developing algorithms to automate the induction process, which provided another alternative to regression analysis. In 1979, Quinlan used the well-known concept learning system developed by Hunt et al. to implement one of the first intelligent systems—ID3—which was succeeded by C4.5 and C5.0 [17, 18]. These algorithms are also considered data mining tools but have not successfully crossed over to the statistical community.

The interface of statistics and ML began in earnest in the 1980s. ML researchers became familiar with the three classical problems facing statisticians: regression (predicting a continuous outcome variable), classification (predicting a categorical outcome variable), and clustering (generating a few composite variables that carry a large percentage of the information in the original variables). They started using their machinery (algorithms and the PC) for a nonstatistical, assumption-free nonparametric approach to the three problem areas. At the same time, statisticians began harnessing the power of the desktop PC to influence the classical problems they know so well, thus relieving themselves from the starchy parametric road.

The ML community has many specialty groups working on data mining: neural networks, support vector machines, fuzzy logic, genetic algorithms and programming, information retrieval, knowledge acquisition, text processing, inductive logic programming, expert systems, and dynamic programming. All areas have the same objective in mind but accomplish it with their own tools and techniques. Unfortunately, the statistics community and the ML subgroups have no real exchanges of ideas or best practices. They create distinctions of no distinction.

1.9 Statistical Data Mining

In the spirit of EDA, it is incumbent on data analysts to try something new and retry something old. They can benefit not only from the computational power of the PC in doing the heavy lifting of big data but also from the ML ability of the PC in uncovering structure nestled in big data. In the spirit of trying something old, statistics still has a lot to offer.

Thus, today's data mining can be defined in terms of three easy concepts:

1. *Statistics with emphasis on EDA proper:* This includes using the descriptive and noninferential parts of classical statistical machinery as indicators. The parts include sum of squares, degrees of freedom, F-ratios, chi-square values, and p values, but exclude inferential conclusions.

2. *Big data:* Big data are given special mention because of today's digital environment. However, because small data are a component of big data, they are not excluded.

3. *Machine learning:* The PC is the learning machine, the *essential processing unit,* having the ability to learn without being explicitly programmed and the intelligence to find structure in the data. Moreover, the PC is essential for big data, as it can always do what it is explicitly programmed to do.

In view of these terms, the following data mining mnemonic can be formed:

Data Mining = Statistics + Big Data + Machine Learning and Lifting

Thus, *data mining* is defined today as a*ll of statistics and EDA for big and small data with the power of PC for the lifting of data and learning the structures within the data.* Explicitly referring to big and small data implies the process works equally well on both.

Again, in the spirit of EDA, it is prudent to parse the mnemonic equation. Lifting and learning require two different aspects of the data table. Lifting focuses on the rows of the data table and uses the capacity of the PC in terms of MIPS (million instructions per second), the speed in which explicitly programmed steps are executed. Calculating the average income of 1 million individuals is an example of PC lifting.

Learning focuses on the columns of the data table and the ability of the PC to find the structure within the columns without being explicitly programmed. Learning is more demanding on the PC than lifting in the same way that learning from books is always more demanding than merely lifting the books. **Identifying structure, such as the square root of ($a^2 + b^2$), is an** example of PC learning.

When there are indicators that the population is not homogeneous (i.e., there are subpopulations or clusters), the PC has to learn the rows and their relationships to each other to identify the row structures. Thus, when lifting and learning of the rows are required in addition to learning within the columns, the PC must work exceptionally hard but can yield extraordinary results.

As presented here, my definition of *statistical data mining* is the EDA/statistics component with PC lifting. Further in the book, I elaborate on *machine-learning data mining,* which I define as PC learning without the EDA/statistics component.

References

1. Tukey, J.W., The future of data analysis, *Annals of Mathematical Statistics,* 33, 1–67, 1962.
2. Roy, S.N., *Some Aspects of Multivariate Analysis,* Wiley, New York, 1957.
3. Tukey, J.W., *Exploratory Data Analysis,* Addison-Wesley, Reading, MA, 1977.
4. Mosteller, F., and Tukey, J.W., *Data Analysis and Regression,* Addison-Wesley, Reading, MA, 1977.
5. Hoaglin, D.C., Mosteller, F., and Tukey, J.W., *Understanding Robust and Exploratory Data Analysis,* Wiley, New York, 1983.
6. Hoaglin, D.C., Mosteller, F., and Tukey, J.W., *Fundamentals of Exploratory Analysis of Variance,* Wiley, New York, 1991.
7. Chambers, M.J., Cleveland, W.S., Kleiner, B., and Tukey, P.A., *Graphical Methods for Data Analysis,* Wadsworth & Brooks/Cole, Pacific Grove, CA, 1983.

8. du Toit, S.H.C., Steyn, A.G.W., and Stumpf, R.H., *Graphical Exploratory Data Analysis*, Springer-Verlag, New York, 1986.

9. Jacoby, W.G., *Statistical Graphics for Visualizing Univariate and Bivariate Data*, Sage, Thousand Oaks, CA, 1997.

10. Jacoby, W.G., *Statistical Graphics for Visualizing Multivariate Data*, Sage, Thousand Oaks, CA, 1998.

11. Box, G.E.P., Science and statistics, *Journal of the American Statistical Association*, 71, 791–799, 1976.

12. Weiss, S.M., and Indurkhya, N., *Predictive Data Mining*, Morgan Kaufman., San Francisco, CA, 1998.

13. Dun, O.J., Multiple comparison among means, *Journal of the American Statistical Association*, 54, 52–64, 1961.

14. Ye, J., On measuring and correcting the effects of data mining and model selection, *Journal of the American Statistical Association*, 93, 120–131, 1998.

15. Samuel, A., Some studies in machine learning using the game of checkers, In Feigenbaum, E., and Feldman, J., Eds., *Computers and Thought*, McGraw-Hill, New York, 14–36, 1963.

16. Morgan, J.N., and Sonquist, J.A., Problems in the analysis of survey data, and a proposal, *Journal of the American Statistical Association*, 58, 415–435, 1963.

17. Hunt, E., Marin, J., and Stone, P., *Experiments in Induction*, Academic Press, New York, 1966.

18. Quinlan, J.R., Discovering rules by induction from large collections of examples, In Mite, D., Ed., *Expert Systems in the Micro Electronic Age*, Edinburgh University Press, Edinburgh, UK, 143–159, 1979.

2

Two Basic Data Mining Methods for Variable Assessment

2.1 Introduction

Assessing the relationship between a predictor variable and a dependent variable is an essential task in the model-building process. If the relationship is identified and tractable, then the predictor variable is reexpressed to reflect the uncovered relationship and consequently tested for inclusion into the model. Most methods of variable assessment are based on the correlation coefficient, which is often misused. The linearity assumption of the correlation coefficient is frequently not subjected to testing, in which case the utility of the coefficient is unknown. The purpose of this chapter is twofold: to present (1) the *smoothed scatterplot* as an easy and effective data mining method and (2) a *general association* nonparametric test for assessing the relationship between two variables. The intent of point 1 is to embolden the data analyst to test the linearity assumption to ensure the proper use of the correlation coefficient. The intent of point 2 is an effectual data mining method for assessing the indicative message of the smoothed scatterplot.

I review the correlation coefficient with a quick tutorial, which includes an illustration of the importance of testing the linearity assumption, and outline the construction of the smoothed scatterplot, which serves as an easy method for testing the linearity assumption. Next, I introduce the *general association test*, as a data mining method for assessing a general association between two variables.

2.2 Correlation Coefficient

The correlation coefficient, denoted by r, is a measure of the strength of the straight-line or linear relationship between two variables. The correlation coefficient takes on values ranging between +1 and -1. The following points are the accepted guidelines for interpreting the correlation coefficient:

1. 0 indicates no linear relationship.

2. +1 indicates a perfect positive linear relationship: As one variable increases in its values, the other variable also increases in its values via an exact linear rule.

3. -1 indicates a perfect negative linear relationship: As one variable increases in its values, the other variable decreases in its values via an exact linear rule.

4. Values between 0 and 0.3 (0 and -0.3) indicate a weak positive (negative) linear relationship via a shaky linear rule.

5. Values between 0.3 and 0.7 (0.3 and -0.7) indicate a moderate positive (negative) linear relationship via a fuzzy-firm linear rule.

6. Values between 0.7 and 1.0 (-0.7 and -1.0) indicate a strong positive (negative) linear relationship via a firm linear rule.

7. The value of r squared is typically taken as the percent of variation in one variable explained by the other variable or the percent of variation shared between the two variables.

8. Linearity assumption: The correlation coefficient requires the underlying relationship between the two variables under consideration to be linear. If the relationship is known to be linear, or the observed pattern between the two variables appears to be linear, then the correlation coefficient provides a reliable measure of the strength of the linear relationship. If the relationship is known to be nonlinear, or the observed pattern appears to be nonlinear, then the correlation coefficient is not useful or is at least questionable.

The calculation of the correlation coefficient for two variables, say X and Y, is simple to understand. Let zX and zY be the standardized versions of X and Y, respectively. That is, zX and zY are both reexpressed to have means equal to zero, and standard deviations (std) equal to one. The reexpressions used to obtain the standardized scores are in Equations (2.1) and (2.2):

$$zX_i = [X_i - mean(X)]/std(X) \qquad (2.1)$$

$$zY_i = [Y_i - mean(Y)]/std(Y) \qquad (2.2)$$

The correlation coefficient is defined as the mean product of the paired standardized scores (zX_i, zY_i) as expressed in Equation (2.3).

$$r_{XY} = sum\ of\ [zX_i * zY_i]/(n - 1) \qquad (2.3)$$

where n is the sample size.

TABLE 2.1

Calculation of Correlation Coefficient

obs	X	Y	zX	zY	zX*zY
1	12	77	−1.14	−0.96	1.11
2	15	98	−0.62	1.07	−0.66
3	17	75	−0.27	−1.16	0.32
4	23	93	0.76	0.58	0.44
5	26	92	1.28	0.48	0.62
mean	18.6	87		sum	1.83
std	5.77	10.32			
n	5			r	0.46

$$r^2 = 0.21\overline{7}$$

For a simple illustration of the calculation of the correlation coefficient, consider the sample of five observations in Table 2.1. Columns zX and zY contain the standardized scores of X and Y, respectively. The last column is the product of the paired standardized scores. The sum of these scores is 1.83. The mean of these scores (using the adjusted divisor n - 1, not n) is 0.46. Thus, $r_{XY} = 0.46$.

See pg 47 for more on correlation coef.

2.3 Scatterplots

The linearity assumption of the correlation coefficient can easily be tested with a *scatterplot*, which is a mapping of the paired points (X_i, Y_i) in a graph with two orthogonal axes; X and Y are typically assigned as the predictor and dependent variables, respectively; index i represents the observations from 1 to n, where n is the sample size. The scatterplot provides a visual display of a discoverable relation between two variables within a framework of a horizontal X-axis perpendicular to a vertical Y-axis graph (without a causality implication suggested by the designation of dependent and predictor variables). If the scatter of points in the scatterplot appears to overlay a straight line, then the assumption has been satisfied, and r_{XY} provides a meaningful measure of the linear relationship between X and Y. If the scatter does not appear to overlay a straight line, then the assumption has not been satisfied, and the r_{XY} value is at best questionable. Thus, when using the correlation coefficient to measure the strength of the linear relationship, it is advisable to construct the scatterplot to test the linearity assumption. Unfortunately, many data analysts do not construct the scatterplot, thus rendering any analysis based on the correlation coefficient as potentially invalid. The following illustration is presented to reinforce the importance of evaluating scatterplots.

TABLE 2.2

Four Pairs of (X, Y) with the Same Correlation
Coefficient (r = 0.82)

obs	X1	Y1	X2	Y2	X3	Y3	X4	Y4
1	10	8.04	10	9.14	10	7.46	8	6.58
2	8	6.95	8	8.14	8	6.77	8	5.76
3	13	7.58	13	8.74	13	12.74	8	7.71
4	9	8.81	9	8.77	9	7.11	8	8.84
5	11	8.33	11	9.26	11	7.81	8	8.47
6	14	9.96	14	8.1	14	8.84	8	7.04
7	6	7.24	6	6.13	6	6.08	8	5.25
8	4	4.26	4	3.1	4	5.39	19	12.5
9	12	10.84	12	9.13	12	8.15	8	5.56
10	7	4.82	7	7.26	7	6.42	8	7.91
11	5	5.68	5	4.74	5	5.73	8	6.89

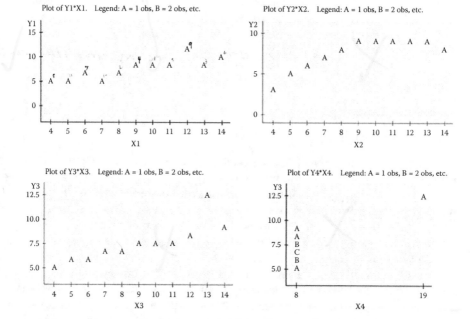

FIGURE 2.1
Four different datasets with the same correlation coefficient.

Consider the four datasets with 11 observations in Table 2.2 [1]. There
are four sets of (X, Y) points, with the same correlation coefficient value of
0.82. However, each X-Y relationship is distinct from one another, reflecting a different underlying structure, as depicted in the scatterplots in
Figure 2.1.

Scatterplot for X1-Y1 (upper left) indicates a linear relationship; thus, the $r_{X1,Y1}$ value of 0.82 correctly suggests a strong positive linear relationship between X1 and Y1. The scatterplot for X2-Y2 (upper right) reveals a curved relationship; $r_{X2,Y2} = 0.82$. The scatterplot for X3-Y3 (lower left) reveals a straight line except for the "outside" observation 3, data point (13, 12.74); $r_{X3,Y3} = 0.82$. A scatterplot for X4-Y4 (lower right) has a "shape of its own," which is clearly not linear; $r_{X4,Y4} = 0.82$. Accordingly, the correlation coefficient value of 0.82 is not a meaningful measure for the last three X-Y relationships.

See Chept 7, Section 7.4 → for more on Correlation coef.

see pg 11

2.4 Data Mining

Data mining—the process of revealing unexpected relationships in data—is needed to unmask the underlying relationships in scatterplots filled with big data. Big data, so much a part of the information world, have rendered the scatterplot overloaded with data points, or information. Paradoxically, scatterplots based on more information are actually less informative. With a quantitative target variable, the scatterplot typically becomes a cloud of points with sample-specific variation, called *rough*, which masks the underlying relationship. With a qualitative target variable, there is *discrete* rough, which masks the underlying relationship. In either case, if the rough can be removed from the big data scatterplot, then the underlying relationship can be revealed. After presenting two examples that illustrate how scatterplots filled with more data can actually provide less information, I outline the construction of the *smoothed scatterplot*, a rough-free scatterplot, which reveals the underlying relationship in big data.

2.4.1 Example 2.1

— see pg 19, 23

Consider the quantitative target variable Toll Calls (TC) in dollars and the predictor variable Household Income (HI) in dollars from a sample of size 102,000. The calculated $r_{TC,HI}$ is 0.09. The TC-HI scatterplot in Figure 2.2 shows a cloud of points obscuring the underlying relationship within the data (assuming a relationship exists). This scatterplot is uninformative regarding an indication for the reliable use of the calculated $r_{TC,HI}$.

2.4.2 Example 2.2

Consider the qualitative target variable Response (RS), which measures the response to a mailing, and the predictor variable HI from a sample of size of 102,000. RS assumes "yes" and "no" values, which are coded as 1 and 0, respectively. The calculated $r_{RS,HI}$ is 0.01. The RS-HI scatterplot in Figure 2.3 shows "train tracks" obscuring the underlying relationship within the data

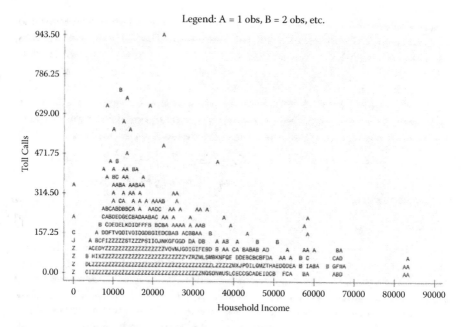

FIGURE 2.2
Scatterplot for TOLL CALLS and HOUSEHOLD INCOME.

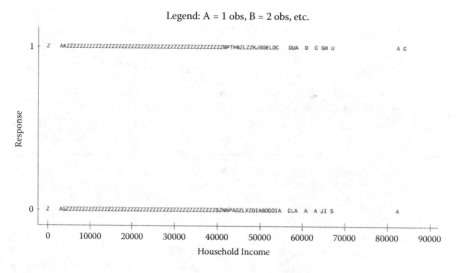

FIGURE 2.3
Scatterplot for RESPONSE and HOUSEHOLD INCOME.

(assuming a relationship exists). The tracks appear because the target variable takes on only two values, 0 and 1. As in the first example, this scatterplot is uninformative regarding an indication for the reliable use of the calculated $r_{RS,HI}$.

2.5 Smoothed Scatterplot

The *smoothed scatterplot* is the desired visual display for revealing a rough-free relationship lying within big data. *Smoothing* is a method of removing the rough and retaining the predictable underlying relationship (the *smooth*) in data by averaging within *neighborhoods* of similar values. Smoothing an X-Y scatterplot involves taking the averages of both the target (dependent) variable Y and the continuous predictor (independent) variable X, within X-based neighborhoods [2]. The six-step procedure to construct a smoothed scatterplot is as follows:

1. Plot the (X_i, Y_i) data points in an X-Y graph.
2. For a continuous X variable, divide the X-axis into distinct and nonoverlapping neighborhoods (slices). A common approach to dividing the X-axis is creating 10 equal-size slices (also known as *deciles*), whose aggregation equals the total sample [3–5]. Each slice accounts for 10% of the sample. For a categorical X variable, slicing per se cannot be performed. The categorical labels (levels) define *single-point* slices. Each single-point slice accounts for a percentage of the sample dependent on the distribution of the categorical levels in the sample.
3. Take the average of X within each slice. The mean or median can be used as the average. The average of X within each slice is known as a smooth X value, or smooth X. Notation for smooth X is sm_X.
4. Take the average of Y within each slice.
    a. For a continuous Y, the mean or median can be used as the average.
    b. For a categorical Y that assumes only two levels, the levels are typically reassigned numeric values 0 and 1. Clearly, only the mean can be calculated. This coding yields Y proportions, or Y rates.
    c. For a multinomial Y that assumes more than two levels, say k, clearly the average cannot be calculated. (Level-specific proportions can be calculated, but they do not fit into any developed procedure for the intended task.)

i. The appropriate procedure, which involves generating all combinations of pairwise-level scatterplots, is cumbersome and rarely used.

ii. The procedure that is often used because it is an easy and efficient technique is discussed in Chapter 15. Notation for smooth Y is sm_Y.

5. Plot the smooth points (smooth Y, smooth X), constructing a *smooth scatterplot*.

6. Connect the smooth points, starting from the most-left smooth point through the most-right smooth point. The resultant *smooth trace* line reveals the underlying relationship between X and Y.

Returning to examples 1 and 2, the HI data are grouped into 10 equal-size slices, each consisting of 10,200 observations. The averages (means) or smooth points for HI with both TC and RS within the slices (numbered from 0 to 9) are presented in Tables 2.3 and 2.4, respectively. The smooth points are plotted and connected.

The TC smooth trace line in Figure 2.4 clearly indicates a linear relationship. Thus, the $r_{TC, HI}$ value of 0.09 is a reliable measure of a weak positive linear relationship between TC and HI. Moreover, the variable HI itself (without any reexpression) can be tested for inclusion in the TC model. Note: The small r value does not preclude testing HI for model inclusion. This point is discussed in Chapter 5, Section 5.5.1.

The RS smooth trace line in Figure 2.5 is clear: The relationship between RS and HI is not linear. Thus, the $r_{RS, HI}$ value of 0.01 is invalid. This raises the following question: Does the RS smooth trace line indicate a general

TABLE 2.3

Smooth Points: Toll Calls and Household Income

Slice	Average Toll Calls	Average Household Income
0	$31.98	$26,157
1	$27.95	$18,697
2	$26.94	$16,271
3	$25.47	$14,712
4	$25.04	$13,493
5	$25.30	$12,474
6	$24.43	$11,644
7	$24.84	$10,803
8	$23.79	$9,796
9	$22.86	$6,748

TABLE 2.4

Smooth Points: Response and Household Income

Slice	Average Response	Average Household Income
0	2.8%	$26,157
1	2.6%	$18,697
2	2.6%	$16,271
3	2.5%	$14,712
4	2.3%	$13,493
5	2.2%	$12,474
6	2.2%	$11,644
7	2.1%	$10,803
8	2.1%	$9,796
9	2.3%	$6,748

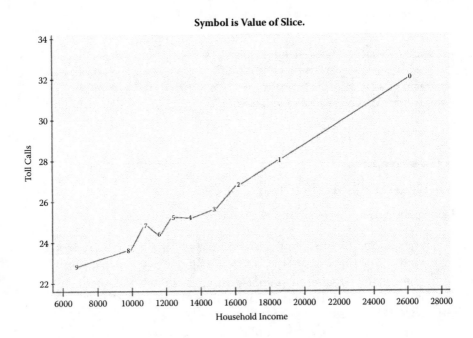

Symbol is Value of Slice.

FIGURE 2.4
Smoothed scatterplot for TOLL CALLS and HOUSEHOLD INCOME.

Symbol is Value of Slice.

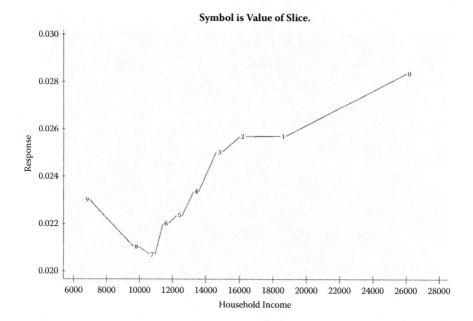

FIGURE 2.5
Smoothed scatterplot for RESPONSE and HOUSEHOLD INCOME.

association between RS and HI, implying a nonlinear relationship, or does the RS smooth trace line indicate a random scatter, implying no relationship between RS and HI? The answer can be readily found with the graphical nonparametric general association test [5].

2.6 General Association Test

Here is the general association test:

1. *Plot* the N smooth points in a scatterplot and draw a horizontal medial line that divides the N points into two equal-size groups.
2. *Connect* the N smooth points starting from the leftmost smooth point. N - 1 line segments result. Count the number m of line segments that cross the medial line.
3. *Test* for significance. The null hypothesis: There is no association between the two variables at hand. The alternative hypothesis: There is an association between the two variables.
4. *Consider* the test statistic TS is N - 1 - m.

Reject the null hypothesis if TS is greater than or equal to the cutoff score in Table 2.5. It is concluded that there is an association between the two variables. The smooth trace line indicates the "shape" or structure of the association.

Fail-to-reject the null hypothesis: If TS is less than the cutoff score in Table 2.5, it is concluded that there is no association between the two variables.

Returning to the smoothed scatterplot of RS and HI, I determine the following:

1. There are 10 smooth points: N = 10.

2. The medial line divides the smooth points such that points 5 to 9 are below the line and points 0 to 4 are above.

3. The line segment formed by points 4 and 5 in Figure 2.6 is the only segment that crosses the medial line. Accordingly, m = 1.

4. TS equals 8 (= 10 - 1 - 1), which is greater than or equal to the 95% and 99% confidence cutoff scores 7 and 8, respectively.

TABLE 2.5

Cutoff Scores for General Association
Test (95% and 99% Confidence Levels)

N	95%	99%
8–9	6	—
10–11	7	8
12–13	9	10
14–15	10	11
16–17	11	12
18–19	12	14
20–21	14	15
22–23	15	16
24–25	16	17
26–27	17	19
28–29	18	20
30–31	19	21
32–33	21	22
34–35	22	24
36–37	23	25
38–39	24	26
40–41	25	27
42–43	26	28
44–45	27	30
46–47	29	31
48–49	30	32
50–51	31	33

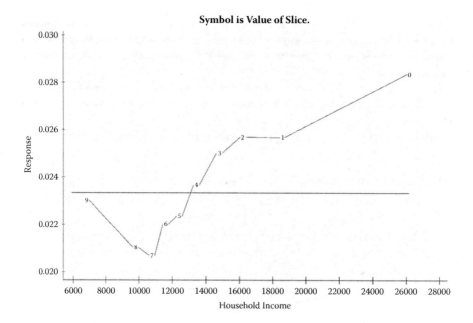

FIGURE 2.6
General association test for smoothed RS-HI scatterplot.

Thus, there is a 99% (and of course 95%) confidence level that there is an association between RS and HI. The RS smooth trace line in Figure 2.5 suggests that the observed relationship between RS and Hi appears to be polynomial to the third power. Accordingly, the linear (HI), quadratic (HI²), and cubed (HI³) Household Income terms should be tested in the Response model.

2.7 Summary

It should be clear that an analysis based on the uncritical use of the coefficient correlation is problematic. The strength of a relationship between two variables cannot simply be taken as the calculated r value itself. The testing for the linearity assumption, which is made easy by the simple scatterplot or smoothed scatterplot, is necessary for a "thorough-and-ready" analysis. If the observed relationship is linear, then the r value can be taken at face value for the strength of the relationship at hand. If the observed relationship is not linear, then the r value must be disregarded or used with extreme caution.

When a smoothed scatterplot for big data does not reveal a linear relationship, its scatter can be tested for randomness or for a noticeable general association by the proposed nonparametric method. If the former is true, then it is concluded there is no association between the variables. If the latter is true, then the predictor variable is reexpressed to reflect the observed relationship and therefore tested for inclusion in the model.

References

1. Anscombe, F.J., Graphs in statistical analysis, *American Statistician*, 27, 17–22, 1973.
2. Tukey, J.W., *Exploratory Data Analysis*, Addison-Wesley, Reading, MA, 1997.
3. Hardle, W., *Smoothing Techniques*, Springer-Verlag, New York, 1990.
4. Simonoff, J.S., *Smoothing Methods in Statistics*, Springer-Verlag, New York, 1996.
5. Quenouille, M.H., *Rapid Statistical Calculations*, Hafner, New York, 1959.

3

CHAID-Based Data Mining for Paired-Variable Assessment

(Machine Learning technique)
a.k.a. Data Mining

3.1 Introduction

Building on the concepts of the scatterplot and the smoothed scatterplot as data mining methods presented in Chapter 2, I introduce a new data mining method: a smoother scatterplot based on CHAID (chi-squared automatic interaction detection). The new method has the potential of exposing a more reliable depiction of the unmasked relationship for paired-variable assessment than that of the scatterplot and the smoothed scatterplot. I use a new dataset to keep an edge on refocusing another illustration of the scatterplot and smoothed scatterplot. Then, I present a primer on CHAID, after which I bring the proposed subject to the attention of data miners, who are buried deep in data, to help exhume themselves along with the patterns and relationships within the data.

3.2 The Scatterplot

verbatim repeat of pg 21

Big data are more than the bulk of today's information world. They contain valuable elements of the streaming current of digital data. Data analysts find themselves in troubled waters while pulling out the valuable elements. One impact of big data is that, on the basic analytical tool, the scatterplot has become overloaded with data points or with information. Paradoxically, the scatterplot based on more information is actually less informative. The scatterplot displays a cloud of data points, of which too many are due to sample variation, namely, the *rough* that masks a presuming existent underlying a relationship [1]. If the rough is removed from the cloudy scatterplot, then the sought-after underlying relationship, namely, the *smooth*, hidden behind the cloud can shine through. I offer to view an exemplar scatterplot filled with too much data and then show the corresponding *smooth scatterplot*, a

31

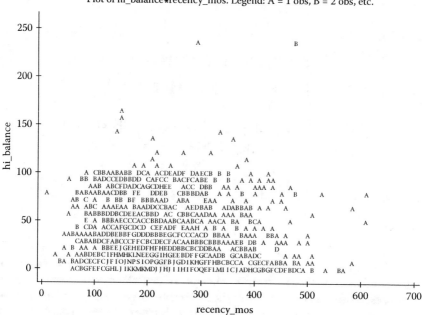

FIGURE 3.1
Scatterplot of HI_BALANCE and RECENCY_MOS.

rough-free, smooth-full scatterplot, which reveals the sunny-side up of the inherent character of a paired-variable assessment.

3.2.1 An Exemplar Scatterplot

I data mine for an honest display of the relationship between two variables from a real study: HI_BALANCE (the highest balance attained among credit card transactions for an individual) and RECENCY_MOS (the number of months since last purchase by an individual). The first step in this data mining process is to produce the scatterplot of HI_BALANCE and RECENCY_MOS. It is clear the relationship between the two variables reflects an irregular and diffuse cloud of data in the scatterplot (Figure 3.1). To lift the cloudiness in the scatterplot (i.e., to remove the rough to expose the smooth in the data), in the next section I create the smooth scatterplot of the present scatterplot.

3.3 The Smooth Scatterplot

The smooth scatterplot is the appropriate visual display for uncovering the sought-after underlying relationship within an X-Y graph of big raw data.

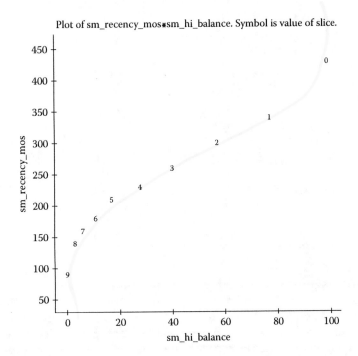

FIGURE 3.2
Smooth scatterplot of HI_BALANCE and RECENCY_MOS.

Smoothing is a method of averaging within *neighborhoods* of similar values—for removing the rough and retaining the smooth in data. Specifically, smoothing of a scatterplot involves taking the averages of both X and Y within X-based neighborhoods [2]. The six-step construction procedure for a smooth scatterplot is detailed in Chapter 2.

The smooth scatterplot of HI_BALANCE and RECENCY_MOS is in Figure 3.2. The *SM_HI_BALANCE* and *SM_RECENCY_MOS* values per slice, which assumes numeric labels from 0 to 9, are in Table 3.1. The uncovered relationship is an S-shape trend. This relationship can be analyzed further using Tukey's bulging rule, which is discussed in Chapter 8.

3.4 Primer on CHAID

Before discussing the CHAID-based smoother scatterplot, I provide a succinct primer on CHAID (the acronym for chi-squared automatic interaction detection, not detector). CHAID is a popular technique, especially among wannabe regression modelers with no significant statistical training because

TABLE 3.1

Smooth Values for HI_BALANCE and
RECENCY_MOS by Slice

slice	SM_HI_BALANCE	SM_RECENCY_MOS
0	99.0092	430.630
1	77.6742	341.485
2	56.7908	296.130
3	39.9817	262.940
4	27.1600	230.605
5	17.2617	205.825
6	10.6875	181.320
7	5.6342	159.870
8	2.5075	135.375
9	0.7550	90.250

(1) CHAID regression tree models are easy to build, understand, and implement; and (2) CHAID underpinnings are quite attractive: CHAID is an assumption-free method, meaning there are no formal theoretical assumptions to meet (traditional regression models are assumption full, which makes them susceptible to risky results). The CHAID model is highlighted by its major feature of handling a "big data" number of many predictor variables (this is problematic for traditional regression models). Take note, I use interchangeably the following CHAID terms: CHAID, CHAID tree, CHAID regression tree, CHAID regression tree model, and CHAID model.[*][†]

CHAID is a recursive technique that splits a population (node 1) into non-overlapping binary (two) subpopulations (nodes, bins, slices), defined by the "most important" predictor variable. Then, CHAID splits the first-level resultant nodes, defined by the "next" most important predictor variables, and continues to split second-level, third-level,..., and nth-level resultant nodes until stopping rules are met or the splitting criterion is *not* achieved. For the splitting criterion,[†] the variance of the dependent variable is minimized within each of the two resultant nodes and maximized between the two resultant nodes. To clarify the recursive splitting process, after the first splitting of the population (virtually always[§]) produces resultant nodes 2 and 3, further splitting is attempted on the resultant nodes. Nodes are split with constraints of two conditions: (1) if the user-defined stopping rules (e.g.,

---

[*] There are many CHAID software packages in the marketplace. The best ones are based on the original AID (automatic interaction detection) algorithm.

[†] See *A Pithy History of CHAID and Its Offspring* on the author's Web site (http://www.geniq.net/res/Reference-Pithy-history-of-CHAID-and-Offspring.html).

[‡] There are many splitting criterion metrics beyond the variance (e.g., gini, entropy, and misclassification cost).

[§] There is no guarantee that the top node can be split, regardless of the number of predictor variables at hand.

minimum size node to split and maximum level of the tree) are not met; and (2) if the splitting criterion bears resultant nodes with significantly different means for the Y variable. Assuming the conditions are in check, node 2 splits into nodes 4 and 5; node 3 splits into nodes 6 and 7. For each of nodes 4–7, splitting is performed if the two conditions are true; otherwise, the splitting is stopped, and the CHAID tree is complete.

3.5 CHAID-Based Data Mining for a Smoother Scatterplot

I illustrate the CHAID model for making a *smoother scatterplot* with the HI_BALANCE and RECENCY_MOS variables of the previously mentioned study. The fundamental characteristic of the CHAID-based smoother scatterplot is as follows: The CHAID model consists of *only one* predictor vari- (X) able. I build a CHAID regression tree model, regressing HI_BALANCE on RECENCY_MOS; the stopping rules used are minimum size node to split is 10, and maximum level of the tree is three. The CHAID regression tree model in Figure 3.3 is read as follows:

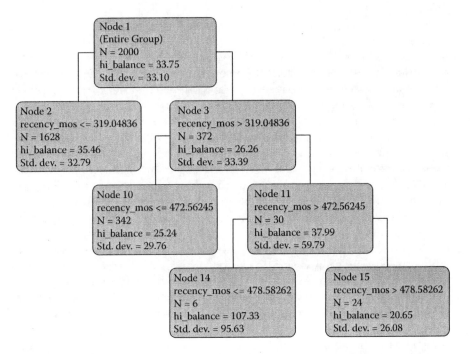

FIGURE 3.3
Illustration of CHAID tree of HI_BALANCE and RECENCY_MOS.

1. Node 1 mean for HI_BALANCE is 33.75 within the sample of size 2,000.
2. The splitting of node 1 yields nodes 2 and 3.
   a. Node 2 includes individuals whose RECENCY_MOS ≤ 319.04836; mean HI_BALANCE is 35.46; and node size is 1,628.
   b. Node 3 includes individuals whose RECENCY_MOS > 319.04836; mean HI_BALANCE is 26.26; and node size is 372.
3. The splitting of node 3 yields nodes 10 and 11; the nodes are read similarly to items 2a and 2b.
4. The splitting of node 11 yields nodes 14 and 15; the nodes are read similarly to items 2a and 2b.
5. The eight nodes 4–9, 12, and 13 cannot be created because the splitting criterion is not satisfied.

The *usual* interpretation of the instructive CHAID tree is moot, as it is a *simple* CHAID model (i.e., it has one predictor variable). Explaining and predicting HI_BALANCE based on the singleton variable RECENCY_MOS needs a fuller explanation and more accurate prediction of HI_BALANCE that would be achieved with more than one predictor variable. However, the simple CHAID model is *not* abstract-academic as it is the vehicle for the proposed method.

The *unique* interpretation of the simple CHAID model is the core of CHAID-based data mining for a smoother scatterplot: CHAID-based smoothing. The end nodes of the CHAID model are *end-node slices*, which are numerous (in the hundreds) by user design. The end-node slices have quite a bit of accuracy because they are predicted (fitted) by a CHAID model *accounting for* the X-axis variable. The end-node slices are aggregated into 10 *CHAID slices*. The CHAID slices of the X-axis variable clearly produce *more accurate (smoother)* values (CHAID-based sm_X) than the smooth X values from the dummy slicing of the X-axis (sm_X from the smooth scatterplot). Ergo, the CHAID slices produce smoother Y values (CHAID-based sm_Y) than the smooth Y values from the dummy slicing of the X-axis (sm_Y from the smooth scatterplot). In sum, CHAID slices produce CHAID-based sm_X that are smoother than sm_X and CHAID-based sm_Y that are smoother than sm_Y.

It is noted that the instructive CHAID tree in Figure 3.3 does not have numerous end-node slices. For this section only, I ask the reader to assume the CHAID tree has numerous end nodes so I can bring forth the exposition of CHAID-based smoothing. The real study illustration of CHAID-based smoothing, in the next section, indeed has end nodes in the hundreds.

I continue with the CHAID tree in Figure 3.3 to assist the understanding of CHAID-based smoothing. The CHAID slices of the RECENCY_MOS X-axis produce smoother HI_BALANCE values, *CHAID-based SM_HI_BALANCE*, than the smooth HI_BALANCE values from the dummy slicing of the RECENCY_MOS X-axis. In other words, CHAID-based SM_HI_BALANCE has less rough and more smooth. The following sequential identities may serve as a mnemonic guide to the concepts of rough and smooth, and how CHAID-based smoothing works:

1. Data point/value = reliable value + *error* value, from theoretical statistics.
2. Data value = predicted/fitted value + residual value, from applied statistics.
    a. The residual is reduced, and
    b. The fitted value is more reliable given a well-built model.
3. Data value = fitted value + residual value, a clearer restatement of number 2.
4. Data = smooth + rough, from Tukey's EDA (exploratory data analysis) [2].
    a. The rough is reduced, and
    b. The smooth is increased/more accurate, given a well-built model.
5. Data = smooth per slice + rough per slice, from the CHAID model.
    a. The rough per slice is reduced, and
    b. The smooth per slice is more accurate, given a well-built CHAID model.

To facilitate the discussion thus far, I have presented the dependent-predictor variable framework as the standard paired-variable notation (X_i, Y_i) suggests. However, when assessing the relationship between two variables, there is no dependent-predictor variable framework, in which case the standard paired-variable notation is $(X1_i, X2_i)$. Thus, analytical logic necessitates building a second CHAID model: Regressing RECENCY_MOS on HI_BALANCE yields the end nodes of the HI_BALANCE X-axis with smooth RECENCY_MOS values, *CHAID-based SM_RECENCY_MOS*. CHAID-based SM_RECENCY_MOS values are smoother than the smooth RECENCY_MOS values from the dummy slicing of HI_BALANCE. CHAID-based SM_RECENCY_MOS has less rough and more smooth.

3.5.1 The Smoother Scatterplot

The CHAID-based smoother scatterplot of HI_BALANCE and RECENCY_MOS is in Figure 3.4. The SM_HI_BALANCE and SM_RECENCY_MOS values per CHAID slice, which assumes numeric labels from 0 to 9, are in Table 3.2. The uncovered relationship in the majority view is linear, except for some *jumpy* slices in the middle (3–6), and for the stickler, slice 0 is slightly under the trace line. Regardless of these last diagnostics, the *smoother*

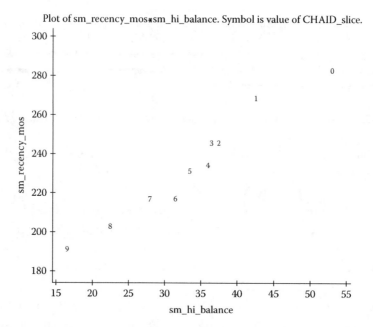

FIGURE 3.4
CHAID-based smoother scatterplot of HI_BALANCE and RECENCY_MOS.

TABLE 3.2

Smoother Values for HI_BALANCE and RECENCY_
MOS by CHAID Slices

CHAID_slice	SM_HI_BALANCE	SM_RECENCY_MOS
0	52.9450	281.550
1	42.3392	268.055
2	37.4108	246.270
3	36.5675	245.545
4	36.1242	233.910
5	33.3158	232.155
6	31.5350	216.750
7	27.8492	215.955
8	22.6783	202.670
9	16.6967	191.570

scatterplot suggests that no reexpressing of HI_BALANCE or RECENCY_
MOS is required: Testing the original variables for model inclusion is the
determinate of what shape either or both variables take in the final model.
Lest one forget, the data analyst compares the smoother scatterplot findings
to the smooth scatterplot findings.

The full, hard-to-read CHAID HI_BALANCE model *accounting for* RECENCY_MOS is in the Appendix (Figure 3.5). The CHAID HI_BALANCE model has 181 end nodes stemming from minimum-size node to split is 10, and the maximum tree level is 10. A readable midsection of the tree with node 1 front and center is in the Appendix (Figure 3.6). (This model's full-size output spans nine pages. So, to save trees *in the forest* I captured the hard-to-read tree as a JPEG image). The full, hard-to-read CHAID RECENCY_MOS model *accounting for* HI_BALANCE is in the Appendix (Figure 3.7). The CHAID RECENCY_MOS model has 121 end nodes stemming from minimum-size node to split is 10, and maximum tree level is 14. A readable midsection of the tree with node 1 front and center is in the Appendix (Figure 3.8). (This model's full-size output spans four pages.)

3.6 Summary

The basic (raw data) scatterplot and smooth scatterplot are the current data mining methods for assessing the relationship between predictor and dependent variables, an essential task in the model-building process. Working with today's big data, containing valuable elements within, data analysts find themselves in troubled waters while pulling out the valuable elements. Big data have rendered the scatterplot overloaded with data points, or information. Paradoxically, scatterplots based on more information are actually less informative. To data mine the information in the data-overloaded scatterplot, I reviewed the smooth scatterplot for unmasking an underlying relationship as depicted in a raw data scatterplot. Then, I proposed a CHAID-based method of data mining for paired-variable assessment, a new technique of obtaining a smoother scatterplot, which exposes a more reliable depiction of the unmasked relationship than that of the smooth scatterplot. The smooth scatterplot uses averages of raw data, and the smoother scatterplot uses averages of fitted values of CHAID end nodes. I illustrated the basic, the smooth, and smoother scatterplots using a real study.

References

1. Scatterplots with the X or Y variables as qualitative dependent variables yield not a cloud, but either two or more parallel lines or a dot matrix corresponding to the number of categorical levels. In the latter situation, the scatterplot displays mostly the rough in the data.
2. Tukey, J.W., *Exploratory Data Analysis*, Addison-Wesley, Reading, MA, 1997.

Appendix

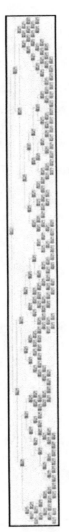

FIGURE 3.5
CHAID regression tree: HI_BALANCE regressed on RECENCY_MOS.

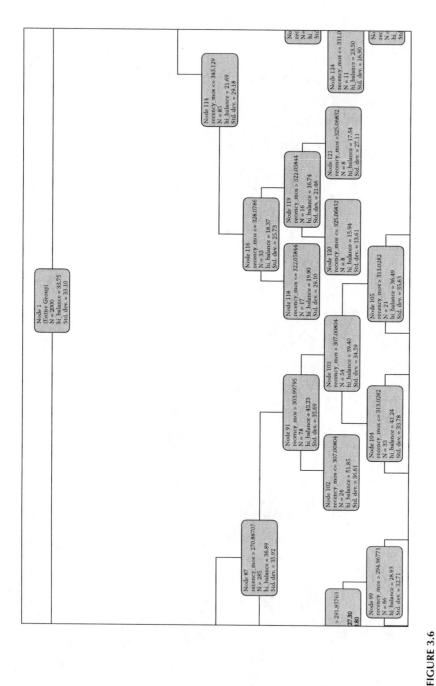

FIGURE 3.6
Midsection of CHAID regression tree: HI_BALANCE regressed on RECENCY_MOS.

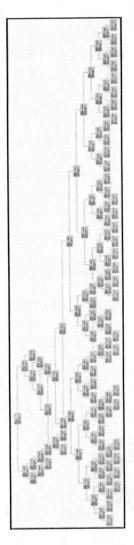

FIGURE 3.7
CHAID regression tree: RECENCY_MOS regressed on HI_BALANCE.

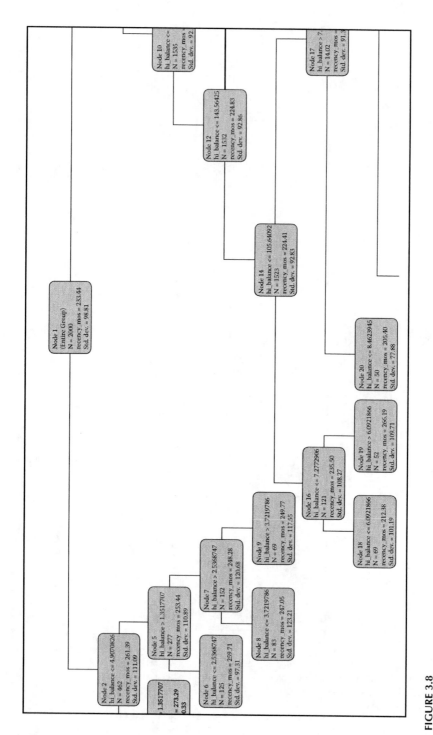

FIGURE 3.8
Midsection of CHAID regression tree: RECENCY_MOS regressed on HI_BALANCE.

4

The Importance of Straight Data: Simplicity and Desirability for Good Model-Building Practice

4.1 Introduction

The purpose of this chapter is to show the importance of straight data for the simplicity and desirability it brings for good model-building practice. I illustrate the content of the chapter title by giving details of what to do when an observed relationship between two variables depicted in a scatterplot is masking an acute underlying relationship. Data mining is employed to unmask and straighten the obtuse relationship. The correlation coefficient is used to quantify the strength of the exposed relationship, which possesses straight-line simplicity.

4.2 Straightness and Symmetry in Data

Today's data mechanics* (DMers), who consist of the lot of statisticians, data analysts, data miners, knowledge discovers, and the like, know that exploratory data analysis, better known as EDA, places special importance on straight data, not in the least for the sake of simplicity itself. The paradigm of life is simplicity (at least for those of us who are older and wiser). In the physical world, Einstein

see pg 20

---

* Data mechanics (DMers) are a group of individuals skilled by the study of understanding data, stressing the throwing of light upon data by offering details or inferences previously unclear or only implicit. According to The Royal Society, founded in 1660, the first statistician was John Graunt, who believed that the bills of mortality could provide information for more than just for gossip. He compiled and analyzed all bills from 1604 through 1661. Graunt's work was published as *Natural and Political Observations Made upon the Bills of Mortality*. Graunt became the first to put forth today's common statistics as: More boys are born than girls; women live longer than men; and, except in epidemics, the number of people dying each year is relatively constant. http://www.answers.com/topic/the-first-statistician.

uncovered one of life's ruling principles using only three letters: $E = mc^2$. In the visual world, however, simplicity is undervalued and overlooked. A smiley face is an unsophisticated, simple shape that nevertheless communicates effectively, clearly, and instantly. Why should DMers accept anything less than simplicity in their life's work? Numbers, as well, should communicate powerfully, unmistakably, and without more ado. Accordingly, DMers should seek two features that reflect simplicity: straightness and symmetry in data.

There are five reasons why it is important to straighten data:

1. The straight-line (linear) relationship between two continuous variables, say X and Y, *is as simple as it gets.* As X increases (decreases) in its values, so does Y increase (decrease) in its values, in which case it is said that X and Y are positively correlated. Or, as X increases (decreases) in its values, so does Y decrease (increase) in its values, in which case it is said that X and Y are negatively correlated. As an example of this setting of simplicity (and everlasting importance), Einstein's E and m have a perfect positive linear relationship.

2. With linear data, the data analyst without difficulty sees *what is going on within the data.* The class of linear data is the desirable element for good model-building practice.

3. Most marketing models, belonging to the class of innumerable varieties of the statistical linear model, *require linear relationships* between a dependent variable and (a) each predictor variable in a model and (b) *all* predictor variables considered jointly, regarding them as an array of predictor variables that have a multivariate normal distribution.

4. It has been shown that *nonlinear models*, which are attributed with yielding good predictions with nonstraight data, in fact *do better with straight data.*

5. I have not ignored the feature of symmetry. Not accidentally, there are theoretical reasons for *symmetry and straightness going hand in hand.* Straightening data often makes data symmetric and vice versa. Recall that symmetric data have values that are in correspondence in size and shape on opposite sides of a dividing line or middle value of the data. The iconic symmetric data profile in statistics is bell shaped.

4.3 Data Mining Is a High Concept

Data mining is a high concept having elements of fast action in its development, glamour as it stirs the imagination for the unconventional, unexpected, and a mystique that appeals to a wide audience that knows curiosity feeds human thought. Conventional wisdom, in the DM space, has it that everyone knows *what data mining is* [1]. Everyone does it—that is what he or she says. I do not believe it. I know that everyone talks about it, but only a small, self-seeking group of data analysts genuinely does data mining. I make this bold and excessively self-confident assertion based on my consulting experience as a statistical modeler, data miner, and computer scientist for the many years that have befallen me.

4.4 The Correlation Coefficient *see pgs 17, 18*

The term *correlation coefficient,* denoted by r, was coined by Karl Pearson in 1896. This statistic, over a century old, is still going strong. It is one of the most used statistics, second to the mean. The correlation coefficient weaknesses and warnings of misuse are well documented. As a practiced consulting statistician and instructor of statistical modeling and data mining for continuing professional studies in the DM Space,* I see too often the weaknesses, and misuses are not heeded, perhaps because they are rarely mentioned. The correlation coefficient, whose values theoretically range within the left- and right-closed interval [-1, +1], is restricted in practice by the individual distributions of the two variables being correlated (see Chapter 7). The misuse of the correlation coefficient is the nontesting of the *linear assumption,* which is discussed in this section. *Pg 18*

Assessing the relationship between dependent and predictor variables is an essential task in statistical linear and nonlinear regression model building. If the relationship is linear, then the modeler tests to determine whether the predictor variable has statistical importance to be included in the model. If the relationship is either nonlinear or indiscernible, then one or both of the two variables are reexpressed, that is, *data mined* to be voguish with terminology, to reshape the observed relationship into a data-mined linear relationship. As a result, the reexpressed variable(s) is (are) tested for inclusion into the model.

The everyday method of assessing a relationship between two variables—lest the data analyst forget: *linear* relationships only—is based on the correlation coefficient. The correlation coefficient is often misused because its linearity assumption *is not tested,* albeit simple to do. (I put forth an obvious practical, but still not acceptable, reason why the nontesting has a long shelf life further in the chapter.) I state the linear assumption, discuss the testing of the assumption, and provide how to interpret the observed correlation coefficient values.

The correlation coefficient requires that the underlying relationship between two variables is linear. If the observed pattern displayed in the

---

* DM Space includes industry sectors such as direct and database marketing, banking, insurance, finance, retail, telecommunications, health care, pharmaceuticals, publication and circulation, mass and direct advertising, catalog marketing, e-commerce, Web-mining, B2B (business to business), human capital management, and risk management, and the like.

scatterplot of two variables has an outward aspect of being linear, then the correlation coefficient provides a reliable measure of the *linear* strength of the relationship. If the observed pattern is either nonlinear or indiscernible, then the correlation coefficient is inutile or offers risky results. If the latter data condition exists, data mining efforts should be attempted to straighten the relationship. In the remote situation when the proposed data mining method is not successful, then *extra*-data mining techniques, like binning, should be explored. The last techniques are not discussed because they are outside the scope of this chapter. Sources for extra-data mining are many [2–4].

If the relationship is deemed linear, then the *strength* of the relationship is quantified by an accompanying value of r. For convenience, I restate the accepted guidelines (from Chapter 2) for interpreting the correlation coefficient:

(You are repeating everything you said in chapt 2, many times!)

1. 0 indicates no linear relationship.

2. +1 indicates a perfect positive linear relationship: As one variable increases in its values, the other variable also increases in its values via an exact linear rule.

3. –1 indicates a perfect negative linear relationship: As one variable increases in its values, the other variable decreases in its values via an exact linear rule.

4. Values between 0 and 0.3 (0 and -0.3) indicate a weak positive (negative) linear relationship via a shaky linear rule.

5. Values between 0.3 and 0.7 (0.3 and -0.7) indicate a moderate positive (negative) linear relationship via a fuzzy-firm linear rule.

6. Values between 0.7 and 1.0 (-0.7 and -1.0) indicate a strong positive (negative) linear relationship via a firm linear rule.

I present the sought-after scatterplot of paired variables (x, y)—a cloud of data points that indicates a silver lining of a straight line. The correlation coefficient $r_{(x,y)}$, corresponding to this scatterplot, ensures that the value of r reliably reflects the strength of the linear relationship between x and y (see Figure 4.1). The cloud of points in Figure 4.1 is not typical due to the small, 11-observation dataset used in the illustration. However, the discussion still holds true, as if the presentation involves, say, 11,000 observations or greater. I take the freedom of writing to refer to the silver lining scatterplot as a thin, wispy cirrus cloud.

4.5 Scatterplot of (xx3, yy3)

Consider the scatterplot of 11 data points of the third paired variables (x3, y3) in Table 4.1 regarding the Ancombe data. I construct the scatterplot of (x3, y3) in Figure 4.2, renaming (xx3, yy3) for a seemingly unnecessary

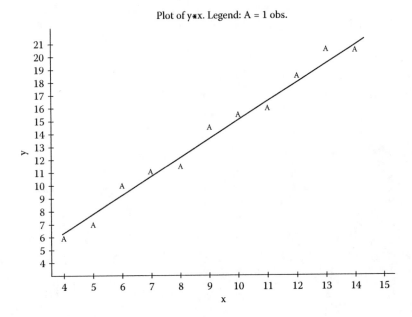

FIGURE 4.1

Sought-after scatterplot of pair (x, y).

TABLE 4.1

Anscombe Data

ID	x1	y1	x2	y2	x3	y3	x4	y4
1	10	8.04	10	9.14	10	7.46	8	6.58
2	8	6.95	8	8.14	8	6.77	8	5.76
3	13	7.58	13	8.74	13	12.74	8	7.71
4	9	8.81	9	8.77	9	7.11	8	8.84
5	11	8.33	11	9.26	11	7.81	8	8.47
6	14	9.96	14	8.10	14	8.84	8	7.04
7	6	7.24	6	6.13	6	6.08	8	5.25
8	4	4.26	4	3.10	4	5.39	19	12.50
9	12	10.84	12	9.13	12	8.15	8	5.56
10	7	4.82	7	7.26	7	6.42	8	7.91
11	5	5.68	5	4.74	5	5.73	8	6.89

Source: Anscombe, F.J., Graphs in statistical analysis, American Statistician, 27, 17–21, 1973.

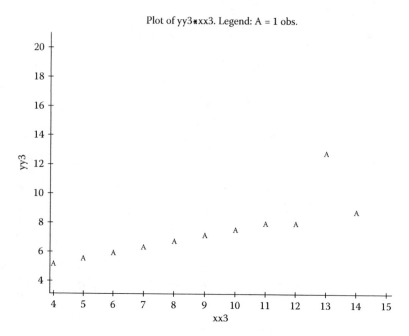

FIGURE 4.2
Scatterplot of (xx3, yy3).

inconvenience. (The reason for the renaming is explained further in the book.) Clearly, the relationship between xx3 and yy3 is problematic: It would be straightaway linear if not for the *far-out* point ID3, (13, 12.74). The scatterplot does not ocularly reflect a linear relationship. The nice large value of $r_{(xx3, yy3)} = 0.8163$ is meaningless and useless. I leave it to the reader to draw his or her underlying straight line (!).

4.6 Data Mining the Relationship of (xx3, yy3)

I data mine for the underlying structure of the paired variables (xx3, yy3) using a machine-learning approach under the discipline of evolutionary computation, specifically *genetic programming* (GP). The fruits of my data mining work yield the scatterplot in Figure 4.3. The data mining work is not an expenditure of time-consuming results or mental effort, as the GP-based data mining (GP-DM) is a machine-learning adaptive intelligence process that is effective and efficient for straightening data. The data mining tool used is the GenIQ Model, which renames the data-mined variable with the

⎿ covered in Chapt; 24-31

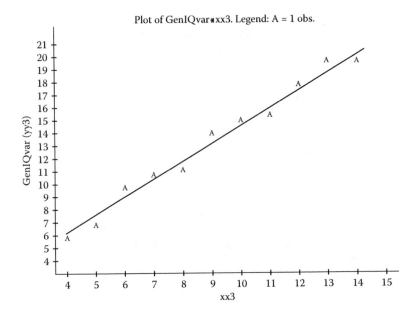

FIGURE 4.3
Scatterplot of (xx3, GenIQvar(yy3)).

prefix GenIQvar. Data-mined (xx3, yy3) is relabeled (xx3, GenIQvar(yy3)). (The GenIQ Model is formally introduced replete with eye-opening examples in Chapter 29.)

The correlation coefficient $r_{(xx3, \text{GenIQvar(yy3)})}$ = 0.9895 and the uncurtained underlying relationship in Figure 4.3 warrant a silver, if not gold, medal straight line. The correlation coefficient is a reliable measure of the linear relationship between xx3 and GenIQvar(yy3). The almost-maximum value of $r_{(xx3, \text{GenIQvar(yy3)})}$ indicates an almost-perfect linear relationship between the original xx3 and the data-mined GenIQvar(yy3). (Note: The scatterplot indicates GenIQvar_yy3 on the Y-axis, not the correct notation GenIQvar(yy3) due to syntax restrictions of the graphics software.)

The values of the variables xx3, yy3, and GenIQvar(yy3) are in Table 4.2. The 11 data points are ordered based on the descending values of GenIQvar(yy3).

4.6.1 Side-by-Side Scatterplot

A side-by-side scatterplot of all the goings-on is best pictured, as it is worth a 1,000 words. The side-by-side scatterplot speaks for itself of a data mining piece of work done well (see Figure 4.4).

TABLE 4.2

Reexpressed yy3, GenIQ(yy3),
Descendingly Ranked by GenIQ(yy3)

xx3	yy3	GenIQ(yy3)
13	12.74	20.4919
14	8.84	20.4089
12	8.15	18.7426
11	7.81	15.7920
10	7.46	15.6735
9	7.11	14.3992
8	6.77	11.2546
7	6.42	10.8225
6	6.08	10.0031
5	5.73	6.7936
4	5.39	5.9607

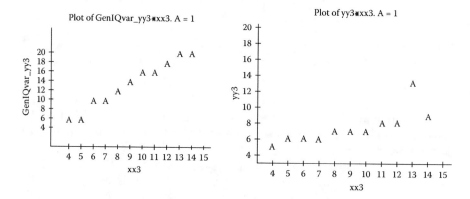

FIGURE 4.4
Side-by-side scatterplot.

4.7 What Is the GP-Based Data Mining Doing to the Data?

As evidenced by the detailed illustration, GP-DM maximizes the straight-line correlation between a genetically reexpressed dependent variable and a single predictor variable. When building a multiple (at least two predictor variables) regression model, GP-DM maximizes the straight-line correlations between a genetically reexpressed dependent variable with each predictor variable and the array of predictor variables considered jointly.

Although the GenIQ Model, as used here, is revisited in Chapter 28, it is not appropriate to leave without the definition of the reexpressed variable. The definition is in Equation (4.1):

$$GenIQvar\_yy3 = \cos(2^*xx3) + (xx3/0.655) \qquad (4.1)$$

The GenIQ model did eminently good data mining that rendered excellent results: $r_{(xx3, GenIQvar(yy3))} = 0.9895$.

4.8 Straightening a Handful of Variables and a Baker's Dozen of Variables

A handful of variables (10 pairs of variables) can be dealt with presumably without difficulty. As for a baker's dozen of variables (78 pairs), they can be a *handful*. However, GP-DM, as outlined with its main features and points of functioning, *reduces* initially, effectively, and efficiently the 78 pairs to a practical number of variables. This reduction operation is examined next.

One requires clearness about a baker's dozen of variables. A data analyst cannot expect to straighten 78 pairs of variables. So many pairs require so many scatterplots. This is the reason for misusing or ignoring the linearity assumption. GP-DM can. In fact, it does so with the speed of a Gatling gun. In practice, it is a common sight to work on a dataset of, say, 400 variables (79, 800 pairs of variables). GP-DM, in its beginning step, deletes variables (single and paired variables) that are deemed to have no predictive power by dint of the probabilistic-selected biological operators of reproduction, mating, and mutation. During the second evolutionary step, GP-DM further decreases the number of variables to only handfuls. The remaining step of GP-DM is the full-fledged evolutionary process of GP proper, by which the data strengthening is carried out in earnest [5].

Ergo, GP-DM can handle efficiently virtually any number of variables as long as the computer being used has the capacity to process initially all the variables of the original dataset. Illustrating GP-DM with an enlarged dataset of many, many pairs of variables is beyond the scope of this chapter, Moreover, demonstrating GP-DM with an enlarged dataset would be spurious injustice and reckless wronging of an exposition demanding that which must go beyond 2D (two dimensions) and even the 3D of the movie *Avatar*.

4.9 Summary

The objective of this chapter was share my personal encounters: Entering the mines of data, going deep to unearth acute underlying relationships, rising from the inner workings of the data mine to show the importance of straight data for the simplicity and desirability it brings for good model-building practice. I discussed an illustration, simple but with an object lesson, of what to do when an observed relationship between two variables in a scatterplot is masking an acute underlying relationship. A GP-DM method was proposed directly toward unmasking and straightening the obtuse relationship. The correlation coefficient was used correctly when the exposed character of the exampled relationship had straight-line simplicity.

References

1. Ratner, B., Data mining: an ill-defined concept. http://www.geniq.net/res/data-mining-is-an-ill-defined-concept.html, 2009.
2. Han, J., and Kamber, M., *Data Mining: Concepts and Techniques,* Morgan Kaufmann, San Francisco, 2001.
3. Bozdogan, H., ed., *Statistical Data Mining and Knowledge Discovery*, CRC Press, Boca Raton, FL, 2004.
4. Refaat, M., *Data Preparation for Data Mining Using SAS*, Morgan Kaufmann Series in Data Management Systems, Morgan Kaufmann, Maryland Heights, MO, 2006.
5. Ratner, B., What is genetic programming? http://www.geniq.net/Koza_GPs.html, 2007.

5

Symmetrizing Ranked Data: A Statistical Data Mining Method for Improving the Predictive Power of Data

5.1 Introduction

The purpose of this chapter is to introduce a new statistical data mining method, the *symmetrizing ranked data* method, and add it to the paradigm of simplicity and desirability for good model-building practice as presented in Chapter 4. The new method carries out the action of two basic statistical tools, symmetrizing and ranking variables, yielding new reexpressed variables with likely improved predictive power. I detail Steven's scales of measurement (nominal, ordinal, interval, and ratio). Then, I define an *approximate-interval* scale that is an offspring of the new statistical data mining method. Next, I provide a quick review of the simplest of EDA elements: (1) the stem-and-leaf display and (2) the box-and-whiskers plot. Both are needed for presenting the new method, which itself falls under EDA (exploratory data analysis) proper. Last, I illustrate the proposed method with two examples, which provide the data miner with a starting point for more applications of this utile statistical data mining tool.

5.2 Scales of Measurement

There are four scales of data measurement due to Steven's scales of measurement [1]:

1. Nominal data are classification *labels*, for example, color (red, white, and blue); there is no ordering of the data values. Clearly, arithmetic operations cannot be performed on nominal data. That is, one cannot add red + blue (= ?).

2. Ordinal data are *ordered* numeric labels in that higher/lower numbers represent higher/lower values on the scale. The intervals between the numbers are not necessarily equal.

a. For example, consider the variables CLASS and AGE as per traveling on a cruise liner. I recode the CLASS labels (first, second, third, and crew) into the ordinal variable CLASS_, implying some measure of income. Also, I recode the AGE labels (adult and child) into the ordinal variable AGE_, implying years old. In addition, I create CLASS interaction variables with both AGE and GENDER, denoted by CLASS_AGE_ and CLASS_GENDER_, respectively. The definitions of the recoded variables are listed next. The definitions of the interaction variables are discussed in the second illustration.

   i. if GENDER=male then GENDER_=0
   ii. if GENDER=female then GENDER_=1
   i. if CLASS=first then CLASS_ =4
   ii. if CLASS=second then CLASS_=3
   iii. if CLASS=third then CLASS_=2
   iv. if CLASS=crew then CLASS_=1
   i. if AGE=adult then AGE_=2
   ii. if AGE=child then AGE_=1
   i. if CLASS=second and age=child then Class_Age_=8
   ii. if CLASS=first and age=child then Class_Age_=7
   iii. if CLASS=first and age=adult then Class_Age_=6
   iv. if CLASS=second and age=adult then Class_Age_=5
   v. if CLASS=third and age=child then Class_Age_=4
   vi. if CLASS=third and age=adult then Class_Age_=3
   vii. if CLASS=crew and age=adult then Class_Age_=2
   viii. if CLASS=crew and age=child then Class_Age_=1
   i. if CLASS=first and GENDER=female then Class_Gender_=8
   ii. if CLASS=second and GENDER=female then Class_Gender_=7
   iii. if CLASS=crew and GENDER=female then Class_Gender_=6
   iv. if CLASS=third and GENDER=female then Class_Gender_=5
   v. if CLASS=first and GENDER=male then Class_Gender_=4
   vi. if CLASS=third and GENDER=male then Class_Gender_=2
   vii. if CLASS=crew and GENDER=male then Class_Gender_=3
   viii. if CLASS=second and GENDER=male then Class_Gender_=1

b. One cannot assume the difference in income between CLASS_=4 and CLASS_=3 equals the difference in income between CLASS_=3 and CLASS_=2.

    c. Arithmetic operations (e.g., subtraction) cannot be performed. With CLASS_ numeric labels, one cannot conclude 4 - 3 = 3 - 2.

    d. Only the logical operators "less than" and "greater than" can be performed.

    e. Another feature of an ordinal scale is that there is no "true" zero. This is so because the CLASS_ scale, which goes from 4 through 1, could have been recorded to go from 3 through 0.

3. Interval data are measured along a scale in which each position is equidistant from one another. This allows for the distance between two pairs to be equivalent in some way.

    a. Consider the HAPPINESS scale of 10 (= most happy) through 1 (= very sad). Four persons rate themselves on HAPPINESS:

        i. Persons A and B state 10 and 8, respectively, and

        ii. Persons C and D state 5 and 3, respectively.

        iii. One can conclude that person-pair A and B (with happiness difference of 2) represents the same difference in happiness as that of person-pair C and D (with happiness difference of 2).

        iv. Interval scales *do not* have a true zero point; therefore, it is not possible to make statements about how many times happier one score is than another.

            1) Interval data cannot be multiplied or divided. The common example of interval data is the Fahrenheit scale for temperature. Equal differences on this scale represent equal differences in temperature, but a temperature of $30°$ is not twice as warm as a temperature of $15°$. That is, $30° - 20° = 20° - 10°$, but $20°/10°$ is not equal to 2. That is, $20°$ is not twice as hot as $10°$.

4. Ratio data are like interval data except they *have* true zero points. The common example is the Kelvin scale of temperature. This scale has an absolute zero. Thus, a temperature of 300 K is twice as high as a temperature of 150 K.

5. What is a true zero? Some scales of measurement have a true or natural zero.

    a. For example, WEIGHT has a natural 0 at no weight. Thus, it makes sense to say that my beagle Matzi weighing 26 pounds is twice as heavy as my dachshund Dappy weighing 13 pounds. WEIGHT is on a ratio scale.

    b. On the other hand, YEAR does not have a natural zero. The YEAR 0 is arbitrary, and it is not sensible to say that the year 2000 is twice as old as the year 1000. Thus, YEAR is measured on an interval scale.

6. Note: Some data analysts unfortunately make no distinction between interval or ratio data, calling them both continuous. Moreover, most data analysts put their heads in the sand to treat ordinal data, which assumes numerical values, as interval data. In both situations, this is not correct technically.

5.3 Stem-and-Leaf Display

The stem-and-leaf display is a graphical presentation of quantitative data to assist in visualizing the relative density and shape of a distribution. A salient feature of the display is that it retains the original data to at least two significant digits and puts the data in order. A basic stem-and-leaf display is drawn with two columns separated by a vertical line. The left column contains the *stems,* and the right column contains the *leaves.* Typically, the leaf contains the last digit of the number, and the stem contains all of the other digits. In the case of very large numbers, the data values may be rounded to a particular place value (e.g., the hundredths place) that will be used for the leaves. The remaining digits to the left of the rounded place value are used as the stem. The stem-and-leaf display is also useful for highlighting outliers and finding the mode. The display is most useful for datasets of moderate "EDA" size (around 250 data points), after which the stem-and-leaf display becomes a histogram, rotated counterclockwise 90°, and all the digits of the leaves are represented by a*.

5.4 Box-and-Whiskers Plot

The box-and-whiskers plot provides a detailed visual summary of various features of a distribution. The box stretches from the bottom horizontal line, the lower hinge, defined as the 25th percentile, to the top horizontal line, the upper hinge, defined as the 75th percentile. The box is completed by adding the vertical lines on both ends of the hinges. The horizontal line within the box is the median. The + represents the mean.

The *H-spread* is defined as the difference between the hinges, and a *step* is defined as 1.5 times the H-spread. *Inner fences* are one step beyond the hinges. *Outer fences* are two steps beyond the hinges. These fences are used to create the *whiskers,* which are vertical lines at either end of the box, extended up to the inner fences. Every value between the inner and outer fences is indicated by an o; a score beyond the outer fences is indicated by a *. A symmetric distribution's stem-and-leaf display and box-and-whiskers plot are in Figure 5.1.

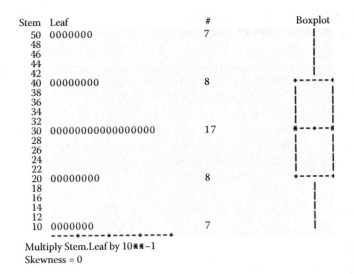

```
Stem  Leaf                        #        Boxplot
  50  0000000                     7           |
  48                                          |
  46                                          |
  44                                          |
  42                                          |
  40  00000000                    8        +-----+
  38                                        |     |
  36                                        |     |
  34                                        |     |
  32                                        |     |
  30  00000000000000000          17        *--+--*
  28                                        |     |
  26                                        |     |
  24                                        |     |
  22                                        |     |
  20  00000000                    8        +-----+
  18                                          |
  16                                          |
  14                                          |
  12                                          |
  10  0000000                     7           |
      ----+----+----+----+
```

Multiply Stem.Leaf by 10**−1

Skewness = 0

FIGURE 5.1

Stem-and-leaf display, box-and whiskers plot of symmetric data.

Note: I added the statistic skewness, which measures the lack of symmetry of a distribution. The skewness scale is interval. Skewness = 0 means the distribution is symmetric.

If the skewness value is positive, then the distribution is said to be right skewed or positive skewed, which means the distribution has a long tail in the positive direction. Similarly, if the skewness value is negative, then the distribution is left skewed or negative skewed, which means the distribution has a long tail in the negative direction.

5.5 Illustration of the Symmetrizing Ranked Data Method

The best way of describing the proposed symmetrizing ranked data (SRD) method is by example. I illustrate the new statistical data mining method with two examples that provide the data miner with a starting point for more applications of the statistical data mining tool.

5.5.1 Illustration 1

Consider the two variables from the real study presented in Chapter 3, HI_ BALANCE (the highest balance attained among credit card transactions for an individual) and RECENCY_MOS (the number of months since last purchase by an individual). The two steps of the SRD data mining process are

1. *Rank* the values of both HI_BALANCE and RECENCY_MOS and create the *rank scores* variables rHI_BALANCE and rRECENCY_MOS, respectively. Use any method for handling tied rank scores.

2. *Symmetrize* the ranked variables, which have the same names as the rank scores variables, namely, rHI_BALANCE and rRECENCY_MOS.

The sequential steps are performed by the SAS procedure RANK. The procedure creates the rank scores variables, and the option "normal = TUKEY" instructs for symmetrizing the rank scores variables. The input data, say, DTReg, and the symmetrized-ranked data are in dataset, say, DTReg_NORMAL. The SAS program is

```
proc rank data = DTReg_data_ normal = TUKEY
out = DTReg_NORMAL;
var
HI_BALANCE RECENCY_MOS;
ranks
rHI_BALANCE rRECENCY_MOS;
run;
```

5.5.1.1 Discussion of Illustration 1

1. The stem-and-leaf displays and the box-and-whiskers plots for HI_BALANCE and rHI_BALANCE are in Figures 5.2 and 5.3, respectively. HI_BALANCE and rHI_BALANCE have skewness values of 1.0888, and 0.0098, respectively.

2. The stem-and-leaf displays and the box-and-whiskers plots for RECENCY_MOS and rRECENCY_MOS are in Figures 5.4 and 5.5, respectively. RECENCY_MOS and rRECENCY_MOS have skewness values of 0.0621 and -0.0001, respectively.

3. Note: The stem-and-leaf displays turn into histograms because the sample size is too large, 2,000. Regardless, information can still be obtained about the shape of the distributions.

I *acknowledge hesitantly* throwing, with an overhand, my head in the sand, only to further the SRD method by treating ordinal data, which assumes recoded numeric values, as interval data.

Recall, symmetrizing data helps in straightening data. Accordingly, without generating scatterplots, the *reliable* correlation coefficients between the two pairs of variables, the raw variables HI_BALANCE and RECENCY_MOS, and the reexpressed variables, via the SRD method, rHI_BALANCE and rRECENCY_MOS are -0.6412 and -0.10063, respectively (see Tables 5.1 and 5.2). Hence, the *SRD method has improved the strength* of the predictive

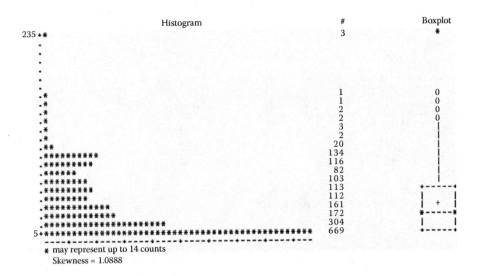

FIGURE 5.2

Histogram and boxplot of HI_BALANCE.

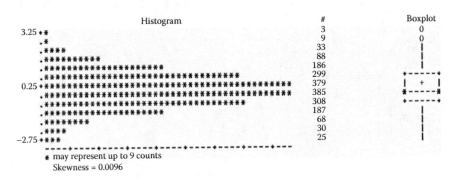

FIGURE 5.3

Histogram and boxplot for rHI_BALANCE.

relationship between the two raw variables by 56.9% (=abs(-0.10063) - abs (-0.06412))/abs(-0.06421)), where abs = the absolute value function, which ignores the negative sign. In sum, the paired-variable (rHI_BALANCE, rRE-CENCY_MOS) is not necessarily the correct one pair; but it has more predictive power than the original pair-variable, and offer more potential in the model building process.

5.5.2 Illustration 2

Since the fatal night of April 15, 1912, when the White Star liner *Titanic* hit an iceberg and went down in the mid-Atlantic, fascination with the disaster has never abated. In recent years, interest in the *Titanic* has increased

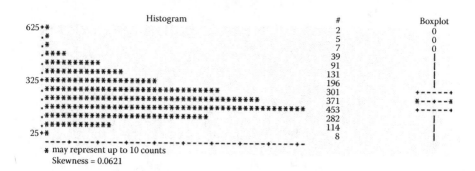

FIGURE 5.4
Histogram and boxplot for RECENCY_MOS.

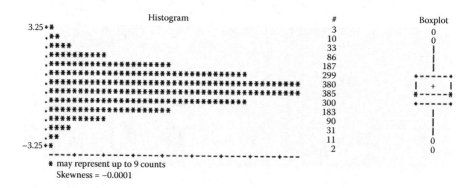

FIGURE 5.5
Histogram and boxplot for rRECENCY_MOS.

TABLE 5.1

Correlation Coefficient between HI_BALANCE and RECENCY_MOS

Pearson Correlation Coefficients, N = 2,000 Prob > r under H0: Rho = 0

	HI_BALANCE	**RECENCY_MOS**
HI_BALANCE	1.00000	−0.06412
		0.0041
RECENCY_MOS	−0.06412	1.00000
	0.0041	

dramatically because of the discovery of the wreck site by Dr. Robert Ballard. The century-old tragedy has become a national obsession. Any fresh morsel of information about the sinking is savored. I believe that the SRD method can satisfy the appetite of the *Titanic* aficionado. I build a *preliminary* Titanic *model* to identify survivors so when *Titanic II* sails it will know beforehand who will be most likely to survive an iceberg hitting with outcome odds

TABLE 5.2

Correlation Coefficient between rHI_BALANCE and rRECENCY_MOS

Pearson Correlation Coefficients, N = 2,000 Prob > r under H0: Rho = 0

	RHI_BALANCE	**RECENCY_MOS**
rRHI_BALANCE	1.00000	-0.10063
Rank for Variable rHI_BALANCE		<.0001
rRECENCY_MOS	-0.10063	1.00000
Rank for Variable RECENCY_MOS	<.0001	

of 2.0408e-12 to 1.* The *Titanic* model application, detailed in the remaining sections of the chapter, shows clearly the power of the SRD data mining technique, worthy of inclusion in every data miner's toolkit.

5.5.2.1 Titanic Dataset

There were 2,201 passengers and crew aboard the *Titanic*. Only 711 persons survived, resulting in a 32.2% survival rate. For all persons, their basic demographic variables are known: GENDER (female, male); CLASS (first, second, third, crew); and AGE (adult, child). The passengers fall into 14 patterns of GENDER-CLASS-AGE (Table 5.3). Also, Table 5.3 includes pattern cell size (N), number of survivors within each cell (S), and the Survival Rate (format is %).

Because there are only three variables and their scales are of the least informative, nominal (GENDER), and ordinal (CLASS and AGE), building a *Titanic* model has been a challenge for the best of academics and practitioners [2–6]. The SRD method is an original and valuable a data mining method that belongs with the *Titanic* modeling literature. In the following sections, I present the building of a *Titanic* model.

5.5.2.2 Looking at the Recoded Titanic Ordinal Variables CLASS_, AGE_, CLASS_AGE_, and CLASS_GENDER_

To see what the data look like, I generate stem-and-leaf displays and box-and-whisker plots for CLASS_, AGE_, and GENDER_ (Figures 5.6, 5.7, and 5.8, respectively). Also, I bring out the interaction variables CLASS_AGE_ and CLASS_GENDER_, which I created in the scales of measurement Section 5.2. The graphics of CLASS_AGE and CLASS_GENDER_ are in Figures 5.9 and 5.10, respectively. Regarding the coding of ordinal values for the interaction variables, I use the commonly known refrain during a crisis, "Women and children, first." Survival rates for women and children are 74.35%, and 52.29%, respectively (Table 5.4), and bear out the refrain.

---

* Source is unknown (actually, I lost it).

TABLE 5.3

Titanic Dataset

Pattern	GENDER	CLASS	AGE	N	S	Survival Rate
1	Male	First	Adult	175	57	32.5
2	Male	First	Child	5	5	100.0
3	Male	Second	Adult	168	14	8.3
4	Male	Second	Child	11	11	100.0
5	Male	Third	Adult	462	75	16.2
6	Male	Third	Child	48	13	27.1
7	Male	Crew	Adult	862	192	22.3
8	Female	First	Adult	144	140	97.2
9	Female	First	Child	1	1	100.0
10	Female	Second	Adult	93	80	86.0
11	Female	Second	Child	13	13	100.0
12	Female	Third	Adult	165	76	46.1
13	Female	Third	Child	31	14	45.2
14	Female	Crew	Adult	23	20	87.0
			Total	2,201	711	32.2

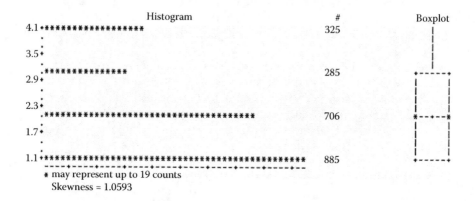

FIGURE 5.6
Histogram and boxplot of CLASS_.

5.5.2.3 Looking at the Symmetrized-Ranked Titanic Ordinal Variables rCLASS_, rAGE_, rCLASS_AGE_, and rCLASS_GENDER_

The stem-and-leaf displays and box-and-whisker plots for the symmetrized-ranked variables rCLASS_, rAGE_, rGENDER_, rCLASS_AGE_, and rCLASS_GENDER_ are in Figures 5.11 through 5.15, respectively.

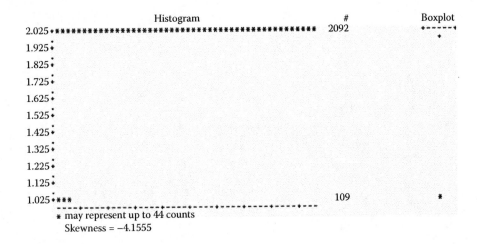

```
                  Histogram                              #        Boxplot
2.025 ★★★★★★★★★★★★★★★★★★★★★★★★★★★★★★★★★★★★★★★★★★★    2092      +------+
      .                                                               +
1.925 +.
      .
1.825 +.
      .
1.725 +.
      .
1.625 +.
      .
1.525 +.
      .
1.425 +.
      .
1.325 +.
      .
1.225 +.
      .
1.125 +.
      .
1.025 +★★★                                            109           *
      ----+----+----+----+----+----+----+----+----+---
      ★ may represent up to 44 counts
        Skewness = −4.1555
```

FIGURE 5.7
Histogram and boxplot for AGE_.

```
                  Histogram                              #        Boxplot
1.025 +★★★★★★★★★★★★★                                 470          *
      .
0.925 +.
      .
0.825 +.
      .
0.725 +.
      .
0.625 +.
      .
0.525 +.
      .
0.425 +.
      .
0.325 +.
      .
0.225 +.
      .
0.125 +.
      .
0.025 +★★★★★★★★★★★★★★★★★★★★★★★★★★★★★★★★★★★★★★★★★★★★★★★   1731      +------+
      ----+----+----+----+----+----+----+----+----+---
      ★ may represent up to 37 counts
        Skewness = 1.3989
```

FIGURE 5.8
Histogram and boxplot for GENDER_.

The results of the SRD method are shown in Table 5.5, a comparison of skewness for the original and SRD variables. The CLASS_, CLASS_AGE_, and CLASS_GENDER_ variables have been reexpressed, rendering significance shifts in the corresponding skewed distributions to almost symmetric distributions: Skewness values decrease significantly in the direction of zero. Although AGE_ and GENDER_ are moot variables with null (useless) graphics because the variables assume only two values, I include them as an object lesson.

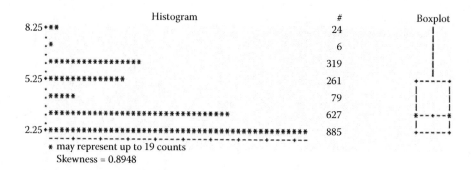

FIGURE 5.9
Histogram and boxplot for CLASS_AGE_.

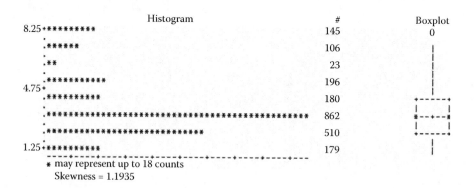

FIGURE 5.10
Histogram and boxplot for CLASS_GENDER_.

TABLE 5.4

Women and Children by SURVIVED

Row Pct Frequency	No	Yes	Total
Child	52	57	109
	47.71	52.29	
Female	109	316	425
	25.65	74.35	
Total	161	373	534

5.5.2.4 Building a Preliminary Titanic Model

Reflecting on the definition of ordinal and interval variables, I know that the symmetrized-ranked variables are not ordinal variables. However, the scale property of the reexpressed variables rCLASS_, rCLASS_AGE_, and rCLASS_GENDER_ is not obvious. The variables are not on a ratio scale

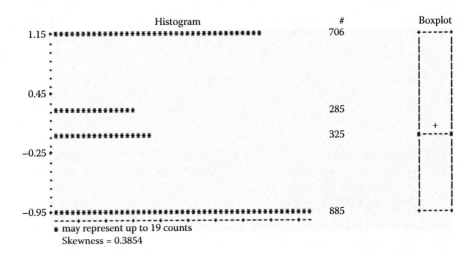

FIGURE 5.11
Histogram and boxplot for rCLASS_.

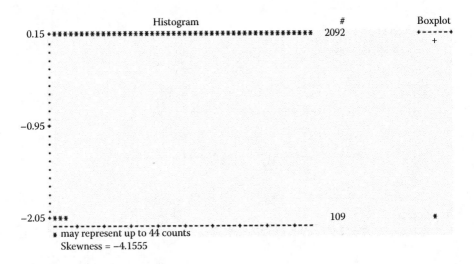

FIGURE 5.12
Histogram and boxplot for rAGE_.

because a true zero value has no interpretation. Accordingly, I define the symmetrized-ranked variable as an *approximate interval* variable.

The preliminary *Titanic* model is a logistic regression model with the dependent variable SURVIVED, which assumes 1 = yes and 0 = no. The preliminary *Titanic* model is built by the SAS procedure LOGISTIC, and

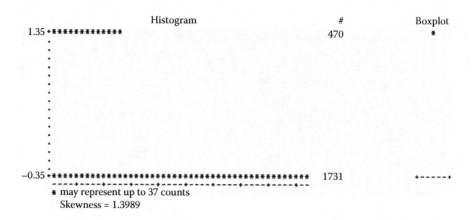

FIGURE 5.13
Histogram and boxplot for rGENDER_.

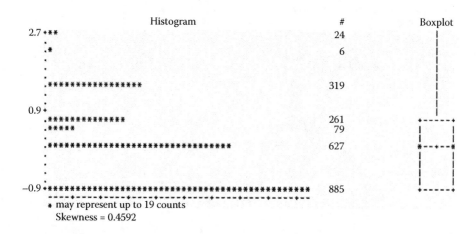

FIGURE 5.14
Histogram and boxplot for rCLASS_AGE_.

its definition, consisting of two interaction symmetrized-ranked variables rCLASS_AGE_ and rCLASS_GENDER_ is detailed in Table 5.6.

The preliminary *Titanic* model results are as follows: 59.1% (= 420/711) survivors are correctly classified among those predicted survivors. The classification matrix for the preliminary model (Table 5.7) I believe yields a clearer display of the predictiveness of a binary (yes/no) classification model.

I present only a preliminary model because there is much work to be tested with respect to the three-way interaction variables, which are required because the *Titanic* data only have two ordinal variables (CLASS and AGE)

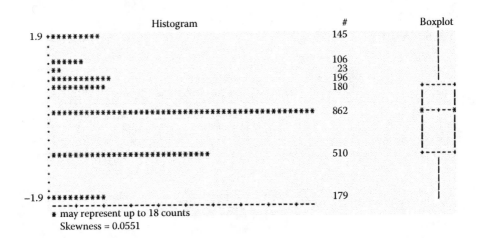

FIGURE 5.15

Histogram and boxplot for rCLASS_GENDER_.

TABLE 5.5

Comparison of Skewness Values for Original and Symmetrized-Ranked Variables

Variable	Skewness Value	Effect of the Symmetrized-Ranked Method (Yes, No, Null)
CLASS_	1.0593	Yes
rCLASS_	0.3854	
AGE_	-4.1555	Null. Binary variables render two parallel lines.
rAGE_	-4.1555	
GENDER_	1.3989	Null. Binary variables render two parallel lines.
rGENDER_	1.3989	
CLASS_AGE_	0.9848	Yes
rCLASS_AGE_	0.4592	
CLASS_ GENDER_	1.1935	Yes
rCLASS_ GENDER_	0.0551	

and one nominal class (GENDER). To build on the preliminary model is going beyond the objectives of introducing the SRD method; the statistical data mining SRD method offers promise in improving the predictive power of the data and providing the data miner with a starting point for more applications of this utile method. Perhaps the data miner will continue with the preliminary model with the SRD data mining approach to finalize the *Titanic* model.

TABLE 5.6

Preliminary Titanic Model

The LOGISTIC Procedure: Analysis of Maximum Likelihood Estimates

Parameter	DF	Estimate	Standard Error	Wald Chi-Square	Pr > ChiSq
Intercept	1	-0.8690	0.0533	265.4989	<.0001
rCLASS_AGE_	1	0.4037	0.0581	48.1935	<.0001
rCLASS_GENDER_	1	1.0104	0.0635	252.9202	<.0001

TABLE 5.7

Classification Table for the Preliminary Titanic Model

	Predicted Deceased	Predicted Survivors	Total
Actual deceased	1,199	291	1,490
Actual survivors	291	420	711
Total	1,490	711	2,201

5.6 Summary

I introduced a new statistical data mining method, the SRD method, and added it to the paradigm of simplicity and desirability for good model-building practice. The method uses two basic statistical tools, symmetrizing and ranking variables, yielding new reexpressed variables with likely improved predictive power. First, I detailed Steven's scales of measurement to provide a framework for the new symmetric reexpressed variables. Accordingly, I defined the newly created SRD variables on an approximate interval scale. Then, I provided a quick review of the simplest of EDA elements, the stem-and-leaf display and the box-and-whiskers plot, as both are needed for presenting the underpinnings of the SRD method. Last, I illustrated the new method with two examples, which showed improved predictive power of the symmetric reexpressed variables over the raw variables. It was my intent that the examples are the starting point for more applications of the SRD method.

References

1. Stevens, S.S., On the theory of scales of measurement, *American Association for the Advancement of Science*, 103(2684), 677–680, 1946.
2. Simonoff, J.S., The "Unusual Episode" and a Second Statistics Course, *Journal of Statistics Education* v.5, n.1, 1997. http://www.amstat.org/publications/jse/v5n1/simonoff.html.

3. Friendly, M., Extending Mosaic Displays: Marginal, Partial, and Conditional Views of Categorical Data, *Journal of Computational and Graphical Statistics*, 8:373—395, 1999. http://www.datavis.ca/papers/drew/.

4. http://www.geniqmodel.com/res/TitanicGenIQModel.html, 999.

5. Young, F.W., *Visualizing Categorical Data in ViSta*, University of North Carolina, 1993. http://forrest.psych.unc.edu/research/vista-frames/pdf/Categorical DataAnalysis.pdf.

6. SAS Institute, http://support.sas.com/publishing/pubcat/chaps/56571.pdf, 2009.

6

Principal Component Analysis: A Statistical Data Mining Method for Many-Variable Assessment

6.1 Introduction

Principal component analysis (PCA), invented in 1901 by Karl Pearson[*][†] as a data reduction technique, uncovers the interrelationship among many variables by creating linear combinations of the many original variables into a few new variables such that most of the variation among the many original variables is accounted for or retained by the few new *uncorrelated* variables. The literature is sparse on PCA used as a reexpression,[‡] not a reduction, technique. I posit the latter distinction for repositioning PCA as an EDA (exploratory data analysis) technique. EDA is a force for identifying structure: PCA is a classical data reduction technique of the 1900s; PCA is a reexpression method of 1997, a very good year for the statistics community as Tukey's seminal EDA book was released then; and PCA is a data mining method of today, as if the voguish term *data mining* can replace EDA proper. In this chapter, I put PCA in its appropriate place in the EDA reexpression paradigm. I illustrate PCA as a statistical data mining technique capable of serving in a *common* application with expected solution and illustrate PCA in an *uncommon* application, yielding a reliable and robust solution. In addition, I provide an original and valuable use of PCA in the construction of quasi-interaction variables, furthering the case of PCA as a powerful data mining method.

---

[*] Pearson, K., On lines and planes of closest fit to systems of points in space, *Philosophical Magazine*, 2(6), 559–572, 1901.

[†] Harold developed independently principal component analysis in 1933.

[‡] Tukey coined the term *reexpression* without defining it. I need a definition of terms that I use. My definition of reexpression is changing the composition, structure, or scale of original variables by applying functions, such as arithmetic, mathematic, and truncation function, to produce reexpressed variables of the original variables. The objective is to bring out or uncover reexpressed variables that have more information than the original variables.

TABLE 6.1

Objective of Reexpression and Method by Number of Variables

	Number of Variables		
	1	**2**	**Many**
Reexpression	Symmetrize	Straighten	Retain variation
Method	Ladder of powers with boxplot	Ladder of powers with bulging rule	PCA

6.2 EDA Reexpression Paradigm

Consider Table 6.1, the EDA reexpression paradigm, which indicates the objective of reexpression and method by number of variables. In Section 4.2, "Straightness and Symmetry in Data," in Chapter 4, I discuss the relationship between the two concepts for one and two variables and provide a machine-learning data mining approach to straighten the data. The methods of ladder of powers with the boxplot* and bulging rule, for one and two variables, are discussed in exacting detail in Chapters 8 and 9. In Table 6.1, PCA is put in its proper place. PCA is used to retain variation among many variables by reexpressing the many original variables into a few new variables such that *most* of the variation among the many original variables is accounted for or retained by the few new uncorrelated variables. The literature (to my knowledge) is sparse on PCA used as a reexpression, not a reduction, technique. PCA, viewed as an EDA technique to identify structure, gives awareness that PCA is a valid (new) data mining tool.

6.3 What Is the Big Deal?

The use of many variables burdens the data miner with two costs:

1. Working with many variables takes more time and space. This is self-evident—just ask a data miner.

2. Modeling dependent variable Y by including many predictor variables yields fitting many coefficients and renders the predicted Y with greater error variance if the fit to Y is adequate with few variables.

---

* Boxplot is also known as box-and-whiskers plot.

Thus, by reexpressing many predictor variables to a few new variables, the data miner saves time and space, and, importantly, reduces error variance of the predicted Y.

6.4 PCA Basics

PCA transforms a set of p variables, X1, X2, ¼ , Xp into p linear combination variables PC1, PC2, PC1, PCp (PC for principal component) such that most of the information (variation) in the original set of variables can be represented in a smaller set of the new variables, which are uncorrelated with each other. That is,

$$PC1 = a11^* X1 + a12 * X2 + \ldots + a1j2 * X1j + \ldots + aip * Xp$$

$$PC2 = a21^* X1 + a22 * X2 + \ldots + a2j2 * X2j + \ldots + a2p * Xp$$

$$\ldots \ldots \ldots \ldots .$$

$$PCi = ai1^* X1 + ai2 * X2 + \ldots + aij2 * Xij + \ldots + aip * Xp$$

$$\ldots \ldots \ldots \ldots .$$
$$\ldots \ldots \ldots \ldots .$$

$$PCp = ai1^* X1 + ai2 * X2 + \ldots + aij2 * Xij + \ldots + aip * Xp$$

where the aij's are constants called PC coefficients.

In addition, the PCs and the aij's have many algebraic and interpretive properties. (For ease of presentation, the X's are assumed to be standardized.)

6.5 Exemplary Detailed Illustration

Consider the correlation matrix of four census EDUCATION variables (X1, X2, X3, X4) and the corresponding PCA output in Tables 6.2 and 6.3, respectively.

6.5.1 Discussion

1. Because there are four variables, it is possible to extract four PCs from the correlation matrix.
2. The basic statistics of PCA are

TABLE 6.2

Correlation Matrix of X1, X2, X3, X4

	X1	X2	X3	X4
% less than high school (X1)	1.000	.2689	−.7532	−.8116
% graduated high school (X2)		1.000	.3823	−.6200
% some college (X3)			1.000	.4311
% college or more (X4)				1.000

TABLE 6.3

Latent Roots (Variance) and Latent Vectors (Coefficients) of Correlation Matrix

Variable	Latent Vector			
	a1	a2	a3	a4
X1	−.5514	−.4222	.2912	.6578
X2	−.4042	.7779	−.3595	.3196
X3	.4844	.4120	.6766	.3710
X4	.5457	.2162	−.5727	.5721
LRi	2.6620	.8238	.5141	.0000
Prop. Var %	66.55	20.59	12.85	.0000
Cum. Var %	66.55	87.14	100	100

- the four variances: Latent Roots (LR1, LR2, LR3, LR4), which are ordered by size, and
- the associated weight (i.e., coefficient) vectors: Latent Vectors (a1, a2, a3 , a4).

3. The total variance in the system or dataset is four—the sum of the variances of the four (standardized) variables.

4. Each Latent Vector contains four elements, one corresponding to each variable.

For a1 there are

$$\left[-.5514, -.4041, .4844, .5457\right]$$

which are the four coefficients associated with the first and largest PC whose variance is 2.6620.

5. The first PC is the linear combination.

$$PC1 = -.5514 * X1 - .4042 * X2 + .4844 * X3 + .5457 * X4$$

6. PC1 explains (100 * 2.6620/4) = 66.55% of the total variance of the four variables.

7. The second PC is the linear combination

$$PC1=-.4222 * X1+.7779 * X2+.4120 * X3-.2162 * X4$$

with next-to-largest variance as 1.202, and explains (100 * .8238/4) = 20.59% of the total variance of the four variables.

8. Together, the first two PCs account for 66.55% + 20.55% or 87.14% of the variance in the four variables.

9. For the first PC, the first two coefficients are negative, and the last two are positive; accordingly, PC1 is interpreted as follows:

It is a *contrast* between persons who at most graduated from high school and persons who at least attended college.

High scores on PC1 are associated with zip codes where the percentage of persons who at least attended college is *greater* than the percentage of persons who at most graduated from high school.

Low scores on PC1 are associated with zip codes where the percentage of persons who at least attended college is *less* than the percentage of persons who at most graduated from high school.

6.6 Algebraic Properties of PCA

Almost always PCA is performed on a correlation matrix; that is, the analysis is done with standardized variables (means = 0, and variances = 1).

1. Each PC,

$$PCi = ai1* X1+ ai2 * X2 +...+ aij2 * Xij + ...+ aip * Xp$$

has a variance also called a latent root or eigenvalue such that
   a. Var(PC1) is maximum
   b. Var(PC1) > Var(PC2) >...> Var(PCp)
      b1. Equality can occur but it is rare.
   c. Mean(PCi) = 0.

2. All PCs are uncorrelated.

3. Associated with each latent root i is a latent vector

$$\left(ai1, ai2,...,aij, ..., aip\right)$$

which together are the weights for the linear combination of the original variables forming PCi.

$$PCi = ai1 * X1 + ai2 * X2 +...+ aip * Xij +...+ aip * Xp$$

4. The sum of the variances of the PCs (i.e., sum of the latent roots) is equal to the sum of the variances of the original variables. Because the variables are standardized, the following identity results: Sum of latent roots = p.

5. The proportion of variance in the original p variables that k PCs account for is

$$= \frac{\text{sum of latent roots for first k PCs}}{P}$$

6. Correlation between Xi and PCj equals

$$aij * sqrt\left[Var\left(PC_1\right)\right]$$

This correlation is called a *PC loading*.

7. The rule of thumb for identifying significant loadings is

$$aij > .5 / sqrt\left[Var\left(PC_1\right)\right]$$

8. The sum of the squares of loading across all the PCs for an original variable indicates how much variance for that variable (communality) is accounted for by the PCs.

9. Var(PC) = small (less than .001) implies high multicollinearity.

10. Var(PC) = 0 implies a perfect collinear relationship exists.

6.7 Uncommon Illustration

The objective of this uncommon illustration is to examine the procedures for considering a categorical predictor variable R_CD, which assumes 64 distinct values, defined by six binary (0, 1) elemental variables (X1, X2, X3, X4,

X5, X6), for inclusion in a binary RESPONSE predictive model. The standard approach is to create 63 dummy variables, which are tested for inclusion into a statistical regression model. The classical approach instructs that the complete set of 63 dummy variables is included in the model regardless of the number of dummy variables that are declared nonsignificant. This approach is problematic: Putting all the dummy variables in the model effectively adds "noise" or unreliability to the model as nonsignificant variables are known to be noisy. Intuitively, a large set of inseparable dummy variables poses a difficulty in model building in that they quickly "fill up" the model, not allowing room for other variables. Even if the dummy variables are not considered as a set and regardless of the variable selection method used, too many dummy variables (typically, dummy variables that reflect 100% and 0% response rates based on a very small number of individuals) are included spuriously in a model. As with the classical approach, this situation yields too many dummy variables that fill up the model, making it difficult for other candidate predictor variables to enter the model. An alternative, viable approach—*smoothing a categorical variable for model inclusion*—is illustrated in a case study in Chapter 8. (It is not discussed here because the required background has not yet been presented.) I prefer the procedure presented in the next section because it is effective, reliable, and (most important) easy to carry out; this is performing a PCA on the six elemental variables (X1, X2, X3, X4, X5, X6).

6.7.1 PCA of R_CD Elements (X1, X2, X3, X4, X5, X6)

The output of the PCA of R_CD six elemental variables is presented in Table 6.4.

6.7.2 Discussion of the PCA of R_CD Elements

1. Six elements of R_CD produce six PCs. PCA factoid: k original variables always produce k PCs.
2. The first two components, R1 and R3, account for 80.642% of the total variation; the first accounts for 50.634% of the total variation.
3. R1 is a contrast of X3 and X6 with X2, X4, and X5. This is the fruit of the PCA data mining method.
4. R3 is a weighted average of all six positive elements. PCA factoid: A *weighted average* component, also known as a generalized component, is often produced. This is also one of the fruits of the data mining PCA method, as the generalized component is often used instead of any one or all of the many original variables.
5. Consider Table 6.5. The variables are ranked from most to least correlated to RESPONSE based on the absolute value of the correlation coefficient.

TABLE 6.4

PCA of the Six Elements of R_CD (X1, X2, X3, X4, X5, X6)

	Eigenvalues of the Correlation Matrix				
	Eigenvalue		Difference	Proportion	Cumulative
R1	3.03807		1.23759	0.506345	0.50634
R2	1.80048		0.89416	0.300079	0.80642
R3	0.90632		0.71641	0.151053	0.95748
R4	0.18991		0.14428	0.031652	0.98913
R5	0.04563		0.02604	0.007605	0.99673
R6	0.01959		•	0.003265	1.00000

	Eigenvectors					
	R1	R2	R3	R4	R5	R6
X1	0.038567	-.700382	0.304779	-.251930	0.592915	0.008353
X2	0.473495	0.177061	0.445763	-.636097	-.323745	0.190570
X3	-.216239	0.674166	0.102083	-.234035	0.658377	0.009403
X4	0.553323	0.084856	0.010638	0.494018	0.260060	0.612237
X5	0.556382	0.112421	0.125057	0.231257	0.141395	-.767261
X6	-.334408	0.061446	0.825973	0.423799	-.150190	0.001190

TABLE 6.5

Correlation Coefficients: RESPONSE with the Original and Principal Component Variables Ranked by Absolute Coefficient Values

	Original and Principal Component Variables					
	R1	R3	R4	R6	X4	X6
RESPONSE with	0.10048**	0.08797**	0.01257*	-0.01109*	0.00973*	-0.00959*
	R2	X5	R5	X3	X2	X1
	0.00082*	0.00753*	0.00 741*	0.00729*	0.00538*	0.00258*

$**p < 0.0001; *0.015 < p < 0.7334.$

a. PCs R1, R3, R4, and R6 have larger correlation coefficients than the original X variables.

b. PCA factoid: It is typical for some PCs to have correlation coefficients larger than the correlation coefficients of some of the original variables.

c. In fact, this is the reason to perform a PCA.

d. Only R1 and R3 are statistically significant with p values less than 0.0001; the other variables have p values between 0.015 and 0.7334.

I build a RESPONSE model with the candidate predictor variable set consisting of the six original and six PC variables. I could only substantiate a two-variable model that includes (not surprisingly) R1 and R3. Regarding the predictive power of the model:

a. The model identifies the top 10% of the most responsive individuals with a response rate 24% greater than chance (i.e., the average response rate of the data file).

b. The model identifies the bottom 10% of the least-responsive individuals with a response rate 68% less than chance.

c. Thus, the model's predictive power index (top 10%/bottom 10%) = 1.8. This index value is relatively okay given that I use only 12 variables. With additional variables in the candidate predictor variable set, I know a stronger predictive model could be built. And, I believe the PCs R1 and R3 will be in the final model.

Lest one thinks that I forgot about the importance of straightness and symmetry mentioned in Chapter 4, I note that PCs are typically normally distributed. Moreover, because straightness and symmetry go hand in hand, there is little concern about checking the straightness of R1 and R3 in the model.

6.8 PCA in the Construction of Quasi-Interaction Variables

I provide an original and valuable use of PCA in the construction of quasi-interaction variables. I use SAS for this task and provide the program after the steps of the construction are detailed. Consider the dataset IN in Table 6.6. There are two categorical variables: GENDER (assumes M for male, F for female, and blank for missing) and MARITAL (assumes M for married, S for single, D for divorced, and blank for missing).

I necessarily recode the variables, replacing the blank with the letter *x*. Thus, GENDER is recoded to GENDER_, MARITAL recoded to MARITAL_ (see Table 6.7).

Then, I use the SAS procedure TRANSREG to create the dummy variables for both GENDER_ and MARITAL_. For each value of the variables, there are corresponding dummy variables. For example, for GENDER_ = M, the dummy variable is GENDER_M. The reference dummy variable is for the missing value of x (see Table 6.8).

I perform a PCA with both GENDER_ and MARITAL_dummy variables. This produces five quasi-interaction variables: GENDER_x_MARITAL_pc1 to GENDER_x_MARITAL_pc5. The output of the PCA is in Table 6.9. I leave the reader to interpret the results; notwithstanding the detailed findings, it is clear that PCA is a powerful data mining method.

TABLE 6.6

Data IN

ID	GENDER	MARITAL
1	M	S
2	M	M
3	M	
4		
5	F	S
6	F	M
7	F	
8		M
9		S
10	M	D

TABLE 6.7

Data In with Necessary Recoding of Variables for Missing Values

ID	GENDER	GENDER_	MARITAL	MARITAL_
1	M	M	S	S
2	M	M	M	M
3	M	M		×
4		×		×
5	F	F	S	S
8	F	F	M	M
7	F	F		×
8		×	M	M
9		×	S	S
10	M	M	D	D

6.8.1 SAS Program for the PCA of the Quasi-Interaction Variable

data IN;

input ID 2.0 GENDER $1. MARITAL $1.;

cards;

01MS

02MM

03M

04

TABLE 6.8

Data IN with Dummy Variables for GENDER_ and MARITAL_Using SAS Proc TRANSREG

ID	GENDER_	MARITAL_	GENDER_F	GENDER_M	MARITAL_D	MARITAL_M	MARITAL_S
1	M	S	0	1	0	0	1
2	M	M	0	1	0	1	0
3	M	×	0	1	0	0	0
4	×	×	0	0	0	0	0
5	F	S	1	0	0	0	1
7	F	×	1	0	0	0	0
8	F	M	1	0	0	1	0
8	×	M	0	0	0	1	0
9	×	S	0	0	0	0	1
10	M	D	1	1	1	0	0

TABLE 6.9

PCA with GENDER_ and MARITAL_Dummy Variables Producing Quasi-GENDER_x_MARITAL Interaction Variables

The PRINCOMP Procedure

Observations	10
Variables	5

Simple Statistics

	GENDER_F	FIGURE	MARITAL_D	MARITAL_M	MARITAL_S
Mean	0.3000000000	0.4000000000	0.1000000000	0.3000000000	0.3000000000
StD	0.4830458915	0.5163977795	0.3162277660	0.4830458915	0.4830458915

Correlation Matrix

	GENDER_F	GENDER_M	MARITAL_D	MARITAL_M	MARITAL_S
Gender_F	1.0000	-.5345	-.2182	0.0476	0.0476
Gender_M	-.5345	1.0000	0.4082	-.0891	.0891
Marital_D	-.2182	0.4082	1.0000	-.2182	.2182
Marital_M	0.0476	-.0891	-.2182	1.0000	-.4286
Marital_S	0.0476	-.0891	-.2182	.4286	1.0000

Eigenvalues of the Correlation Matrix

	Eigenvalue	Difference	Proportion	Cumulative
1	1.84840072	0.41982929	0.3697	0.3697
2	1.42857143	0.53896598	0.2857	0.6554
3	0.88960545	0.43723188	0.1779	0.8333
4	0.45237357	0.07132474	0.0905	0.9238
5	0.38104883		0.0762	1.0000

Eigenvectors

	GENDER_x_ MARITAL_pc1	GENDER_x_ MARITAL_pc2	GENDER_x_ MARITAL_pc3	GENDER_x_ MARITAL_pc4	GENDER_x_ MARITAL_pc5
Gender_F	-.543563	0.000000	0.567380	0.551467	-.280184
Gender_M	0.623943	0.000000	-.209190	0.597326	-.458405
Marital_D	0.518445	0.000000	0.663472	0.097928	0.530499
Marital_M	-.152394	0.707107	-.311547	0.405892	0.463644
Marital_S	-.152394	-.707107	-.311547	0.405892	0.463644

```
05FS
08FM
07F
08 M
09 S
10MD
;
run;
proc print noobs data=IN;
title2 ' Data IN ';
run;
data IN;
set IN;
GENDER_ = GENDER; if GENDER =' ' then GENDER_ ='x';
MARITAL_= MARITAL; if MARITAL=' ' then MARITAL_='x';
run;
proc print noobs; var ID GENDER GENDER_ MARITAL MARITAL_ ;
title2 ' ';
title3 ' Data IN with necessary Recoding of Vars. for Missing Values ';
title4 ' GENDER now Recoded to GENDER_, MARITAL now Recoded
   to MARITAL_';
title5 ' Missing Values are replaced with letter x ';
run;
/* Using proc TRANSREG to create Dummy Variables for GENDER_ */
proc transreg data=IN DESIGN;
model class (GENDER_ / ZERO='x');
output out = GENDER_ (drop = Intercept _NAME_ _TYPE_);
id ID;
run;
/* Appending GENDER_ Dummy Variables */
proc sort data=GENDER_ ;by ID;
proc sort data=IN ;by ID;
run;
data IN;
```

```
merge IN GENDER_ ;
by ID;
run;
/* Using proc TRANSREG to create Dummy Variables for GENDER_ */
proc transreg data=IN DESIGN;
model class (MARITAL_ / ZERO='x');
output out=MARITAL_ (drop= Intercept _NAME_ _TYPE_);
id ID;
run;
/* Appending MARITAL_ Dummy Variables */
proc sort data=MARITAL_;by ID;
proc sort data=IN; by ID; run;
data IN;
merge IN MARITAL_;
by ID;
run;
proc print data=IN (drop= GENDER MARITAL)noobs;
title2' Using proc TRANSREG to create Dummy Vars. for both
    GENDER_ and MARITAL_ ';
run;
/* Running PCA with GENDER_ and MARITAL_ Variables Together */
/* This PCA of a Quasi-GENDER_x_MARITAL Interaction */
proc princomp data= IN n=4 outstat=coef
out=IN_pcs prefix=GENDER_x_MARITAL_pc std;
var
GENDER_F GENDER_M MARITAL_D MARITAL_M MARITAL_S;
title2 ' PCA with both GENDER_ and MARITAL_ Dummy Variables ';
title3 ' This is PCA of a Quasi-GENDER_x_MARITAL Interaction ';
run;
proc print data=IN_pcs noobs;
title2 ' Data appended with the PCs for Quasi-GENDER_x_MARITAL
    Interaction ';
title3 ' ';
run;
```

6.9 Summary

I repositioned the classical data reduction technique of PCA as a reexpression method of EDA (1997). Then, I relabeled PCA as a voguish data mining method of today. I laid PCA in its appropriate place in the EDA reexpression paradigm. I illustrated PCA as a statistical data mining technique capable of serving in common applications with expected solutions; specifically, PCA was illustrated in an exemplary detailed presentation of a set of census EDUCATION variables. And, I illustrated PCA in an uncommon application of finding a structural approach for preparing a categorical predictive variable for possible model inclusion. The results were compelling as they highlighted the power of the PCA data mining tool. In addition, I provided an original and valuable use of PCA in the construction of quasi-interaction variables, along with the SAS program, to perform this novel PCA application.

7

The Correlation Coefficient: Its Values Range between Plus/Minus 1, or Do They?

7.1 Introduction

The correlation coefficient was invented by Karl Pearson in 1896. This century-old statistic is still going strong today, second to the mean statistic in frequency of use. The correlation coefficient's weaknesses and warnings of misuse are well documented. Based on my consulting experience as a statistical modeler, data miner, and instructor of continuing professional studies in statistics for many years, I see too often that the weaknesses and warnings are not heeded. Among the weaknesses, one is rarely mentioned: The correlation coefficient interval [–1, +1] is restricted by the distributions of the two variables being correlated. The purposes of this chapter are (1) to discuss the effects that the distributions of the two variables have on the correlation coefficient interval and (2) thus to provide a procedure for calculating an *adjusted correlation coefficient,* whose realized correlation coefficient interval is often shorter than the original correlation coefficient.

7.2 Basics of the Correlation Coefficient

The correlation coefficient, denoted by r, is a measure of the strength of the straight-line or linear relationship between two variables. The correlation coefficient—by definition—assumes theoretically any value in the interval between +1 and –1, including the end values ±1, namely, the closed interval denoted by [+1, –1].

See chpt 2, and pg 48

The following points are the accepted guidelines for interpreting the correlation coefficient values:

1. 0 indicates no linear relationship.

2. +1 indicates a perfect positive linear relationship: As one variable increases in its values, the other variable also increases in its values via an exact linear rule.

3. −1 indicates a perfect negative linear relationship: As one variable increases in its values, the other variable decreases in its values via an exact linear rule.

4. Values between 0 and 0.3 (0 and -0.3) indicate a weak positive (negative) linear relationship via a shaky linear rule.

5. Values between 0.3 and 0.7 (0.3 and -0.7) indicate a moderate positive (negative) linear relationship via a fuzzy-firm linear rule.

6. Values between 0.7 and 1.0 (-0.7 and -1.0) indicate a strong positive (negative) linear relationship via a firm linear rule.

7. The value of r squared, called the coefficient of determination, and denoted R-squared, is typically interpreted as the percent of variation in one variable explained by the other variable or the percent of variation shared between the two variables. These are good things to know about R-squared:

    a. R-squared is the correlation coefficient between the observed and modeled (predicted) data values.

    b. R-squared can increase as the number of predictor variables in the model increases; R-squared does not decrease. Modelers unwittingly may think a "better" model is being built, as they have a tendency to include more (unnecessary) predictor variables in the model. Accordingly, an adjustment of R-squared was developed, appropriately called adjusted R-squared. The explanation of this statistic is the same as for R-squared, but it penalizes the statistic when unnecessary variables are included in the model.

    c. Specifically, the adjusted R-squared adjusts the R-squared for the sample size and the number of variables in the regression model. Therefore, the adjusted R-squared allows for an "apples-to-apples" comparison between models with different numbers of variables and different sample sizes. Unlike R-squared, adjusted R-squared does not necessarily increase if a predictor variable is added to a model.

    d. R-squared is a first-blush indicator of a good model. R-squared is often *misused* as the measure to assess which model produces better predictions. The RMSE (root mean squared error) is the measure for determining the better model. The smaller the

RMSE value, the better the model is (viz., the more precise are the predictions). It is usually best to report the RMSE rather than mean squared error (MSE) because the RMSE is measured in the *same units as the data*, rather than in squared units, and is representative of the size of a "typical" error. The RMSE is a valid indicator of relative model quality only if it is *well fitted* (e.g., if the model is neither overfitted nor underfitted).

8. Linearity assumption: The correlation coefficient requires that the underlying relationship between the two variables under consideration is linear. If the relationship is known to be linear, or the observed pattern between the two variables appears to be linear, then the correlation coefficient provides a reliable measure of the strength of the linear relationship. If the relationship is known to be nonlinear or the observed pattern appears to be nonlinear, then the correlation coefficient is not useful or at least is questionable.

I see too often that the correlation coefficient is often *misused* because the linearity assumption is not subjected to testing.

7.3 Calculation of the Correlation Coefficient

The calculation of the correlation coefficient for two variables, say X and Y, is simple to understand. Let zX and zY be the standardized versions of X and Y, respectively. That is, zX and zY are both reexpressed to have means equal to zero and standard deviations (std) equal to one. The reexpressions used to obtain the standardized scores and $r_{x,y}$ are in Equations (7.1), (7.2), and (7.3), respectively:

$$zX_i = [X_i - \text{mean }(X)]/\text{std}(X) \tag{7.1}$$

$$zY_i = [Y_i - \text{mean }(Y)]/\text{std }(Y) \tag{7.2}$$

The correlation coefficient is defined as the mean production of the paired standardized scores (zX_i, zY_i) as expressed in Equation (7.3).

$$r_{x,y} = \text{sum of } [zX_i * zY_i]/(n - 1) \tag{7.3}$$

where n is the sample size.

For a simple illustration of the calculation, consider the sample of five observations in Table 7.1. Columns zX and zY contain the standardized scores of X and Y, respectively. The rightmost column is the product of the paired standardized scores. The sum of these scores is 1.83. The mean of these scores (using the adjusted divisor n - 1, not n) is 0.46. Thus, $r_{X,Y} = 0.46$.

TABLE 7.1

Calculation of the Correlation Coefficient

obs	X	Y	zX	zY	zX*zY
1	12	77	-1.14	-0.96	1.11
2	15	98	-0.62	1.07	-0.66
3	17	75	-0.27	-1.16	0.32
4	23	93	0.76	0.58	0.44
5	26	92	1.28	0.48	0.62
mean	18.6	87.0		Sum	1.83
std	5.77	10.32			
n	5			r	0.46

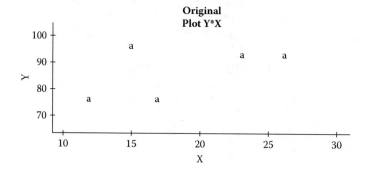

FIGURE 7.1
Original plot of Y with X.

For the sake of completeness, I provide the plot of the original data, Plot Y and X, in Figure 7.1. Unfortunately, the small sample size renders the plot visually unhelpful.

7.4 Rematching

As mentioned, the correlation coefficient theoretically assumes values in the closed interval [+1, –1]. However, it is *not well known* that the correlation coefficient closed interval is *restricted** by the shapes (distributions) of the individual X and Y data. Specifically, the extent to which the shapes of the individual X and Y data are not the same; the length of the realized correlation coefficient closed interval is shorter than the theoretical closed

---

* My first sighting of the term restricted is in Tukey's EDA, 1977. However, I have known about it many years before: I guess from my days in graduate school. I cannot provide any pre-EDA reference.

interval. Clearly, a shorter realized correlation coefficient closed interval necessitates the calculation of the *adjusted correlation coefficient* (discussed in the following).

The length of the realized correlation coefficient closed interval is determined by the process of *rematching*. Rematching takes the original (X, Y) paired data to create new (X, Y) "rematched-paired" data such that the rematched-paired data produce the strongest positive and strongest negative relationships. The correlation coefficients of the strongest positive and strongest negative relationships yield the length of the realized correlation coefficient closed interval. The rematching process is as follows:

1. The strongest positive relationship comes about when the highest X value is paired with the highest Y value, the second highest X is paired with the second-highest Y value, and so on until the lowest X value is paired with the lowest Y value.

2. The strongest negative relationship comes about when the highest, say, X value is paired with the lowest Y value, the second highest X is paired with the second-lowest Y value, and so on until the highest X value is paired with the lowest Y value.

Continuing with the data in Table 7.1, I rematch the X-Y data in Table 7.2. The rematching produces

$$r_{X,Y}(\text{negative rematch}) = -0.99$$

and

$$r_{X,Y}(\text{positive rematch}) = +0.90.$$

For sake of completeness, I provide the "rematched" plots. Unfortunately, the small sample size renders the plots visually unhelpful. The plots of the

TABLE 7.2

Rematched (X, Y) Data from Table 7.1

	Original (X, Y)		Positive Rematch		Negative Rematch	
Obs	X	Y	X	Y	X	Y
1	12	77	26	98	26	75
2	15	98	23	93	23	77
3	17	75	17	92	17	92
4	23	93	15	77	15	93
5	26	92	12	75	12	98
r		0.46		+0.90		-0.99

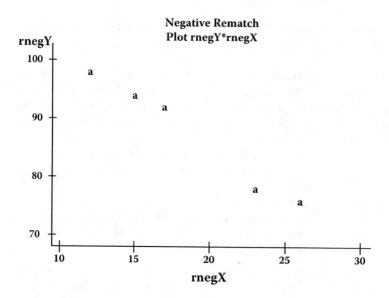

FIGURE 7.2
Negative rematch plot of Y with X.

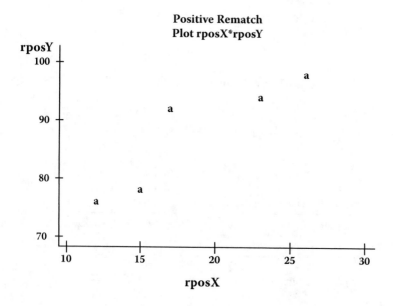

FIGURE 7.3
Positive rematch plot of Y with X.

negative and positive rematched data, Plot rnegY and rnegX and Plot rposY and rposX are in Figures 7.2 and 7.3, respectively.

So, just as there is an adjustment for R-squared, there is an adjustment for the correlation coefficient due to the individual shapes of the X and Y data. Thus, the restricted, realized correlation coefficient closed interval is [-0.99, +0.90], and the adjusted correlation coefficient can now be calculated.

[neg. rematch , pos. rematch]

7.5 Calculation of the Adjusted Correlation Coefficient

The adjusted correlation coefficient is obtained by dividing the original correlation coefficient by the rematched correlation coefficient, and the sign of the adjusted correlation coefficient is the sign of the original correlation coefficient. If the sign of the original r is *negative*, then the sign of the adjusted r is *negative*, even though the arithmetic of dividing two negative numbers yields a positive number. Equation (7.4) provides only the numerical value of the adjusted correlation coefficient. In this example, the adjusted correlation coefficient between X and Y is defined in Equation (7.4): The original correlation coefficient with a positive sign is divided by the positive-rematched original correlation.

$$r_{x,y}(\text{adjusted}) = r_{x,y}(\text{original})/r_{x,y}(\text{positive rematch}) \qquad (7.4)$$

Thus, $r_{x,y}(\text{adjusted}) = 0.51$ (= 0.45/0.90), a 10.9% increase over the original correlation coefficient.

7.6 Implication of Rematching

The correlation coefficient is restricted by the observed shapes of the X and Y data. The shape of the data has the following effects:

1. Regardless of the shape of either variable, symmetric or otherwise, if one variable's shape is different from the other variable's shape, the correlation coefficient is restricted.
2. The restriction is indicated by the rematch.
3. It is not possible to obtain perfect correlation unless the variables have the same shape, symmetric or otherwise.
4. A condition that is necessary for a perfect correlation is that the shapes must be the same, but it does not guarantee a perfect correlation.

7.7 Summary

The everyday correlation coefficient is still going strong after its introduction over 100 years ago. The statistic is well studied, and its weakness and warnings of misuse, unfortunately, at least from my observations, have not been heeded. Among the weaknesses, one is rarely mentioned: the restriction on the values that the correlation coefficient assumes: Namely, the correlation coefficient interval [-1, +1] is restricted by the distributions of the two variables being correlated. I discussed with a simple, yet compelling, illustration the effects that the distributions of the two variables have on the correlation coefficient interval. And, I provided and illustrated a procedure for calculating an adjusted correlation coefficient, whose realized correlation coefficient interval is often shorter than the original correlation coefficient.

8

Logistic Regression: The Workhorse of Response Modeling

8.1 Introduction

Logistic regression is a popular technique for classifying individuals into two mutually exclusive and exhaustive categories, for example, buyer-nonbuyer and responder-nonresponder. Logistic regression is the workhorse of *response* modeling as its results are considered the gold standard. Accordingly, it is used as the benchmark for assessing the superiority of new techniques, such as the machine-learning GenIQ Model. In addition, it is used to determine the advantage of popular techniques, such as the regression-tree CHAID (chi-squared automatic interaction detection) model. In a database marketing application, response to a prior solicitation is the binary dependent variable (defined by responder and nonresponder), and a logistic regression model (LRM) is built to classify an individual as either most likely or least likely to respond to a future solicitation.

To explain logistic regression, I first provide a brief overview of the technique and include a SAS©* program for building and scoring[†] an LRM. The program is a welcome addition to the tool kit of techniques used by model builders working on the two-group classification problem. Next, I provide a case study to demonstrate the building of a response model for an investment product solicitation. The case study presentation illustrates a host of statistical data mining techniques that include the following:

Logit plotting

Reexpressing variables with the ladder of powers and the bulging rule

Measuring the straightness of data

---

* The widely used SAS system (SAS Institute Inc., Cary, NC) is the choice of professional model builders.
† My program is for SAS 8 and SAS 6. My program has to be modified (by deleting the procedure SCORE) if the model builder is using SAS 9.2 or 9.3 (under Windows 7), as the newer versions allow for both building and scoring within the procedure LOGISTIC itself.

Assessing the importance of individual predictor variables

Assessing the importance of a subset of predictor variables

Comparing the importance between two subsets of predictor variables

Assessing the relative importance of individual predictor variables

Selecting the best subset of predictor variables

Assessing goodness of model predictions

Smoothing a categorical variable for model inclusion

The data mining techniques are basic skills that model builders, who are effectively acting like data miners, need to acquire; they are easy to understand, execute, and interpret and should be mastered by anyone who wants control of the data and his or her findings. At this point, in keeping within the context of this chapter, which emphasizes data mining, I more often use the term *data miner* than model builder. However, I do acknowledge that an astute model builder is a well-informed data miner.

Used for prediction (odds), not post-analysis. e.g. is this patient likely to develop cancer? yes/no

8.2 Logistic Regression Model

Should be called Binary! It is called logistic because formula 8.1 is derived from natural logarithms.

Let Y be a binary dependent variable that assumes two outcomes or classes (typically labeled 0 and 1). The LRM classifies an individual into one of the classes based on the values of predictor (independent) variables $X_1, X_2, \ldots, X_n$ for that individual. *(Based on what we know, does this belong in class A or B?)*

dependent variable Y is dichotomous
LRM estimates the logit of Y—a log of the odds of an individual belonging to class 1; the logit is defined in Equation (8.1). The logit, which takes on values between -7 and +7, is a virtually abstract measure for all but the experienced model builder. (The logit theoretically assumes values between plus and minus infinity. However, in practice, it rarely goes outside the range of plus and minus 7.) Fortunately, the logit can easily be converted into the probability of an individual belonging to class 1, Prob(Y = 1), which is defined in Equation (8.2).

where O_x = probability of x

$$\ln \left(\frac{O_x}{1-O_x} \right) = \text{Logit } Y = b_0 + b_1 {*} X_1 + b_2 {*} X_2 + \ldots + b_n {*} X_n \qquad \text{see pg 152} \qquad (8.1)$$

$$\text{Prop}(Y=1) = \exp(\text{Logit } Y / 1 + \exp(\text{Logit } y)) \qquad (8.2)$$

An individual's estimated (predicted) probability of belonging to class 1 is calculated by "plugging in" the values of the predictor variables for that individual in Equations (8.1) and (8.2). The b's are the logistic regression coefficients, which are determined by the calculus-based method of maximum likelihood. Note that, unlike the other coefficients, b_0 (referred to as the intercept) has no predictor variable with which it is multiplied.

As presented, the LRM is readily seen as the workhorse of response modeling, as the Yes-No response variable is an exemplary binary class variable.

$b_0 - b_n$ are the regression coefficients
$X_1 - X_n$ are the covariants (e.g. age, gender, weight, etc.)

The illustration in the next section shows the rudiments of logistic regression response modeling.

8.2.1 Illustration

Consider dataset A, which consists of 10 individuals and three variables in Table 8.1: the binary-class variable RESPONSE (Y), INCOME in thousands of dollars (X_1), and Age in years (X_2). I perform a logistic analysis regressing response on Income and Age using dataset A.

The standard LRM output in Table 8.2 includes the logistic regression coefficients and other "columns" of information (a discussion of these is beyond the scope of this chapter). The "Parameter Estimate" column contains the coefficients for Income and variables, Age and the intercept. The intercept variable is a mathematical device; it is defined implicitly as X_0, which is always equal to one (i.e., intercept = X_0 = 1). The coefficient b_0 is used as a "start" value given to all individuals regardless of their specific values of predictor variables in the model.

The estimated LRM is defined by Equation (8.3):

$$\text{Logit of Response} = -0.9367 + 0.0179*\text{INCOME} - 0.0042*\text{AGE} \qquad (8.3)$$

TABLE 8.1

Dataset A

Response (1 = yes, 0 = no)	Income ($000)	Age (years)
1	96	22
1	86	33
1	64	55
1	60	47
1	26	27
0	98	48
0	62	23
0	54	48
0	38	24
0	26	42

TABLE 8.2

LRM Output

Variable	df	Parameter Estimate	Standard Error	Wald Chi-Square	Pr > Chi-Square
INTERCEPT	1	-0.9367	2.5737	0.1325	0.7159
INCOME	1	0.0179	0.0265	0.4570	0.4990
AGE	1	-0.0042	0.0547	0.0059	0.9389

Do not forget that the LRM predicts the logit of Response, not the probability of Response.

8.2.2 Scoring an LRM

The SAS program in Figure 8.1 produces the LRM built with dataset A and scores an external dataset B in Table 8.3. The SAS procedure LOGISTIC produces logistic regression coefficients and puts them in the "coeff" file, as indicated by the code "outest = coeff." The coeff files produced by SAS versions 6 and 8 (SAS 6, SAS 8) are in Tables 8.4 and 8.5, respectively. (The latest versions of SAS are 9.2 and 9.3. SAS currently supports versions 8 and 9 but not version 6. Even though SAS 6 is not supported, there are still die-hard SAS 6 users, as an SAS technical support person informed me.) The procedure

```
/****** Building the LRM on dataset A ***********/
PROC LOGISTIC data = A nosimple des outest = coeff;
model Response =
Income Age;
run;
/****** Scoring the LRM on dataset B ***********/
PROC SCORE data = B predict type = parms score = coeff
out = B_scored;
var Income Age;
run;
/******* Converting Logits into Probabilities ********/
                        SAS version 6
data B_scored;
set B_scored;
Prob_Resp = exp(Estimate)/(1 + exp(Estimate));
run;
                        SAS version 8
data B_scored;
set B_scored;
Prob_Resp = exp(Response)/(1 + exp(Response));
run;
```

FIGURE 8.1
SAS Code for Building and Score LRM.

TABLE 8.3

Dataset B

Income ($000)	Age (years)
148	37
141	43
97	70
90	62
49	42

TABLE 8.4

Coeff File (SAS 6)

OBS	_LINK_	_TYPE_	_NAME_	Intercept	Income	Age	_LNLIKE_
1	LOGIT	PARMS	Estimate	-0.93671	0.017915	-0.0041991	-6.69218

TABLE 8.5

Coeff File (SAS 8)

OBS	_LINK_	_TYPE_	_STATUS_	_NAME_	Intercept	Income	Age	_LNLIKE_
1	LOGIT	PARMS	0 Converged	Response	-0.93671	0.017915	-0.0041991	-6.69218

LOGISTIC, as presented here, holds true for SAS 9. Regardless, I present the SAS 6 program as it is instructive in that it highlights the difference between the uncomfortable-for-most logit and the always-desired probability.) The coeff files differ in two ways:

1. An additional column _STATUS_ in the SAS 8 coeff file, which does not affect the scoring of the model, is this column.
2. The naming of the predicted logit is "Response" in SAS 8, which is indicated by _NAME_ = Response.

Although it is unexpected, the naming of the predicted logit in SAS 8 is the class variable used in the PROC LOGISTIC statement, as indicated by the code "model Response =." In this illustration, the predicted logit is called "Response," which is indicated by _NAME_ = Response, in Table 8.4. The SAS 8 naming convention is unfortunate as it may cause the model builder to think that the Response variable is a binary class variable and not a logit.

The SAS procedure SCORE scores the five individuals in dataset B using the LRM coefficients, as indicated by the code "score = coeff." This effectively appends the predicted logit variable (called Estimate when using SAS 6 and Response when using SAS 8), to the output file B_scored, as indicated by the code "out = B_scored," in Table 8.6. The probability of response (Prob_Resp) is easily obtained with the code at the end of the SAS program in Figure 8.1

8.3 Case Study

By examining the following case study about building a response model for a solicitation for investment products, I illustrate a host of *data mining techniques*. To make the discussion of the techniques manageable, I use small data (a handful of variables, some of which take on few values, and a sample

TABLE 8.6

Dataset B_scored

Income ($000)	Age (years)	Predicted Logit of Response: Estimate (SAS6), Response (SAS8)	Predicted Probability of Response: Prob_Resp
148	37	1.55930	0.82625
141	43	1.40870	0.80356
97	70	0.50708	0.62412
90	62	0.41527	0.60235
49	42	-0.23525	0.44146

of size "petite grande") drawn from the original direct mail solicitation database. Reported results replicate the original findings obtained with slightly bigger data.

I allude to the issue of data size here in anticipation of data miners who subscribe to the idea that big data are better for analysis and modeling. Currently, there is a trend, especially in related statistical communities such as computer science, knowledge discovery, and Web mining, to use extra big data based on the notion that bigger is better. A statistical factoid states if the true model can be built with small data, then the model built with extra big data produces large prediction error variance. Model builders are never aware of the true model, but when building a model they are guided by the principle of simplicity. Therefore, it is wisest to build the model with small data. If the predictions are good, then the model is a good approximation of the true model; if predictions are not acceptable, then EDA (exploratory data analysis) procedures prescribe an increase in data size (by adding predictor variables and individuals) until the model produces good predictions. The data size with which the model produces good predictions is big enough. If extra big data are used, unnecessary variables tend to creep into the model, thereby increasing the prediction error variance.

8.3.1 Candidate Predictor and Dependent Variables

Let TXN_ADD be the Yes-No response dependent variable, which records the activity of existing customers who received a mailing intended to motivate them to purchase additional investment products. Yes-No response, which is coded 1-0, respectively, corresponds to customers who have/have not added at least one new product fund to their investment portfolio. The TXN_ADD response rate is 11.9%, which is typically large for a direct mail campaign, and is usual for solicitations intended to stimulate purchases among existing customers.

The five candidate predictor variables for predicting TXN_ADD whose values reflect measurement prior to the mailing are as follows:

1. FD1_OPEN reflects the number of different types of accounts the customer has.
2. FD2_OPEN reflects the number of total accounts the customer has.
3. INVESTMENT reflects the customer's investment dollars in ordinal values: 1 = $25 to $499, 2 = $500 to $999, 3 = $1,000 to $2,999, 4 = $3,000 to $4,999, 5 = $5,000 to $9,999, and 6 = $10,000+.
4. MOS_OPEN reflects the number of months the account is opened in ordinal values: 1 = 0 to 6 months, 2 = 7 to 12 months, 3 = 13 to 18 months, 4 = 19 to 24 months, 5 = 25 to 36 months, and 6 = 37 months+.
5. FD_TYPE is the product type of the customer's most recent investment purchase: A, B, C, ... , N.

8.4 Logits and Logit Plots

The LRM belongs to the family of linear models that advance the implied assumption that the underlying relationship between a given predictor variable and the logit is linear or straight line. Bear in mind that, to model builders, the adjective *linear* refers to the explicit fact that the logit is expressed as the sum of weighted predictor variables, where the weights are the regression coefficients. In practice, however, the term refers to the implied assumption. To check this assumption, the *logit plot* is needed. A logit plot is the plot of the binary-dependent variable (hereafter, response variable) against the values of the predictor variable. Three steps are required to generate the logit plot:

1. Calculate the mean of the response variable corresponding to each value of the predictor variable. If the predictor variable takes on more than 10 distinct values, then use *typical* values, such as smooth decile values, as defined in Chapter 2.
2. Calculate the logit of response using the formula that converts the mean of response to logit of response: Logit = ln(mean/(1 - mean)), where ln is the natural logarithm.
3. Plot the logit-of-response values against the original distinct or the smooth decile values of the predictor variable.

One point worth noting: The logit plot is an aggregate-level, not individual-level, plot. The logit is an aggregate measure based on the mean of individual response values. Moreover, if smooth decile values are used, the plot is further aggregated as each decile value represents 10% of the sample.

8.4.1 Logits for Case Study

For the case study, the response variable is TXN_ADD, and the logit of TXN_ADD is named LGT_TXN. For no particular reason, I start with candidate predictor variable FD1_OPEN, which takes on the distinct values 1, 2, and 3 in Table 8.7. Following the three-step construction for each FD1_OPEN value, I generate the LGT_TXN logit plot in Figure 8.2. I calculate the mean of TXN_ADD and use the mean-to-logit conversion formula. For example, for FD1_OPEN = 1, the mean of TXN_ADD is 0.07, and the logit LGT_TXN is -2.4 (= ln(0.07/(1 - 0.07)). Last, I plot the LGT_TXN logit values against the FD1_OPEN values.

The LGT_TXN logit plot for FD1_OPEN does not suggest an underlying straight-line relationship between LGT_TXN and FD1_OPEN. To use the LRM correctly, I need to straighten the relationship. A very effective and simple technique for straightening data is reexpressing the variables, which uses Tukey's ladder of powers and the bulging rule. Before presenting the details of the technique, it is worth discussing the importance of straight-line relationships or straight data.

TABLE 8.7

FD1_OPEN

FD1_OPEN	Mean TXN_ADD	LGT_TXN
1	0.07	-2.4
2	0.18	-1.5
3	0.20	-1.4

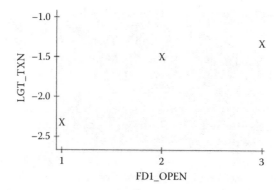

FIGURE 8.2
Logit plot for FD1_OPEN.

8.5 The Importance of Straight Data

EDA places special importance on straight data, not in the least for the sake of simplicity itself. The paradigm of life is simplicity (at least for those of us who are older and wiser). In the physical world, Einstein uncovered one of life's ruling principles using only three letters: $E = mc^2$. In the visual world, however, simplicity is undervalued and overlooked. A smiley face is an unsophisticated, simple shape that nevertheless communicates effectively, clearly, and efficiently. Why should the data miner accept anything less than simplicity in his or her life's work? Numbers, as well, should communicate clearly, effectively, and immediately. In the data miner's world, there are two features that reflect simplicity: symmetry and straightness in the data. The data miner should insist that the numbers be symmetric and straight.

The straight-line relationship between two continuous variables, say X and Y, is as simple as it gets. As X increases or decreases in its values, so does Y increase or decrease in its values, in which case it is said that X and Y are positively or negatively correlated, respectively. Or, as X increases (decreases) in its values, so does Y decrease (increase) in its values, in which case it is said that X and Y are negatively correlated. As further demonstration of its simplicity, Einstein's E and m have a perfect positively correlated straight-line relationship.

The second reason for the importance of straight data is that most response models require it, as they belong to the class of innumerable varieties of the linear model. Moreover, it has been shown that nonlinear models, which pride themselves on making better predictions with nonstraight data, in fact do better with straight data.

I have not ignored the feature of symmetry. Not accidentally, as there are theoretical reasons, symmetry and straightness go hand in hand. Straightening data often makes data symmetric and vice versa. You may recall that icon symmetric data have the profile of the bell-shaped curve. However, symmetric data are defined such that data values are distributed in the same shape (identical on both sides) above and below the middle value of the entire data distribution.

8.6 Reexpressing for Straight Data

The ladder of powers is a method of reexpressing variables to straighten a bulging relationship between two continuous variables, say X and Y. Bulges in the data can be depicted as one of four shapes, as displayed in Figure 8.3. When the X-Y relationship has a bulge similar to any one of the four shapes,

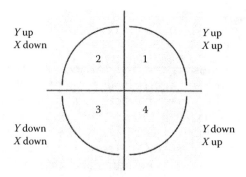

FIGURE 8.3
The bulging rule.

both the ladder of powers and the bulging rule, which guides the choice of "rung" in the ladder, are used to straighten out the bulge. Most data have bulges. However, when kinks or elbows characterize the data, then another approach is required, which is discussed further in the chapter.

8.6.1 Ladder of Powers

Going up-ladder of powers means reexpressing a variable by raising it to a power p greater than 1. (Remember that a variable raised to the power of 1 is still that variable; $X^1 = X$, and $Y^1 = Y$). The most common p values used are 2 and 3. Sometimes values higher up-ladder and in-between values like 1.33 are used. Accordingly, starting at $p = 1$, the data miner goes up-ladder, resulting in reexpressed variables, for X and Y, as follows:

$$\text{Starting at } X^1: X^2, X^3, X^4, X^5, \ldots$$

$$\text{Starting at } Y^1: Y^2, Y^3, Y^4, Y^5, \ldots$$

Some variables reexpressed going up-ladder have special names. Corresponding to power values 2 and 3, they are called X squared and X cubed, respectively. Similarly, for the Y variables, they are called Y squared and Y cubed, respectively.

Going down-ladder of powers means reexpressing a variable by raising it to a power p that is less than 1. The most common p-values are ½, 0, -½, and -1. Sometimes, values lower down-ladder and in-between values like 0.33 are used. Also, for negative powers, the reexpressed variable now sports a negative sign (i.e., is multiplied by -1); the reason for this is theoretical and beyond the scope of this chapter. Accordingly, starting at $p = 1$, the data miner goes down-ladder, resulting in reexpressed variables for X and Y, as follows:

$$\text{Starting at } X^1: X^{1/2}, X^0, X^{-1/2}, -X^1,$$

$$\text{Starting at } Y^1: Y^{1/2}, Y^0, Y_{-1/2}, -Y^1,$$

Some reexpressed variables going down-ladder have special names. Corresponding to values ½, -½, and -1, they are called the square root of X, negative reciprocal square root of X, and negative reciprocal of X, respectively. Similarly, for the Y variables, they are called square root of Y, negative reciprocal square root of Y, and negative reciprocal of Y, respectively. The reexpression for p = 0 is not mathematically defined and is conveniently defined as log to base 10. Thus, $X^0 = \log X$, and $Y^0 = \log Y$.

8.6.2 Bulging Rule

The bulging rule states the following:

1. If the data have a shape similar to that in the first quadrant, then the data miner tries reexpressing by going up-ladder for X, Y, or both.
2. If the data have a shape similar to that shown in the second quadrant, then the data miner tries reexpressing by going down-ladder for X or up-ladder for Y.
3. If the data have a shape similar to that in the third quadrant, then the data miner tries reexpressing by going down-ladder for X, Y, or both.
4. If the data have a shape similar to that in the fourth quadrant, then the data miner tries reexpressing by going up-ladder for X or down-ladder for Y.

Reexpressing is an important, yet fallible, part of EDA detective work. While it will typically result in straightening the data, it might result in a deterioration of information. Here is why: Reexpression (going down too far) has the potential to squeeze the data so much that its values become indistinguishable, resulting in a loss of information. Expansion (going up too far) can potentially pull apart the data so much that the new far-apart values lie within an artificial range, resulting in a spurious gain of information.

Thus, reexpressing requires a careful balance between straightness and soundness. Data miners can always go to the extremes of the ladder by exerting their will to obtain a little more straightness, but they must be mindful of a consequential loss of information. Sometimes, it is evident when one has gone too far up/down on the ladder; there is power p, after which the relationship either does not improve noticeably or inexplicably bulges in the opposite direction due to a corruption of information. I recommend using discretion to avoid overstraightening and its potential deterioration of information. In addition, I caution that extreme reexpressions are sometimes due to the extreme values of the original variables. Thus, always check the maximum and minimum values of the original variables to make sure they are reasonable before reexpressing the variables.

8.6.3 Measuring Straight Data

The correlation coefficient measures the strength of the straight-line or linear relationship between two variables, X and Y, discussed in detail in Chapter 2. However, there is an additional assumption to consider.

In Chapter 2, I referred to a "linear assumption," in that the underlying relationship between X and Y is linear. The second assumption is an implicit one: The (X, Y) data points are at the individual level. When the (X, Y) points are analyzed at an aggregate level, such as in the logit plot and other plots presented in this chapter, the correlation coefficient based on "big" points tends to produce a "big" r value, which serves as a gross estimate of the individual-level r value. The aggregation of data diminishes the idiosyncrasies of the individual (X, Y) points, thereby increasing the resolution of the relationship, for which the r value also increases. Thus, the correlation coefficient on aggregated data serves as a gross indicator of the strength of the original X-Y relationship at hand. There is a drawback of aggregation: It often produces r values without noticeable differences because the power of the distinguishing individual-level information is lost.

8.7 Straight Data for Case Study

Returning to the LGT_TXN logit plot for FD1_OPEN, whose bulging relationship is in need of straightening, I identify its bulge as the type in quadrant 2 in Figure 8.3. According to the bulging rule, I should try going up-ladder for LGT_TXN or down-ladder for FD1_OPEN. LGT_TXN cannot be reexpressed because it is the explicit dependent variable as defined by the logistic regression framework. Reexpressing it would produce grossly illogical results. Thus, I do not go up-ladder for LGT_TXN.

To go down-ladder for FD1_OPEN, I use the powers ½, 0, -½, -1, and -2. This results in the square root of FD1_OPEN, labeled FD1_SQRT; the log to base 10 of FD1_OPEN, labeled FD1_LOG; the negative reciprocal root of FD1_OPEN, labeled FD1_RPRT; the negative reciprocal of FD1_OPEN, labeled FD1_RCP; and the negative reciprocal square of FD1_OPEN, labeled FD1_RSQ. The corresponding LGT_TXN logit plots for these reexpressed variables and the original FD1_OPEN (repeated here for convenience) are in Figure 8.4.

Visually, it appears that reexpressed variables FD1_RSQ, FD1_RCP, and FD1_RPRT do an equal job of straightening the data. I could choose any of them but decide to do a little more detective work by looking at the numerical

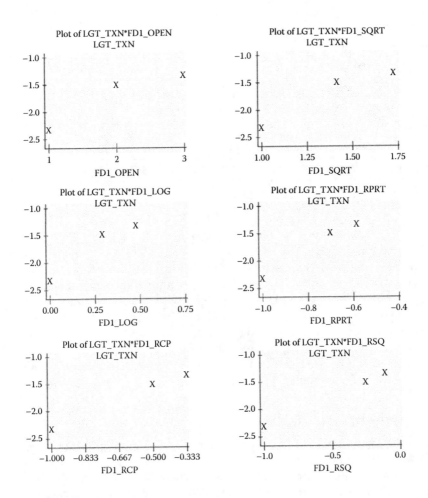

FIGURE 8.4
Logit plots for FD1_OPEN and its reexpressed variables.

indicator—the correlation coefficient between LGT_TXN and the reexpressed variable—to support my choice of the best reexpressed variable. The larger the correlation coefficient, the more effective the reexpressed variable is in straightening the data. Thus, the reexpressed variable with the largest correlation coefficient is declared the best reexpressed variable, with exceptions guided by the data miner's own experience with these visual and numerical indictors in the context of the problem domain.

The correlation coefficients for LGT_TXN with FD1_OPEN and with each reexpressed variable are ranked in descending order in Table 8.8. The correlation coefficients of the reexpressed variables represent noticeable improvements in straightening the data over the correlation coefficient for the original variable FD1_OPEN (r = 0.907). FD1_RSQ has the largest correlation

TABLE 8.8

Correlation Coefficients between LGT_TXN and Reexpressed FD1_OPEN

FD1_RSQ	FD1_RCP	FD1_RPRT	FD1_LOG	FD1_SQRT	FD1_OPEN
0.998	0.988	0.979	0.960	0.937	0.907

TABLE 8.9

Correlation Coefficients between LGT_TXN and Reexpressed FD2_OPEN

FD2_RSQ	FD2_RCP	FD2_RPRT	FD2_LOG	FD2_SQRT	FD2_OPEN
0.995	0.982	0.968	0.949	0.923	0.891

coefficient (r = 0.998), but it is slightly greater than that for FD1_RCP (r = 0.988) and therefore not worthy of notice.

My choice of the best reexpressed variable is FD1_RCP, which represents an 8.9% (= (0.988 - 0.907)/0.907) improvement in straightening the data over the original relationship with FD1_OPEN. I prefer FD1_RCP over FD1_RSQ and other extreme reexpressions down-ladder (defined by power p less than -2) because I do not want to unwittingly select a reexpression that might be too far down-ladder, resulting in loss of information. Thus, I go back one rung to power -1, hoping to get the right balance between straightness and minimal loss of information.

8.7.1 Reexpressing FD2_OPEN

The scenario for FD2_OPEN is virtually identical to the one presented for FD1_OPEN. This is not surprising as FD1_OPEN and FD2_OPEN share a large amount of information. The correlation coefficient between the two variables is 0.97, meaning the two variables share 94.1% of their variation. Thus, I prefer FD2_RCP as the best reexpressed variable for FD2_OPEN (see Table 8.9).

8.7.2 Reexpressing INVESTMENT

The relationship between the LGT_TXN and investment, depicted in the plot in Figure 8.5, is somewhat straight with a negative slope and a slight bulge in the middle for investment values 3, 4 and 5. I identify the bulge of the type in quadrant 3 in Figure 8.3. Thus, going down-ladder for powers ½, 0, -½, -1, and -2 results in the square root of investment, labeled INVEST_SQRT; the log to base 10 of investment, labeled INVEST_LOG; the negative reciprocal root of investment, labeled INVEST_RPRT; the negative reciprocal of investment, labeled INVEST_RCP; and the negative reciprocal square of INVESTMENT, labeled INVEST_RSQ. The corresponding LGT_TXN logit plots for these reexpressed variables and the original INVESTMENT are in Figure 8.5.

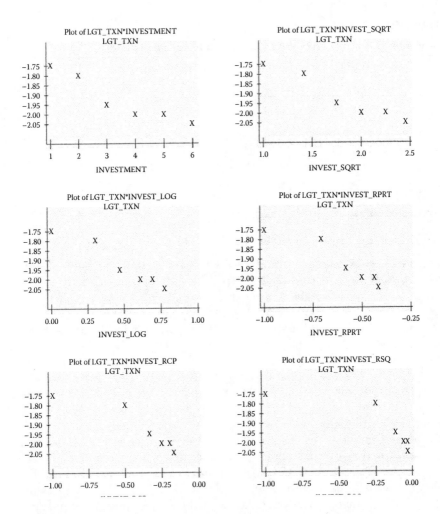

FIGURE 8.5
Logit plots for INVESTMENT and its reexpressed variables.

Visually, I like the straight-line produced by INVEST_SQRT. Quantitatively, INVEST_LOG has the largest correlation coefficient, which supports the statistical factoid that claims a variable in dollar units should be reexpressed with the log function. The correlation coefficients for INVEST_LOG and INVEST_SQRT in Table 8.10 are -0.978 and -0.966, respectively; admittedly, the correlation coefficients do not reflect a noticeable difference. My choice of the best reexpression for investment is INVEST_LOG because I prefer the statistical factoid to my visual choice. Only if a noticeable difference between correlation coefficients for INVEST_LOG and INVEST_SQRT existed would I sway from being guided by the factoid. INVEST_LOG represents an improvement of 3.4% (= (0.978 - 0.946)/0.946; disregarding the negative sign)

TABLE 8.10

Correlation Coefficients between LGT_TXN and Reexpressed INVESTMENT

INVEST_log	INVEST_sqrt	INVEST_rprt	INVESTMENT	INVEST_rcp	INVEST_rsq
−0.978	−0.966	−0.950	0.946	−0.917	−0.840

in straightening the data over the relationship with the original variable INVESTMENT ($r = -0.946$).

8.8 Technique ts When Bulging Rule Does Not Apply

I describe two plotting techniques for uncovering the correct reexpression when the bulging rule does not apply. After discussing the techniques, I return to the next variable for reexpression, MOS_OPEN. The relationship between LGT_TXN and MOS_OPEN is interesting and offers an excellent opportunity to illustrate the data mining flexibility of the EDA methodology.

It behooves the data miner to perform due diligence either to explain qualitatively or to account quantitatively for the relationship in a logit plot. Typically, the latter is easier than the former as the data miner is at best a scientist of data, not a psychologist of data. The data miner seeks to investigate the plotted relationship to uncover the correct representation or structure of the given predictor variable. Briefly, structure is an organization of variables and functions. In this context, variables are broadly defined as both raw variables (e.g., X_1, X_2, ¼ , X_i, ...) and numerical constants, which can be thought as variables assuming any single value k, that is, $X_i = k$. Functions include the arithmetic operators (addition, subtraction, multiplication, and division); comparison operators (e.g., equal to, not equal, greater than); and logical operators (e.g., and, or, not, if ¼ then). For example, $X_1 + X_2/X_1$ is a structure.

By definition, any raw variable X_i is considered a structure as it can be defined by $X_i = X_i + 0$, or $X_i = X_i*1$. A dummy variable (X_dum)—a variable that assumes two numerical values, typically 1 and 0, which indicate the presence and absence of a condition, respectively—is structure. For example, X_dum = 1 if X equals 6; X_dum = 0 if X does not equal 6. The condition is "equals 6."

8.8.1 Fitted Logit Plot

The *fitted logit* plot is a valuable visual aid in uncovering and confirming structure. The fitted logit plot is defined as a plot of the *predicted* logit against a given structure. The steps required to construct the plot and its interpretation are as follows:

1. *Perform* a logistic regression analysis on the response variable with the given structure, obtaining the predicted logit of response, as outlined in Section 8.2.2.
2. *Identify* the values of the structure to use in the plot. Identify the distinct values of the given structure. If the structure has more than 10 values, identify its smooth decile values.
3. *Plot* the predicted (fitted) logit values against the identified values of the structure. Label the points by the identified values.
4. *Infer* that if the fitted logit plot reflects the shape in the original logit plot, the structure is the correct one. This further implies that the structure has some importance in predicting response. The extent to which the fitted logit plot is different from the original logit plot indicates the structure is a poor predictor of response.

8.8.2 Smooth Predicted-versus-Actual Plot

Another valuable plot for exposing the detail of the strength or weakness of a structure is the *smooth predicted-versus-actual* plot, which is defined as the plot of *mean* predicted response against *mean* actual response for values of a reference variable. The steps required to construct the plot and its interpretation are as follows:

1. Calculate the mean predicted response by averaging the individual predicted probabilities of response from the appropriate LRM for each value of the reference variable. Similarly, calculate the mean actual response by averaging the individual actual responses for each value of the reference variable.
2. The paired points (mean predicted response, mean actual response) are called *smooth points.*
3. Plot the smooth points and label them by the values of the reference variable. If the reference variable assumes more than 10 distinct values, then use smooth decile values.
4. Insert the 45° line in the plot. The line serves as a reference for visual assessment of the importance of a structure for predicting response, thereby confirming that the structure under consideration is the correct one. Smooth points on the line imply that the mean predicted response and the mean actual response are equal, and there is great certainty that the structure is the correct one. The tighter the smooth points "hug" the 45° line, the greater the certainty of the structure is. Conversely, the greater the scatter about the line, the lesser the certainty of the structure.

8.9 Reexpressing MOS_OPEN

The relationship between LGT_TXN and MOS_OPEN in Figure 8.6 is not straight in the full range of MOS_OPEN values from one to six but is straight between values one and five. The LGT_TXN logit plot for MOS_OPEN shows a check mark shape with vertex at MOS_OPEN = 5 as LGT_TXN jumps at MOS_OPEN = 6. Clearly, the bulging rule does not apply.

Accordingly, in uncovering the MOS_OPEN structure, I am looking for an organization of variables and functions that renders the ideal straight-line relationship between LGT_TXN and the MOS_OPEN structure. It will implicitly account for the jump in logit of in LGT_TXN at MOS_OPEN = 6. Once the correct MOS_OPEN structure is identified, I can include it in the TXN_ADD response model.

Exploring the structure of MOS_OPEN itself, I generate the LGT_TXN fitted logit plot (Figure 8.7), which is based on the logistic regression analysis on TXN_ADD with MOS_OPEN. The LRM, from which the predicted logits are obtained, is defined in Equation (8.4):

$$\text{Logit(TXN\_ADD)} = -1.24 - 0.17 * \text{MOS\_OPEN} \qquad (8.4)$$

It is acknowledged that MOS_OPEN has six distinct values. The fitted logit plot does not reflect the shape of the relationship in the original LGT_TXN logit plot in Figure 8.6. The predicted point at mos_open = 6 is way too low. The implication and confirmation is that MOS_OPEN alone is not the correct structure as it does not produce the shape in the original logit plot.

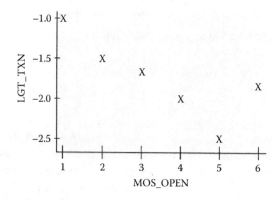

FIGURE 8.6
Logit plot of MOS_OPEN.

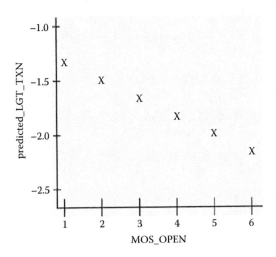

FIGURE 8.7
Fitted logit plot for MOS_OPEN.

8.9.1 Plot of Smooth Predicted versus Actual for MOS_OPEN

I generate the TXN_ADD smooth predicted-versus-actual plot for MOS_OPEN (Figure 8.8), which depicts both the structure under consideration and the reference variable. The smooth predicted values are based on the LRM previously defined in Equation (8.4) and restated here in Equation (8.5) for convenience.

$$\text{logit(TXN\_ADD)} = -1.24 - 0.17*\text{MOS \_OPEN} \qquad (8.5)$$

There are six smooth points, each labeled by the corresponding six values of MOS_OPEN. The scatter about the 45° line is wild, implying MOS_OPEN is not a good predictive structure, especially when MOS_OPEN equals 1, 5, 6, and 4, as their corresponding smooth points are not close to the 45° line. Point MOS_OPEN = 5 is understandable as it can be considered the springboard to jump into LGT_TXN at MOS_OPEN = 6. MOS_OPEN = 1, as the farthest point from the line, strikes me as inexplicable. MOS_OPEN = 4 may be within an acceptable distance from the line.

When MOS_OPEN equals 2 and 3, the prediction appears to be good as the corresponding smooth points are close to the line. Two good predictions of a possible six predictions results in a poor 33% accuracy rate. Thus, MOS_OPEN is not good structure for predicting TXN_ADD. As before, the implication is that MOS_OPEN alone is not the correct structure to reflect the original relationship between LGT_TXN and MOS_OPEN in Figure 8.6. More detective work is needed.

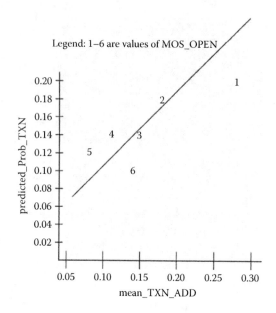

FIGURE 8.8
Plot of smooth predicted versus actual for MOS_OPEN.

It occurs to me that the major problem with MOS_OPEN is the jump point. To account explicitly for the jump, I create a MOS_OPEN dummy variable structure, defined as

$$\text{MOS\_DUM} = 1 \text{ if MOS\_OPEN} = 6;$$

$$\text{MOS\_DUM} = 0 \text{ if MOS\_OPEN not equal to 6.}$$

I generate a second LGT_TXN fitted logit plot in Figure 8.9, this time consisting of the predicted logits from regressing TXN_ADD on the structure consisting of MOS_OPEN and MOS_DUM. The LRM is defined in Equation (8.6):

$$\text{Logit(TXN\_ADD)} = -0.62 - 0.38*\text{MOS\_OPEN} + 1.16*\text{MOS\_DUM} \quad (8.6)$$

This fitted plot accurately reflects the shape of the original relationship between TXN_ADD and MOS_OPEN in Figure 8.6. The implication is that MOS_OPEN and MOS_DUM make up the correct structure of the information carried in MOS_OPEN. The definition of the structure is the right side of the equation itself.

To complete my detective work, I create the second TXN_ADD smooth predicted-versus-actual plot in Figure 8.10, consisting of mean predicted logits of TXN_ADD against mean MOS_OPEN. The predicted logits come

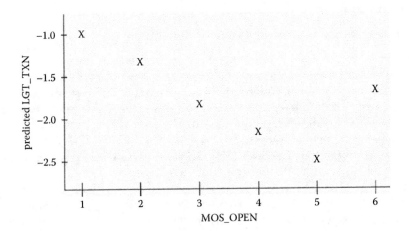

FIGURE 8.9
Fitted logit plot for MOS_OPEN and MOS_DUM.

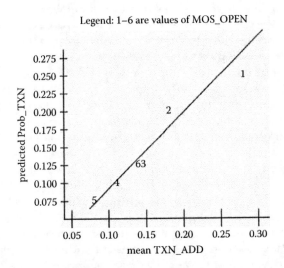

FIGURE 8.10
Plot of smooth predicted versus actual for MOS_OPEN and MOS_DUM.

from the logistic regression Equation (8.6), which includes the predictor variable pair MOS_OPEN and MOS_DUM. MOS_OPEN is used as the reference variable. The smooth points hug the 45° line nicely. The implication is that the MOS_OPEN structure defined by MOS_OPEN and MOS_DUM is again confirmed, and the two-piece structure is an important predictive of TXN_ADD.

8.10 Assessing the Importance of Variables

The classic approach for assessing the statistical significance of a variable considered for model inclusion is the well-known null hypothesis significance testing procedure, which is based on the reduction in prediction error (actual response minus predicted response) associated with the variable in question. The statistical apparatus of the formal testing procedure for logistic regression analysis consists of: The log likelihood (LL) function, the G statistic, degrees of freedom (df), and the p value. The procedure uses the apparatus within a theoretical framework with weighty and untenable assumptions. From a purist point of view, this could cast doubt on findings that actually have statistical significance. Even if findings of statistical significance are accepted as correct, they may not be of practical importance or have *noticeable* value to the study at hand. For the data miner with a pragmatic slant, the limitations and lack of scalability inherent in the classic system cannot be overlooked, especially within big data settings. In contrast, the data mining approach uses the LL units, the G statistic, and degrees of freedom in an informal data-guided search for variables that suggest a noticeable reduction in prediction error. One point worth noting is that the informality of the data mining approach calls for suitable change in terminology, from declaring a result as statistically significant to one worthy of notice or *noticeably important*.

Before I describe the data mining approach of variable assessment, I would like to comment on the objectivity of the classic approach as well as degrees of freedom. The classic approach is so ingrained in the analytic community that no viable alternative occurs to practitioners, especially an alternative based on an informal and sometimes highly individualized series of steps. Declaring a variable statistically significant appears to be purely objective as it is based on sound probability theory and statistical mathematical machinery. However, the settings of the testing machinery defined by model builders could affect the results. The settings include the levels of rejecting a variable as significant when, in fact, it is not, or accepting a variable as not significant when, in fact, it is. Determining the proper sample size is also a subjective setting as it depends on the amount budgeted for the study. Last, the allowable deviation of violations of test assumptions is set by the model builder's experience. Therefore, by acknowledging the subjective nature of the classic approach, the model builder can be receptive to the alternative data mining approach, which is free of theoretical ostentation and mathematical elegance.

A word about degrees of freedom clarifies the discussion. This concept is typically described as a generic measure of the number of independent pieces of information available for analysis. To ensure accurate results, this concept is accompanied by the mathematical adjustment "replace N with N -1." The concept of degrees of freedom gives a deceptive impression of simplicity in counting the pieces of information. However, the principles used in counting are not easy for all but the mathematical statistician. To date, there is no generalized calculus for counting degrees of freedom. Fortunately, the counting already exists for many analytical routines. Therefore, the correct degrees of freedom are readily available; computer output automatically provides them, and there are lookup tables in older statistics textbooks. For the analyses in the following discussions, the counting of degrees of freedom is provided.

8.10.1 Computing the G Statistic

In data mining, the assessment of the importance of a subset of variables for predicting response involves the notion of a noticeable reduction in prediction error due to the subset of variables and is based on the ratio of the G statistic to the degrees of freedom, G/df. The degrees of freedom is defined as the number of variables in the subset. The G statistic is defined, in Equation (8.7), as the difference between two LL quantities, one corresponding to a model *without* the subset of variables and the other corresponding to a model *with* the subset of variables.

$$G = -2LL(\text{model without variables}) - -2\,LL(\text{model with variables}) \quad (8.7)$$

There are two points worth noting: First, the LL units are multiplied by a factor of -2, a mathematical necessity; second, the term *subset* is used to imply there is always a large set of variables available from which the model builder considers the smaller subset, which can include a single variable.

In the following sections, I detail the decision rules in three scenarios for assessing the likelihood that the variables have some predictive power. In brief, the larger the average G value per degrees of freedom (G/df), the more important the variables are in predicting response.

8.10.2 Importance of a Single Variable

If X is the only variable considered for inclusion into the model, the G statistic is defined in Equation (8.8):

$$G = -2LL(\text{model with intercept only}) - -2LL(\text{model with X}) \quad (8.8)$$

The decision rule for declaring X an important variable in predicting response is as follows: If G/df* is greater than the standard G/df value 4, then X is an important predictor variable and should be considered for inclusion in the model. Note that the decision rule only indicates that the variable

---

* Obviously, G/df equals G for a single-predictor variable with df = 1.

has some importance, not how much importance. The decision rule implies that a variable with a greater G/df value has a greater likelihood of some importance than a variable with a smaller G/df value, not that it has greater importance.

8.10.3 Importance of a Subset of Variables

When subset A consisting of k variables is the only subset considered for model inclusion, the G statistic is defined in Equation (8.9):

$$G = -2LL(\text{model with intercept}) - -2LL(\text{model with A(k) variables}) \quad (8.9)$$

The decision rule for declaring subset A important in predicting response is as follows: If G/k is greater than the standard G/df value 4, then subset A is an important subset of the predictor variable and should be considered for inclusion in the model. As before, the decision rule only indicates that the subset has some importance, not how much importance.

8.10.4 Comparing the Importance of Different Subsets of Variables

Let subsets A and B consist of k and p variables, respectively. The number of variables in each subset does not have to be equal. If they are equal, then all but one variable can be the same in both subsets. The G statistics for A and B are defined in Equations (8.10) and (8.11), respectively:

$$G(k) = -2LL(\text{model with intercept}) - -2LL(\text{model with "A" variables}) \quad (8.10)$$

$$G(p) = -2LL(\text{model with intercept}) - -2LL(\text{model with "B" variables}) \quad (8.11)$$

The decision rule for declaring which of the two subsets is more important (i.e., greater likelihood of having some predictive power) in predicting response is as follows:

1. If $G(k)/k$ is greater than $G(p)/p$, then subset A is the more important predictor variable subset; otherwise, B is the more important subset.

2. If $G(k)/k$ and $G(p)/p$ are equal or have comparable values, then both subsets are to be regarded tentatively of comparable importance. The model builder should consider additional indicators to assist in the decision about which subset is better.

It follows clearly from the decision rule that the better model is defined by the more important subset. Of course, this rule assumes that $G(k)/k$ and $G(p)/p$ are greater than the standard G/df value 4.

8.11 Important Variables for Case Study

The first step in variable assessment is to determine the baseline LL value for the data under study. The LRM for TXN_ADD without variables produces two essential bits of information in Table 8.11:

1. The baseline for this case study is -2LL equals 3606.488.
2. The LRM is defined in Equation (8.12):

$$\text{Logit(TNX\_ADD = 1)} = -1.9965 \tag{8.12}$$

There are interesting bits of information in Table 8.11 that illustrate two useful statistical identities:

1. Exponentiation of both sides of Equation (8.12) produces odds of response equal to 0.1358. Recall that exponentiation is the mathematical operation of raising a quantity to a power. The exponentiation of a logit is the odds; consequently, the exponentiation of -1.9965 is 0.1358. See Equations (8.13) to (8.15).

$$\text{Exp(Logit(TNX\_ADD = 1))} = \text{Exp}(-1.9965) \tag{8.13}$$

$$\text{Odds(TNX\_ADD = 1)} = \text{Exp}(-1.9965) \tag{8.14}$$

$$\text{Odds (TNX\_ADD = 1)} = 0.1358 \tag{8.15}$$

TABLE 8.11

The LOGISTIC Procedure for TXN_ADD

	Response Profile	
TXN_ADD	COUNT	
1	589	
0	4,337	
	−2LL = 3,606.488	

Variable	Parameter Estimate	Standard Error	Wald Chi-Square	Pr > Chi-Square
INTERCEPT	−1.9965	0.0439	2,067.050	0.0

$$\text{LOGIT} = 1.9965$$
$$\text{ODDS} = \text{EXP} (.19965) = .1358$$

$$\text{PROB(TXN\_ADD} = 1) = \frac{\text{ODDS}}{1 + \text{ODDS}} = \frac{0.1358}{1 + 0.1358} = 0.119$$

2. Probability of (TNX_ADD = 1), hereafter the probability of RESPONSE, is easily obtained as the ratio of odds divided by 1 + odds. The implication is that the best estimate of RESPONSE—when no information is known or no variables are used—is 11.9%, namely, the average response of the mailing.

8.11.1 Importance of the Predictor Variables

With the LL baseline value 3,606.488, I assess the importance of the five variables: MOS_OPEN and MOS_DUM, FD1_RCP, FD2_RCP, and INVEST_LOG. Starting with MOS_OPEN and MOS_DUM, as they must be together in the model, I perform a logistic regression analysis on TXN_ADD with MOS_OPEN and MOS_DUM; the output is in Table 8.12. From Equation (8.9), the G value is 107.022 (= 3,606.488 - 3,499.466). The degree of freedom is equal to the number of variables; df is 2. Accordingly, G/df equals 53.511, which is greater than the standard G/df value of 4. Thus, MOS_OPEN and MOS_DUM as a pair are declared important predictor variables of TXN_ADD.

From Equation (8.8), the G/df value for each remaining variable in Table 8.12 is greater than 4. Thus, these five variables, each important predictors of TXN_ADD, form a starter subset for predicting TXN_ADD. I have not forgotten about FD_TYPE; it is discussed in another section.

I build a preliminary model by regressing TXN_ADD on the starter subset; the output is in Table 8.13. From Equation (8.9), the five-variable subset has a G/df value of 40.21(= 201.031/5), which is greater than 4. Thus, this is an incipient subset of important variables for predicting TXN_ADD.

8.12 Relative Importance of the Variables

The "mystery" in building a statistical model is that the *true* subset of variables defining the true model is not known. The model builder can be most productive by seeking to find the *best* subset of variables that defines the final model as an "intelliguess" of the true model. The final model reflects more of the model builder's effort given the data at hand than an estimate of the

TABLE 8.12

G and df for Predictor Variables

Variable	-2LL	G	df	p-value
INTERCEPT	3,606.488			
MOS_OPEN + MOS_DUM	3,499.466	107.023	2	0.0001
FD1_RCP	3,511.510	94.978	1	0.0001
FD2_RCP	3,503.993	102.495	1	0.0001
INV_LOG	3,601.881	4.607	1	0.0001

TABLE 8.13

Preliminary Logistic Model for TXN_ADD with Starter Subset

	Intercept Only	Intercept and All Variables	All Variables
-2LL	3,606.488	3,405.457	201.031 with 5 df (p = 0.0001)

Variable	Parameter Estimate	Standard Error	Wald Chi-Square	Pr > Chi-Square
INTERCEPT	0.9948	0.2462	16.3228	0.0001
FD2_RCP	3.6075	0.9679	13.8911	0.0002
MOS_OPEN	-0.3355	0.0383	76.8313	0.0001
MOS_DUM	0.9335	0.1332	49.0856	0.0001
INV_LOG	-0.7820	0.2291	11.6557	0.0006
FD1_RCP	-2.0269	0.9698	4.3686	0.0366

true model itself. The model builder's attention has been drawn to the most noticeable, unavoidable collection of predictor variables, whose behavior is known to the extent the logit plots uncover their shapes and their relationships to response.

There is also magic in building a statistical model, in that the best subset of predictor variables consists of variables whose contributions to the predictions of the model are often unpredictable and unexplainable. Sometimes, the most important variable in the mix drops from the top, in that its contribution in the model is no longer as strong as it was individually. Other times, the least-unlikely variable rises from the bottom, in that its contribution in the model is stronger than it was individually. In the best of times, the variables interact with each other such that their total effect on the predictions of the model is greater than the sum of their individual effects.

Unless the variables are not correlated with each other (the rarest of possibilities), it is impossible for the model builder to assess the unique contribution of a variable. In practice, the model builder can assess the *relative importance* of a variable, specifically, its importance with respect to the presence of the other variables in the model. The Wald chi-square—as posted in logistic regression analysis output—serves as an indicator of the relative importance of a variable, as well as for selecting the best subset. This is discussed in the next section.

8.12.1 Selecting the Best Subset

The decision rules for finding the best subset of important variables consist of the following steps:

1. *Select an initial subset of important variables.* Variables that are thought to be important are probably important; let experience (the model

builders and others) in the problem domain be the rule. If there are many variables from which to choose, rank the variables based on the correlation coefficient r (between response variable and candidate predictor variable). One to two handfuls of the experience-based variables, the largest r-valued variables, and some small r-valued variables form the initial subset. The last variables are included because small r values may falsely exclude important nonlinear variables. (Recall that the correlation coefficient is an indicator of linear relationship.) Categorical variables require special treatment as the correlation coefficient cannot be calculated. (I illustrate with FD_TYPE how to include a categorical variable in a model in the last section.)

2. For the variables in the initial subset, *generate logit plots and straighten the variables as required.* The most noticeable handfuls of original and reexpressed variables form the starter subset.

3. *Perform the preliminary logistic regression analysis on the starter subset.* Delete one or two variables with Wald chi-square values less than the *Wald cutoff value of 4* from the model. This results in the first incipient subset of important variables.

4. *Perform another logistic regression analysis on the incipient subset.* Delete one or two variables with Wald chi-square values less than the Wald cutoff value of 4 from the model. The model builder can create an illusion of important variables appearing and disappearing with the deletion of different variables. The Wald chi-square values can exhibit "bouncing" above and below the Wald cutoff value as the variables are deleted. The bouncing effect is due to the correlation between the "included" variables and the "deleted" variables. A greater correlation implies greater bouncing (unreliability) of Wald chi-square values. Consequently, the greater the bouncing of Wald chi-square values implies the greater the uncertainty of declaring important variables.

5. *Repeat step 4 until all retained predictor variables have comparable Wald chi-square values.* This step often results in different subsets as the model builder deletes judicially different pairings of variables.

6. *Declare the best subset by comparing the relative importance of the different subsets using the decision rule in Section 8.10.4.*

8.13 Best Subset of Variables for Case Study

I perform a logistic regression on TXN_ADD with the five-variable subset MOS_OPEN and MOS_DUM, FD1_RCP, FD2_RCP, and INVEST_LOG; the output is in Table 8.13. FD1_RCP has the smallest Wald chi-square value, 4.3686. FD2_RCP, which has a Wald chi-square of 13.8911, is highly correlated with FD1_RCP ($r_{FD1\_RCP, FD2\_RCP}$ = 0.97), thus rendering their Wald chi-square values unreliable. However, without additional indicators for either variable, I accept their "face" values as an indirect message and delete FD1_RCP, the variable with the lesser value.

INVEST_LOG has the second smallest Wald chi-square value, 11.6557. With no apparent reason other than it just appears to have a less-relative importance given MOS_OPEN, MOS_DUM, FD1_RCP, and FD2_RCP in the model, I also delete INVEST_LOG from the model. Thus, the incipiently best subset consists of FD2_RCP, MOS_OPEN, and MOS_DUM.

I perform another logistic regression on TXN_ADD with the three-variable subset (FD2_RCP, MOS_OPEN, and MOS_DUM); the output is in Table 8.14. MOS_OPEN and FD2_RCP have comparable Wald chi-square values, 81.8072 and 85.7923, respectively, which are obviously greater than the Wald cutoff value 4. The Wald chi-square value for MOD_DUM is half of that of MOS_OPEN, and not comparable to the other values. However, MOS_DUM is staying in the model because it is empirically needed (recall Figures 8.9 and 8.10). I acknowledge that MOS_DUM and MOS_OPEN share information, which could be affecting the reliability of their Wald chi-square values. The actual amount of shared information is 42%, which indicates there is a minimal effect on the reliability of their Wald chi-square values.

I compare the importance of the current three-variable subset (FD2_RCP, MOS_OPEN, MOS_DUM) and the starter five-variable subset (MOS_OPEN, MOS_DUM, FD1_RCP, FD2_RCP, INVEST_LOG). The G/df values are 62.02 (= 186.058/3 from Table 8.14) and 40.21 (= 201.031/5 from Table 8.13) for the former and latter subsets, respectively. Based on the decision rule in Section 8.10.4, I declare the three-variable subset is better than the five-variable subset. Thus, I expect good predictions of TXN_ADD based on the three-variable model defined in Equation (8.16):

TABLE 8.14

Logistic Model for TXN_ADD with Best Incipient Subset

	Intercept Only	Intercept and All Variables	All Variables
-2LL	3,606.488	3,420.430	186.058 with 3 df (p = 0.0001)

Variable	Parameter Estimate	Standard Error	Wald Chi-Square	Pr > Chi-Square
INTERCEPT	0.5164	0.1935	7.1254	0.0076
FD2_RCP	1.4942	0.1652	81.8072	0.0001
MOS_OPEN	-0.3507	0.0379	85.7923	0.0001
MOS_DUM	0.9249	0.1329	48.4654	0.0001

Predicted logit of TXN_ADD =

Predicted LGT_TXN =

0.5164 + 1.4942*FD2_RCP - 0.3507*MOS_OPEN + 0.9249*MOS_DUM   (8.16)

8.14 Visual Indicators of Goodness of Model Predictions

In this section, I provide visual indicators of the quality of model predictions. The LRM itself is a variable as it is a sum of weighted variables with the logistic regression coefficients serving as the weights. As such, the logit model prediction (e.g., the predicted LGT_TXN) is a variable that has a mean, a variance, and all the other descriptive measures afforded any variable. Also, the logit model prediction can be graphically displayed as afforded any variable. Accordingly, I present three valuable plotting techniques, which reflect the EDA-prescribed "graphic detective work" for assessing the goodness of model predictions.

8.14.1 Plot of Smooth Residual by Score Groups

The plot of *smooth residual by score groups* is defined as the plot consisting of the mean residual against the mean predicted response by *score groups*, which are identified by the unique values created by preselected variables, typically the predictor variables in the model under consideration. For example, for the three-variable model, there are 18 score groups: three values of FD2_RCP multiplied by six values of MOS_OPEN. The two values of MOS_DUM are not unique as they are part of the values of MOS_OPEN.

The steps required to construct the plot of the smooth residual by score groups and its interpretation are as follows:

1. *Score* the data by appending the predicted logit as outlined in Section 8.2.2.
2. *Convert* the predicted logit to the predicted probability of response as outlined in Section 8.2.2.
3. *Calculate* the residual (error) for each individual: Residual = actual response minus predicted probability of response.
4. *Determine* the score groups by the unique values created by the preselected variables.
5. For each score group, *calculate* the mean (smooth) residual and mean (smooth) predicted response, producing a set of paired smooth points (smooth residual, smooth predicted response).
6. *Plot* the smooth points by score group.

7. *Draw* a straight line through mean residual = 0. This zero line serves as a reference line for determining whether a general trend exists in the scatter of smooth points. If the smooth residual plot looks like the ideal or *null* plot (i.e., has a random scatter about the zero line with about half of the points above the line and the remaining points below), then it is concluded that there is no general trend in the smooth residuals. Thus, the predictions aggregated at the score group level are considered good. The desired implication is that on average the predictions at the individual level are also good.

8. *Examine* the smooth residual plot for noticeable deviations from random scatter. This is at best a subjective task as it depends on the model builder's unwitting nature to see what is desired. To aid in an objective examination of the smooth residual plot, use the general association test discussed in Chapter 2 to determine whether the smooth residual plot is equivalent to the null plot.

9. When the smooth residual plot is declared null, *look* for a *local pattern*. It is not unusual for a small wave of smooth points to form a local pattern, which has no ripple effect to create a general trend in an otherwise-null plot. A local pattern indicates a weakness or *weak spot* in the model in that there is a prediction bias for the score groups identified by the pattern.

8.14.1.1 Plot of the Smooth Residual by Score Groups for Case Study

I construct the plot of smooth residual by score groups to determine the quality of the predictions of the three-variable (FD2_RCP, MOS_OPEN, MOS_DUM) model. The smooth residual plot in Figure 8.11 is declared to be equivalent to the null-plot-based general association test. Thus, the overall quality of prediction is considered good. That is, on average, the predicted TXN_ADD is equal to the actual TXN_ADD.

Easily seen, but not easily understood (at this point in the analysis), is the local pattern defined by four score groups (labeled 1 through 4) in the lower right-hand side of the plot. The local pattern explicitly shows that the smooth residuals are noticeably negative. The local pattern indicates a weak spot in the model as its predictions for the individuals in the four score groups have, on average, a positive bias, that is, their predicted TXN_ADD tends to be larger than their actual TXN_ADD.

If implementation of the model can afford "exception rules" for individuals in a weak spot, then model performance can be enhanced. For example, response models typically have a weak spot as prediction bias stems from limited information on new customers and outdated information on expired customers. Thus, if model implementation on a solicitation database can include exception rules (e.g., new customers are always targeted—assigned to the top decile) and expired customers are placed in the middle deciles, then the overall quality of prediction is improved.

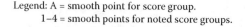

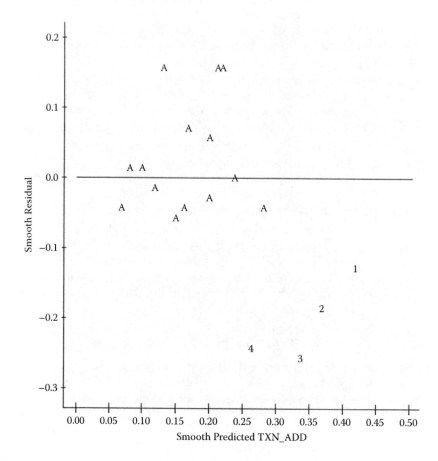

FIGURE 8.11
Plot of smooth residual by score group for three-variable (FD2_RCP, MOS_OPEN, and MOS_DUM) model.

For use in a further discussion, the descriptive statistics for the plot of the smooth residual by score groups/three-variable model are as follows: (1) For the smooth residuals, the minimum and maximum values and the range are -0.26, 0.16, and 0.42, respectively; and (2) the standard deviation of the smooth residuals is 0.124.

8.14.2 Plot of Smooth Actual versus Predicted by Decile Groups

The plot of *smooth actual versus predicted by decile groups* is defined as the plot consisting of the mean actual response against the mean predicted response by *decile groups*. Decile groups are 10 equal-size classes, which are based on the predicted response values from the LRM under consideration. Decile groupings are not an arbitrary partitioning of the data as most database models are implemented at the decile level and consequently are built and validated at the decile level.

The steps required to construct the plot of the smooth actual versus predicted by decile groups and its interpretation are as follows:

1. *Score* the data by appending the predicted logit as outlined in Section 8.2.2.

2. *Convert* the predicted logit to the predicted probability of response as outlined in Section 8.2.2.

3. *Determine* the decile groups. Rank in descending order the scored data by the predicted response values. Then, divide the scored-ranked data into 10 equal-size classes. The first class has the largest mean predicted response, labeled "top"; the next class is labeled "2," and so on. The last class has the smallest mean predicted response, labeled "bottom."

4. For each decile group, *calculate* the mean (smooth) actual response and mean (smooth) predicted response, producing a set of 10 smooth points, (smooth actual response, smooth predicted response).

5. *Plot* the smooth points by decile group, labeling the points by decile group.

6. *Draw* the 45° line on the plot. This line serves as a reference for assessing the quality of predictions at the decile group level. If the smooth points are either on or *hug the 45° line* in their proper order (top to bottom, or bottom to top), then predictions, on average, are considered *good*.

7. *Determine* the "tightness" of the hug of the smooth points about the 45° line. To aid in an objective examination of the smooth plot, use the correlation coefficient between the smooth actual and predicted response points. The correlation coefficient serves as an indicator of the amount of scatter about the 45° straight line. The larger the correlation coefficient is, the less scatter there will be and the better the overall quality of prediction.

8. As discussed in Section 8.6.3, the correlation coefficient based on big points tends to produce a big r value, which serves as a gross estimate of the individual-level r value. The correlation coefficient based on smooth actual and predicted response points is a gross measure of the individual-level predictions of the model. It is best served as a *comparative indicator* in choosing the better model.

8.14.2.1 Plot of Smooth Actual versus Predicted by Decile Groups for Case Study

I construct plot of the smooth actual versus predicted by decile groups based on Table 8.15 to determine the quality of the three-variable model

TABLE 8.15

Smooth Points by Deciles from Model Based on FD_RCP,
MOS_OPEN, and MOS_DUM

	TXN_ADD		Predicted TXN_ADD		
DECILE	**N**	**MEAN**	**MEAN**	**MIN**	**MAX**
Top	492	0.069	0.061	0.061	0.061
2	493	0.047	0.061	0.061	0.061
3	493	0.037	0.061	0.061	0.061
4	492	0.089	0.080	0.061	0.085
5	493	0.116	0.094	0.085	0.104
6	493	0.085	0.104	0.104	0.104
7	492	0.142	0.118	0.104	0.121
8	493	0.156	0.156	0.121	0.196
9	493	0.185	0.198	0.196	0.209
Bottom	492	0.270	0.263	0.209	0.418
Total	4,926	0.119	0.119	0.061	0.418

predictions. The smooth plot in Figure 8.12 has a minimal scatter of the
10 smooth points about the 45° lines, with two noted exceptions. Decile
groups 4 and 6 appear to be the farthest away from the line (in terms of
perpendicular distance). Decile groups 8, 9, and bot are on top of each
other, which indicate the predictions are the same for these groups. The
indication is that the model cannot discriminate among the least-respond-
ing individuals. But, because implementation of response models typically excludes the lower three or four decile groups, their spread about the
45° line and their (lack of) order are not as critical a feature in assessing
the quality of the prediction. Thus, the overall quality of prediction is considered good.

The descriptive statistic for the plot of smooth actual versus predicted by
decile groups/three-variable model is the correlation coefficient between the
smooth points, rsm. actual, sm. predicted: decile group, is 0.972.

8.14.3 Plot of Smooth Actual versus Predicted by Score Groups

The plot of *smooth actual versus predicted by score groups* is defined as the plot
consisting of the mean actual against the mean predicted response by the
score groups. Its construction and interpretation are virtually identical to
the plot for smooth actual versus predicted by decile groups. The painlessly
obvious difference is that decile groups are replaced by score groups, which
are defined in the discussion in Section 8.14.1 on the plot of the smooth residual by score groups.

I outline compactly the steps for the construction and interpretation of the
plot of the smooth actual versus predicted by score groups:

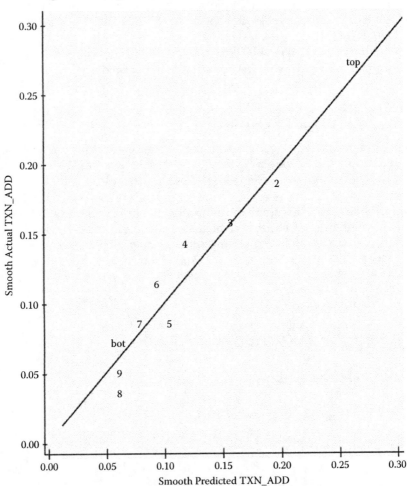

Legend: Values top, 2, ..., 9, bot are decile groups.

FIGURE 8.12
Plot of smooth actual versus predicted by decile group for three-variable (FD2_RCP, MOS_OPEN, and MOS_DUM) model.

1. *Score* the data by appending the predicted logit and convert the predicted logit to the predicted probability of response.
2. *Determine* the score groups and calculate their smooth values for actual response and predicted response.
3. *Plot* the smooth actual and predicted points by score group.
4. *Draw* a 45° line on the plot. If the smooth plot looks like the null plot, then it is concluded that the model predictions aggregated at the score group level are considered good.

5. *Use* the correlation coefficient between the smooth points to aid in an objective examination of the smooth plot. This serves as an indicator of the amount of scatter about the 45° line. The larger the correlation coefficient is, the less scatter there is, and the better the overall quality of predictions is. The correlation coefficient is best served as a comparative measure in choosing the better model.

8.14.3.1 *Plot of Smooth Actual versus Predicted by Score Groups for Case Study*

I construct the plot of the smooth actual versus predicted by score groups based on Table 8.16 to determine the quality of the three-variable model predictions. The smooth plot in Figure 8.13 indicates the scatter of the 18 smooth points about the 45° line is good, except for the four points on the right-hand side of the line, labeled numbers 1 through 4. These points correspond to the four score groups, which became noticeable in the smooth residual plot in Figure 8.11. The indication is the same as that of the smooth residual plot: The overall quality of the prediction is considered good. However, if

TABLE 8.16

Smooth Points by Score Groups from Model Based on FD2_RCP, MOS_OPEN, and MOS_DUM

		TXN_ADD		PROB_HAT
MOS_OPEN	FD2_OPEN	N	MEAN	MEAN
1	1	161	0.267	0.209
	2	56	0.268	0.359
	3	20	0.350	0.418
2	1	186	0.145	0.157
	2	60	0.267	0.282
	3	28	0.214	0.336
3	1	211	0.114	0.116
	2	62	0.274	0.217
	3	19	0.158	0.262
4	1	635	0.087	0.085
	2	141	0.191	0.163
	3	50	0.220	0.200
5	1	1,584	0.052	0.061
	2	293	0.167	0.121
	3	102	0.127	0.150
6	1	769	0.109	0.104
	2	393	0.186	0.196
	3	156	0.237	0.238
Total		4,926	0.119	0.119

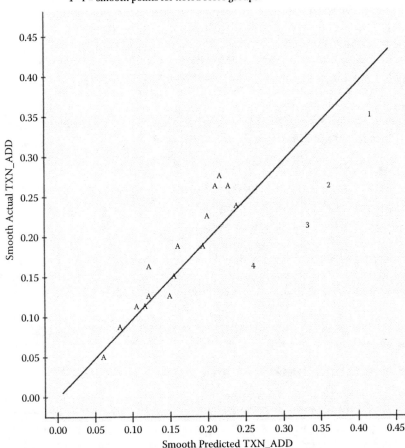

FIGURE 8.13
Plot of smooth actual versus predicted by score group for three-variable (FD2_RCP, MOS_OPEN and MOS_DUM) model.

implementation of the model can afford exception rules for individuals who look like the four score groups, then the model performance can be improved.

The profiling of the individuals in the score groups is immediate from Table 8.16. The original predictor variables, instead of the reexpressed versions, are used to make the interpretation of the profile easier. The sizes (20, 56, 28, 19) of the four noticeable groups are quite small for groups 1 to 4, respectively, which may account for the undesirable spread about the 45° line. However, there are three other groups of small size (60, 62, and 50) that do not have noticeable spread about the 45° line. So, perhaps group size is not the reason for the undesirable spread. Regardless of why the unwanted spread exists, the four noticeable groups indicate that the three-variable model reflects a small weak spot, a segment that accounts for only 2.5% (= (20 + 56 + 28 + 19)/4,926)) of the database population from which the sample was drawn. Thus, implementation of the three-variable model is expected to yield good predictions, even if exception rules cannot be afforded to the weak-spot segment, as its effects on model performance should hardly be noticed.

The descriptive profile of the weak-spot segment is as follows: newly opened (less than 6 months) accounts of customers with two or three accounts; recently opened (between 6 months and 1 year) accounts of customers with three accounts; and older (between 1 and 1½ years) of customers with three accounts. The actual profile cells are

1. MOS_OPEN = 1 and FD2_OPEN = 3
2. MOS_OPEN = 1 and FD2_OPEN = 2
3. MOS_OPEN = 2 and FD2_OPEN = 3
4. MOS_OPEN = 3 and FD2_OPEN = 3

For the descriptive statistic for the plot of smooth actual versus predicted by score groups/three-variable model, the correlation coefficient between the smooth points, rsm. actual, sm. predicted: score group, is 0.848.

8.15 Evaluating the Data Mining Work

To appreciate the data mining analysis that produced the three-variable EDA-model, I build a non-EDA model for comparison. I use the stepwise logistic regression variable selection process, which is a "good" choice for a non-EDA variable selection process as its serious weaknesses are revealed in Chapter 10. The stepwise and other statistics-based variable selection procedures can be questionably classified as minimal data mining techniques as they only find the "best" subset among the original variables without generating potentially important variables. They do *not* generate structure in the search for the best subset of variables. Specifically, they do not create new variables like reexpressed versions of the original variables or derivative variables like dummy variables defined by the original variables. In contrast, the most productive data mining techniques generate structure from the original variables and determine the best combination of those structures along with the original variables. More about variable selection is presented in Chapters 10 and 30.

I perform a stepwise logistic regression analysis on TXN_ADD with the original five variables. The analysis identifies the best non-EDA subset consisting of only two variables: FD2_OPEN and MOS_OPEN; the output is Table 3.17. The G/df value is 61.3 (= 122.631), which is comparable to the G/df

TABLE 8.17

Best Non-EDA Model Criteria for Assessing Model Fit

	Intercept Only	Intercept and All Variables	All Variables	
-2LL	3,606.488	3,483.857	122.631 with 2 df (p = 0.0001)	
Variable	Parameter Estimate	Standard Error	Wald Chi-Square	Pr > Chi-Square
INTERCEPT	-2.0825	0.1634	162.3490	0.0001
FD2_OPEN	0.6162	0.0615	100.5229	0.0001
MOS_OPEN	-0.1790	0.0299	35.8033	0.0001

value (62.02) of the three-variable (FD2_RCP, MOS_OPEN, MOS_DUM) EDA model. Based on the G/df indicator of Section 8.10.4, I cannot declare that the three-variable EDA model is better than the two-variable non-EDA model.

Could it be that all the EDA detective work was for naught—that the "quick-and-dirty" non-EDA model was the obvious one to build? The answer is no. Remember that an indicator is sometimes just an indicator that serves as a pointer to the next thing, such as moving on to the ladder of powers. Sometimes, it is an instrument for making automatically a decision based on visual impulse, such as determining if the relationship is straight enough or the scatter in a smooth residual plot is random. And sometimes, a lowly indicator does not have the force of its own to send a message until it is in the company of other indicators (e.g., smooth plots and their aggregate-level correlation coefficients).

I perform a simple comparative analysis of the descriptive statistics stemming from the EDA and non-EDA models to determine the better model. I need only to construct the three smooth plots—smooth residuals at the score group level, smooth actuals at the decile group level, and smooth actuals at the score group level—from which I can obtain the descriptive statistics for the latter model as I already have the descriptive statistics for the former model.

8.15.1 Comparison of Plots of Smooth Residual by Score Groups: EDA versus Non-EDA Models

I construct the plot of the smooth residual by score groups in Figure 8.14 for the non-EDA model. The plot is not equivalent to the null plot based on the general association test. Thus, the overall quality of the predictions of the non-EDA model is not considered good. There is a local pattern of five smooth points in the lower right-hand corner below the zero line. The five smooth points, labeled I, II, III, IV, and V, indicate that the predictions for the

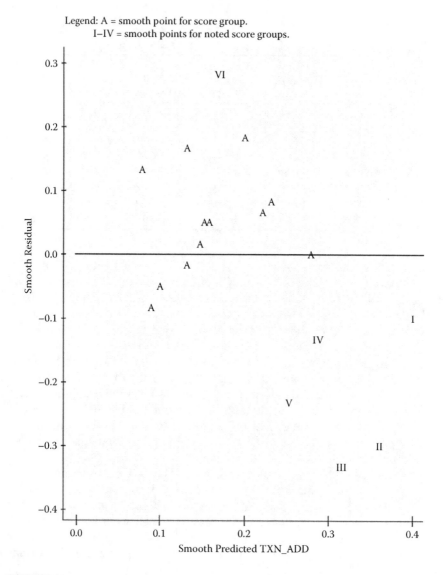

FIGURE 8.14
Smooth residual by score group plot for NonEDA (FD2_OPEN, MOS_OPEN) model.

individuals in the five score groups have, on average, a positive bias; that is, their predicted TXN_ADD tends to be larger than their actual TXN_ADD. There is a smooth point, labeled VI, at the top of the plot that indicates a group of the individuals with an average negative bias. That is, their predicted TXN_ADD tends to be smaller than their actual TXN_ADD.

The descriptive statistics for the plot of the smooth residual by score groups/ non-EDA model are as follows: For the smooth residual, the minimum and maximum values and the range are -0.33, 0.29, and 0.62, respectively. The standard deviation of the smooth residuals is 0.167.

The comparison of the EDA and non-EDA smooth residuals indicates the EDA model produces smaller smooth residuals (prediction errors). The EDA smooth residual range is noticeably smaller than that of the non-EDA: 32.3% (= (0.62 0.42)/0.62) smaller. The EDA smooth residual standard deviation is noticeably smaller than that of the non-EDA: 25.7% (= (0.167 - 0.124)/0.167) smaller. The implication is that the EDA model has a better quality of prediction.

8.15.2 Comparison of the Plots of Smooth Actual versus Predicted by Decile Groups: EDA versus Non-EDA Models

I construct the plot of smooth actual versus predicted by decile groups in Figure 8.15 for the non-EDA model. The plot clearly indicates a scatter that does not hug the 45° line well, as decile groups top and 2 are far from the line, and 8, 9, and bot are out of order, especially bot. The decile-based correlation coefficient between smooth points, rsm. actual, sm. predicted: decile group, is 0.759.

The comparison of the decile-based correlation coefficients of the EDA and non-EDA indicates that the EDA model produces a larger, tighter hug about the 45° line. The EDA correlation coefficient is noticeably larger than that of the non-EDA: 28.1% (= (0.972 - 0.759)/0.759) larger. The implication is that the EDA model has a better quality of prediction at the decile level.

8.15.3 Comparison of Plots of Smooth Actual versus Predicted by Score Groups: EDA versus Non-EDA Models

I construct the plot of smooth actual versus predicted by score groups in Figure 8.16 for the non-EDA model. The plot clearly indicates a scatter that does not hug the 45 line well, as score groups II, III, V, and VI are far from the line. The score-group-based correlation coefficient between smooth points, rsm. actual, sm. predicted: score group, is 0.635.

The comparison of the plots of the score-group-based correlation coefficient of the EDA and non-EDA smooth actual indicate that the EDA model produces a tighter hug about the 45° line. The EDA correlation coefficient is noticeably larger than that of the non-EDA: 33.5% (= (0.848 - 0.635)/0.635) larger. The implication is that the EDA model has a better quality of prediction at the score group level.

8.15.4 Summary of the Data Mining Work

From the comparative analysis, I have the following:

1. The overall quality of the predictions of the EDA model is better than that of the non-EDA as the smooth residual plot of the former is null, and that of the non-EDA model is not.

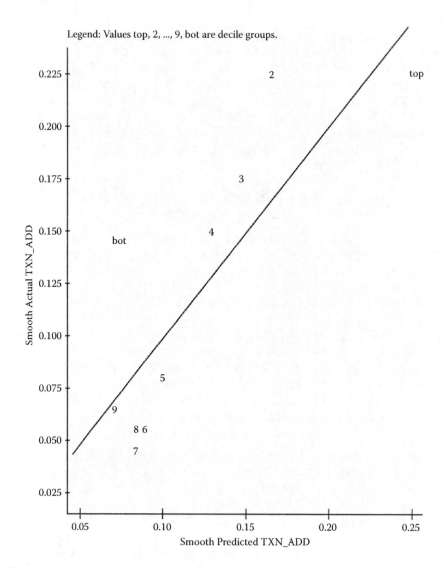

FIGURE 8.15
Smooth actual versus predicted by decile group plot for NonEDA (FD2_OPEN, MOS_OPEN) model.

2. The prediction errors of the EDA model are smaller than those of the non-EDA as the smooth residuals of the former have less spread (smaller range and standard deviation). In addition, the EDA model has better aggregate-level predictions than that of the non-EDA as the former model has less prediction bias (larger correlations between smooth actual and predicted values at decile and score group levels).

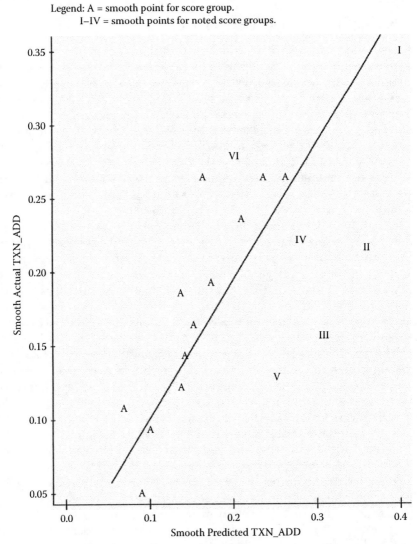

FIGURE 8.16
Smooth actual versus predicted by score group plot for NonEDA (FD2_OPEN, MOS_OPEN) model.

3. I conclude that the three-variable EDA model consisting of FD2_ RCP, MOS_OPEN, and MOS_DUM is better than the two-variable non-EDA model consisting of FD2_OPEN and MOS_OPEN.

As the last effort to improve the EDA model, I consider the last candidate predictor variable FD_TYPE for data mining in the next section.

8.16 Smoothing a Categorical Variable

The classic approach to include a categorical variable into the modeling process involves *dummy variable coding*. A categorical variable with k classes of qualitative (nonnumeric) information is replaced by a set of k - 1 quantitative dummy variables. The dummy variable is defined by the presence or absence of the class values. The class left out is called the *reference class*, to which the other classes are compared when interpreting the effects of dummy variables on response. The classic approach instructs that the complete set of k - 1 dummy variables is included in the model regardless of the number of dummy variables that are declared nonsignificant. This approach is problematic when the number of classes is large, which is typically the case in big data applications. By chance alone, as the number of class values increases, the probability of one or more dummy variables being declared nonsignificant increases. To put all the dummy variables in the model effectively adds "noise" or unreliability to the model, as nonsignificant variables are known to be noisy. Intuitively, a large set of inseparable dummy variables poses difficulty in model building in that they quickly "fill up" the model, not allowing room for other variables.

The EDA approach of treating a categorical variable for model inclusion is a viable alternative to the classic approach as it explicitly addresses the problems associated with a large set of dummy variables. It reduces the number of classes by merging (smoothing or averaging) the classes with comparable values of the dependent variable under study, which for the application of response modeling is the response rate. The smoothed categorical variable, now with fewer classes, is less likely to add noise in the model and allows more room for other variables to get into the model.

There is an additional benefit offered by smoothing of a categorical variable. The information captured by the smoothed categorical variable tends to be more reliable than that of the complete set of dummy variables. The reliability of information of the categorical variable is only as good as the aggregate reliability of information of the individual classes. Classes of small size tend to provide unreliable information. Consider the extreme situation of a class of size one. The estimated response rate for this class is either 100% or 0% because the sole individual either responds or does not respond, respectively. It is unlikely that the estimated response rate is the true response rate for this class. This class is considered to provide unreliable information regarding its true response rate. Thus, the reliability of information for the categorical variable itself decreases as the number of small class values increases. The smoothed categorical variable tends to have greater reliability than the set of dummy variables because it intrinsically has fewer classes and consequently has larger class sizes due to the merging process. The rule of thumb of EDA for small class size is that less than 200 is considered small.

CHAID is often the preferred EDA technique for smoothing a categorical variable. In essence, CHAID is an excellent EDA technique as it involves the three main elements of statistical detective work: "numerical, counting, and graphical." CHAID forms new larger classes based on a numerical merging, or averaging, of response rates and counts the reduction in the number of classes as it determines the best set of merged classes. Last, the output of CHAID is conveniently presented in an easy-to-read and -understand graphical display, a treelike box diagram with leaf boxes representing the merged classes.

The technical details of the merging process of CHAID are beyond the scope of this chapter. CHAID is covered in detail in subsequent chapters, so here I briefly discuss and illustrate it with the smoothing of the last variable to be considered for predicting TXN_ADD response, namely, FD_TYPE.

8.16.1 Smoothing FD_TYPE with CHAID

Remember that FD_TYPE is a categorical variable that represents the product type of the customer's most recent investment purchase. It assumes 14 products (classes) coded A, B, C, ..., N. The TXN_ADD response rate by FD_TYPE values are in Table 8.18.

There are seven small classes (F, G, J, K, L, M, and N) with sizes 42, 45, 57, 94, 126, 19, and 131, respectively. Their response rates—0.26, 0.24, 0.19, 0.20, 0.22, 0.42, and 0.16, respectively—can be considered potentially unreliable. Class B has the largest size, 2,828, with a surely reliable 0.06 response rate.

TABLE 8.18

FD_TYPE

FD_TYPE	TXN_ADD	
	N	MEAN
A	267	0.251
B	2,828	0.066
C	250	0.156
D	219	0.128
E	368	0.261
F	42	0.262
G	45	0.244
H	225	0.138
I	255	0.122
J	57	0.193
K	94	0.202
L	126	0.222
M	19	0.421
N	131	0.160
Total	**4,926**	**0.119**

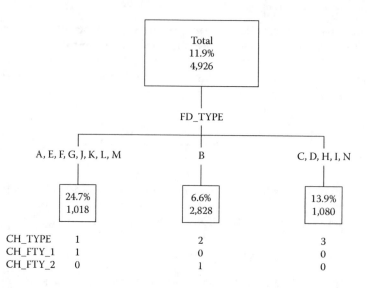

FIGURE 8.17
Double smoothing of FD_TYPE with CHAID.

The remaining six presumably reliable classes (A, C, D, E, and H) have sizes between 219 and 368.

The CHAID tree for FD_TYPE in Figure 8.17 is read and interpreted as follows:

1. The top box, the root of the tree, represents the sample of 4,926 with response rate 11.9%.

2. The CHAID technique smoothes FD_TYPE by way of merging the original 14 classes into 3 merged (smoothed) classes, as displayed in the CHAID tree with three leaf boxes.

3. The leftmost leaf, which consists of the seven small unreliable classes and the two reliable classes A and E, represents a newly merged class with a reliable response rate of 24.7% based on a class size of 1,018. In this situation, the smoothing process increases the reliability of the small classes with two-step averaging. The first step combines all the small classes into a temporary class, which by itself produces a reliable average response rate of 22.7% based on a class size of 383. In the second step, which does not always occur in smoothing, the temporary class is further united with the already-reliable classes A and E because the latter classes have comparable response rates to the temporary class response rate. The *double-smoothed* newly merged class represents the average response rate of the seven small classes and classes A and E. When double smoothing does not occur, the temporary class is the final class.

4. The increased reliability that smoothing of a categorical variable offers can now be clearly illustrated. Consider class M with its unreliable estimated response rate of 42% based on class size 19. The smoothing process puts class M in the larger, more reliable leftmost leaf with a response rate of 24.7%. The implication is that class M now has a more reliable estimate of response rate, namely, the response rate of its newly assigned class, 24.7%. Thus, the smoothing has effectively adjusted the original estimated response rate of class M downward, from a positively biased 42% to a reliable 24.7%. In contrast, within the same smoothing process, the adjustment of class J is upward, from a negatively biased 19% to 24.7%. It is not surprising that the two reliable classes, A and E, remain noticeably unchanged, from 25% and 26% to 24.7%, respectively.

5. The middle leaf consists of only class B, defined by a large class size of 2,828 with a reliable response rate of 6.6%. Apparently, the low response rate of class B is not comparable to any class (original, temporary, or newly merged) response rate to warrant a merging. Thus, the original estimated response rate of class B is unchanged after the smoothing process. This presents no concern over the reliability of class B because its class size is largest from the outset.

6. The rightmost leaf consists of large classes C, D, H, and I and the small class N for an average reliable response rate of 13.9% with class size 1,080. The smoothing process adjusts the response rate of class N downward, from 16% to a smooth 13.9%. The same adjustment occurs for class C. The remaining classes D, H, and I experience an upward adjustment.

I call the smoothed categorical variable CH_TYPE. Its three classes are labeled 1, 2, and 3, corresponding to the leaves from left to right, respectively (see bottom of Figure 8.17). I also create two dummy variables for CH_TYPE:

1. CH_FTY_1 = 1 if FD_TYPE = A, E, F, G, J, K, L, or M; otherwise, CH_FTY_1 = 0;

2. CH_FTY_2 = 1 if FD_TYPE = B; otherwise, CH_FTY_2 = 0.

3. This dummy variable construction uses class CH_TYPE = 3 as the reference class. If an individual has values CH_FTY_1 = 0 and CH_FTY_2 = 0, then the individual has implicitly CH_TYPE = 3 and one of the original classes (C, D, H, I, or N).

8.16.2 Importance of CH_FTY_1 and CH_FTY_2

I assess the importance of the CHAID-based smoothed variable CH_TYPE by performing a logistic regression analysis on TXN_ADD with both CH_FTY_1 and CH_FTY_2, as the set dummy variable must be together in the model; the output is in Table 8.19. The G/df value is 108.234 (= 216.468/2), which is greater than the standard G/df value of 4. Thus, CHFTY_1 and CH_FTY_2 together are declared important predictor variables of TXN_ADD.

8.17 Additional Data Mining Work for Case Study

I try to improve the predictions of the three-variable (MOS_OPEN, MOS_DUM, and FD2_RCP) model with the inclusion of the smoothed variable CH_TYPE. I perform the LRM on TXN_ADD with MOS_OPEN, MOS_DUM, FD2_RCP, and CH_FTY_1 and CH_FTY_2; the output is in Table 8.20. The Wald chi-square value for FD2_RCP is less than 4. Thus, I delete FD2_RCP from the model and rerun the model with the remaining four variables.

The four-variable (MOS_OPEN, MOS_DUM, CH_FTY_1, and CH_FTY_2) model produces comparable Wald chi-square values for the four variables; the output is in Table 8.21. The G/df value equals 64.348 (= 257.395/4), which

TABLE 8.19

G and df for CHAID-Smoothed FD_TYPE

Variable	−2LL	G	df	p-value
INTERCEPT	3,606.488			
CH_FTY_1 & CH_FTY_2	3,390.021	216.468	2	0.0001

TABLE 8.20

Logistic Model: EDA Model Variables plus CH_TYPE Variables

	Intercept Only	Intercept and All Variables	All Variables	
−2LL	3,606.488	3,347.932	258.556 with 5 df (p = 0.0001)	
Variable	Parameter Estimate	Standard Error	Wald Chi-Square	Pr > Chi-Square
INTERCEPT	−0.7497	0.2464	9.253	0.0024
CH_FTY_1	0.6264	0.1175	28.4238	0.0001
CH_FTY_2	−0.6104	0.1376	19.6737	0.0001
FD2_RCP	0.2377	0.2212	1.1546	0.2826
MOS_OPEN	−0.2581	0.0398	42.0054	0.0001
MOS_DUM	0.7051	0.1365	26.6804	0.0001

TABLE 8.21

Logistic Model: Four-Variable EDA Model

	Intercept Only	Intercept and All Variables	All Variables
−2LL	3,606.488	3,349.094	257.395 with 4 df (p = 0.0001)

Variable	Parameter Estimate	Standard Error	Wald Chi-Square	Pr > Chi-Square
INTERCEPT	−0.9446	0.1679	31.6436	0.0001
CH_FTY_1	0.6518	0.1152	32.0362	0.0001
CH_FTY_2	−0.6843	0.1185	33.3517	0.0001
MOS_OPEN	−0.2510	0.0393	40.8141	0.0001
MOS_DUM	0.7005	0.1364	26.3592	0.0001

is slightly larger than the G/df (62.02) of the three-variable (MOS_OPEN, MOS_DUM, FD2_RCP) model. This is not a strong indication that the four-variable model has more predictive power than the three-variable model.

In Sections 8.17.1 to 8.17.4, I perform the comparative analysis, similar to the analysis of EDA versus non-EDA in Section 8.15, to determine whether the four-variable (4var-) EDA model is better than the three-variable (3var-) EDA model. I need the smooth plot descriptive statistics for the latter model as I already have the descriptive statistics for the former model.

8.17.1 Comparison of Plots of Smooth Residual by Score Group: 4var- versus 3var-EDA Models

The plot of smooth residual by score group for the 4var-EDA model in Figure 8.18 is equivalent to the null plot based on the general association test. Thus, the overall quality of the predictions of the model is considered good. It is worthy of notice that there is a far-out smooth point, labeled FO, in the middle of the top of the plot. This smooth residual point corresponds to a score group consisting of 56 individuals (accounting for 1.1% of the data), indicating a weak spot.

The descriptive statistics for the smooth residual by score groups/4var-model plot are as follows: For the smooth residual, the minimum and maximum values and the range are -0.198, 0.560, and 0.758, respectively; the standard deviation of the smooth residual is 0.163.

The descriptive statistics based on all the smooth points excluding the FO smooth point are worthy of notice because such statistics are known to be sensitive to far-out points, especially when they are smoothed and account for a very small percentage of the data. For the FO-adjusted smooth residual, the minimum and maximum values and the range are -0.198, 0.150, and 0.348, respectively; the standard deviation of the smooth residuals is 0.093.

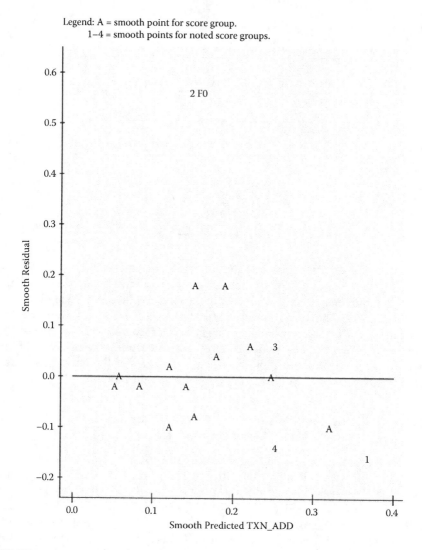

FIGURE 8.18
Smooth residual by score group plot for 4var-EDA (MOS_OPEN, MOS_DUM, CH_FTY_1, CH_FTY_2) model.

The comparison of the 3var- and 4var-EDA model smooth residuals indicates that the former model produces smaller smooth residuals. The 3var-EDA model smooth residual range is noticeably smaller than that of the latter model: 44.6% (= (0.758 - 0.42)/0.758) smaller. The 3var-EDA model smooth residual standard deviation is noticeably smaller than that of the 4var-EDA model: 23.9% (= (0.163 - 0.124/0.163) smaller. The implication is that the CHAID-based dummy variables carrying the information of FD_TYPE are not important enough to produce better predictions than that of the 3var-EDA model. In other words, the 3var-EDA model has a better quality of prediction.

However, if implementation of the TXN_ADD model permits an exception rule for the FO score group/weak spot, the implication is the 4var-EDA model has a better quality of predictions as the model produces smaller smooth residuals. The 4var-EDA model FO-adjusted smooth residual range is noticeably smaller than that of the 3var-EDA model: 17.1% (= (0.42 - 0.348)/0.42) smaller. The 4var-EDA model FO-adjusted smooth residual standard deviation is noticeably smaller than that of the 3var-EDA model: 25.0% (= (0.124 - 0.093)/0.124).

8.17.2 Comparison of the Plots of Smooth Actual versus Predicted by Decile Groups: 4var- versus 3var-EDA Models

The plot of smooth actual versus predicted by decile groups for the 4var-EDA model in Figure 8.19 indicates good hugging of scatter about the 45° lines, despite the following two exceptions. First, there are two pairs of decile groups (6 and 7, 8 and 9), where the decile groups in each pair are adjacent to each other. This indicates that the predictions are different for decile groups within each pair, which should have the same response rate. Second, the bot decile group is very close to the line, but out of order. Because implementation of response models typically exclude the lower three or four decile groups, their spread about the 45° line and their (lack of) order is not as critical a feature in assessing the quality of predictions. Thus, overall the plot is considered very good. The coefficient correlation between smooth actual and predicted points, rsm. actual, sm. predicted: decile group, is 0.989.

The comparison of the decile-based correlation coefficient of the 3var- and 4var-EDA model smooth actual plots indicates the latter model produces a meagerly tighter hug about the 45° line. The 4var-EDA model correlation coefficient is hardly noticeably larger than that of the three-variable model: 1.76% (= (0.989-0.972)/0.972) smaller. The implication is that both models have equivalent quality of prediction at the decile level.

8.17.3 Comparison of Plots of Smooth Actual versus Predicted by Score Groups: 4var- versus 3var-EDA Models

The score groups for plot of the smooth actual versus predicted by score groups for the 4var-EDA model in Figure 8.20 are defined by the variables in the 3var-EDA model to make an uncomplicated comparison. The plot indicates a very nice hugging about the 45° line, except for one far-out smooth point, labeled FO, which was initially uncovered by the smooth residual plot in Figure 8.18. The score-group-based correlation coefficient between all smooth points, rsm. actual, sm. predicted:score group, is 0.784; the score-group-based correlation without the far-out score group FO, $r_{sm. actual, sm. predicted:}$

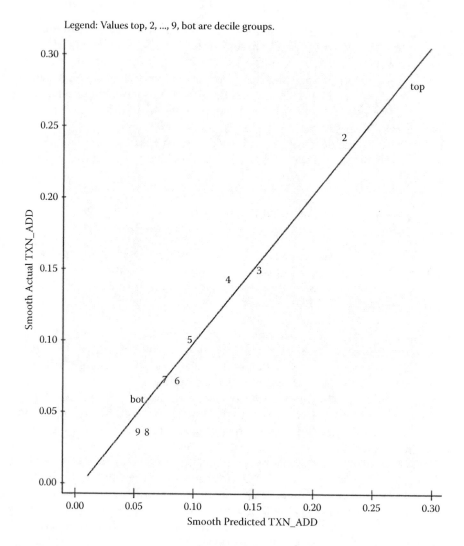

FIGURE 8.19

Smooth actual versus predicted by decile group plot for 4var-EDA (MOS_OPEN, MOS_DUM, CH_FTY_1, CH_FTY_2) model.

score group-FO, is 0.915. The comparison of the score-group-based correlation coefficient of the 3var- and 4var-smooth actual plots indicates that the 3var-EDA model produces a somewhat noticeably tighter hug about the 45 line. The 3var-EDA model score-group-based correlation coefficient is somewhat noticeably larger that of the 4var-EDA model: 8.17% (= (0.848 − 0.784)/0.784) larger. The implication is the 3var-EDA model has a somewhat better quality of prediction at the score group level.

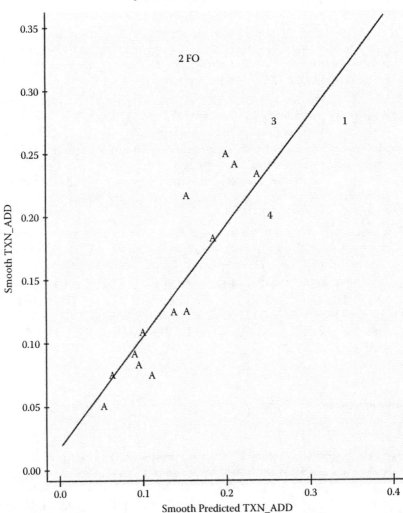

FIGURE 8.20
Smooth Actual vs. Predicted by Score Group Plot for 4var-EDA (MOS_OPEN, MOS_DUM, CH_FTY_1, CH_FTY_2) Model

However, the comparison of the score-group-based correlation coefficient of the 3var- and 4var-smooth actual plots without the FO score group produces a reverse implication. The 4var-EDA model score-group-based correlation coefficient is somewhat noticeably larger than that of the three-variable model: 7.85% (= (0.915 − 0.848)/0.848) larger. The implication is the 4var-EDA model, without the FO score group, has a somewhat better quality of prediction at the score group level.

8.17.4 Final Summary of the Additional Data Mining Work

The comparative analysis offers the following:

1. The overall quality of the 3var- and 4var-EDA models is considered good as both models have a null smooth residual plot. Worthy of notice, there is a very small weak spot (FO score group accounting for 1.1% of the data) in the latter model.

2. The 3var-EDA model prediction errors are smaller than the prediction errors of the 4var-EDA model as the former model smooth residuals have less spread (smaller range and standard deviation). The 3var-EDA model has equivalent or somewhat better aggregate-level predictions as the former model has equivalent or somewhat less prediction bias (equivalent/larger correlation between smooth actual and predicted values at the decile level/score group level).

3. If model implementation can accommodate an exception rule for the FO weak spot, the indicators suggest that the 4var-EDA model has less spread and somewhat better aggregate-level predictions.

4. In sum, I prefer the 3var-EDA model, consisting of MOS_OPEN, MOS_DUM, and FD2_RCP. If exception rules for the far-out score group can be effectively developed and reliably used, I prefer the 4var-EDA model, consisting of MOS_OPEN, MOS_DUM, CH_FTY_1, and CH_FTY_2.

8.18 Summary

I presented the LRM as the workhorse of response modeling. As such, I demonstrated how it fulfills the desired analysis of a Yes-No response variable in that it provides individual probabilities of response as well as yields a meaningful aggregation of individual-level probabilities into decile-level probabilities of response. The decile-level probabilities of response are often required in database implementation of response models. Moreover, I showed the durability and usefulness of the 60-plus-year-old logistic regression analysis and modeling technique as it works well within the EDA/data mining paradigm of today.

I first illustrated the rudiments of the LRM by discussing the SAS program for building and scoring an LRM. Working through a small data set, I pointed out—and ultimately clarified—an often-vexing relationship between the actual and predicted response variables: The former assumes two nominal values, typically 1-0 for yes-no responses, respectively, yet the latter assumes logits, which are continuous values between -7 and +7. This disquieting connection is currently nursed by SAS 8. It is typical for the model builder to name the actual response variable RESPONSE. In SAS procedure LOGISTIC, the naming convention of the predicted response variable is the name that the model builder assigns to the actual response variable, typically RESPONSE. Thus, the predicted response variable labeled RESPONSE may cause the model builder to think that the predicted response variable is a binary variable, not the logit it actually is.

Next, I presented a case study that serves as the vehicle for introducing a host of data mining techniques, which are mostly specific to the LRM. The LRM has the implied assumption that the underlying relationship between a given predictor variable and the logit of response is straight line. I outlined the steps for logit plotting to test the assumption. If the test results are negative (i.e., a straight line exists), then no further attention is required for that variable, which can now be included in the model. If the test result is positive, the variable needs to be straightened before it can justifiably be included in the model.

I discussed and illustrated two straightening methods—reexpressing with the ladder of powers and the bulging rule—both of which are applicable to all linear models, like logistic regression and ordinary regression. The efficacy of these methods is determined by the well-known correlation coefficient. At this point, I reinstated the frequently overlooked assumption of the correlation coefficient, that is, the underlying relationship at hand is a straight line. In addition, I introduced an implied assumption of the correlation coefficient specific to its proper use, which is to acknowledge that the correlation coefficient for reexpressed variables with smoothed big data is a gross measure of straightness of reexpressed variables.

Continuing with the case study, I demonstrated a logistic regression-specific data mining alternative approach to the classical method of assessing. This alternative approach involves the importance of individual predictor variables as well as the importance of a subset of predictor variables and the relative importance of individual predictor variables, in addition to the goodness of model predictions. Additional methods specific to logistic regression set out included selecting the best subset of predictor variables and comparing the importance between two subsets of predictor variables.

My illustration within the case study includes one last alternative method, applicable to all models, linear and nonlinear: smoothing a categorical variable for model inclusion. The method reexpresses a set of dummy variables, which is traditionally used to include a categorical variable in a model, into a new parsimonious set of dummy variables that is more reliable than the original set and easier to include in a model than the original set.

The case study produced a final comparison of a data-guided EDA model and a non-EDA model, based on the stepwise logistic regression variable selection process. The EDA model was the preferred model with a better quality of prediction.

$$\text{Probability of } x \quad O_x = \frac{e^{(\alpha + \beta_1 X_1 + \beta_2 X_2 + \ldots \beta_n X_n)}}{1 + e^{(\alpha + \beta_1 X_1 + \beta_2 X_2 + \ldots \beta_n X_n)}}$$

$$\text{logit}\left[O(x)\right] = \log\left[\frac{O_x}{1 - O_x}\right] = \sum_{k=0}^{k} X_{ik}\beta_k \quad i = 1 \ldots n \; = \alpha + \beta_1 X_1 + \beta_2 X_2 + \ldots \beta_n X_n$$

or $\ln$?∨ — I've seen it both ways!

$\alpha = y$ intercept

$\beta_n = $ regression coefficients

$X_n = $ predictor variables values (maximum likelihoods)

Maximum likelihood method entails finding the set of parameters for which the probability of the observed data is greatest.

9

Ordinary Regression: The Workhorse of Profit Modeling

9.1 Introduction

Ordinary regression is the popular technique for predicting a quantitative outcome, such as profit and sales. It is considered the workhorse of *profit modeling* as its results are taken as the gold standard. Moreover, the ordinary regression model is used as the benchmark for assessing the superiority of new and improved techniques. In a database marketing application, an individual's profit* to a prior solicitation is the quantitative dependent variable, and an ordinary regression model is built to predict the individual's profit to a future solicitation.

I provide a brief overview of ordinary regression and include the SAS program for building and scoring an ordinary regression model. Then, I present a mini case study to illustrate that the data mining techniques presented in Chapter 8 carry over with minor modification to ordinary regression. Model builders, who are called on to provide statistical support to managers monitoring expected revenue from marketing campaigns, will find this chapter an excellent reference for profit modeling.

9.2 Ordinary Regression Model

Let Y be a quantitative dependent variable that assumes a continuum of values. The ordinary regression model, formally known as the ordinary least squares (OLS) regression model, predicts the Y value for an individual based on the values of the predictor (independent) variables $X_1, X_2, \dots, X_n$ for that individual. The OLS model is defined in Equation (9.1):

---

* Profit is variously defined as any measure of an individual's valuable contribution to the bottom line of a business.

$$Y = b_0 + b_1{}^*X_1 + b_2{}^*X_2 + \ldots + b_n{}^*X_n \qquad (9.1)$$

An individual's predicted Y value is calculated by "plugging in" the values of the predictor variables for that individual in Equation (9.1). The b's are the OLS regression coefficients, which are determined by the calculus-based method of least squares estimation; the lead coefficient b_0 is referred to as the intercept.

In practice, the quantitative dependent variable does not have to assume a progression of values that vary by minute degrees. It can assume just several dozens of discrete values and work quite well within the OLS methodology. When the dependent variable assumes only two values, the logistic regression model, not the ordinary regression model, is the appropriate technique. Even though logistic regression has been around for 60-plus years, there is some misunderstanding over the practical (and theoretical) weakness of using the OLS model for a binary response dependent variable. Briefly, an OLS model, with a binary dependent variable, produces typically some probabilities of response greater than 100% and less than 0% and does not typically include some important predictor variables.

9.2.1 Illustration

Consider dataset A, which consists of 10 individuals and three variables (Table 9.1): the quantitative variable PROFIT in dollars (Y), INCOME in thousands of dollars (X1), and AGE in years (X2). I regress PROFIT on INCOME and AGE using dataset A. The OLS output in Table 9.2 includes the ordinary regression coefficients and other "columns" of information. The "Parameter Estimate" column contains the coefficients for INCOME and AGE variables, and the intercept. The coefficient b0 for the intercept variable is used as a

TABLE 9.1

Dataset A

Profit ($)	Income ($000)	Age (years)
78	96	22
74	86	33
66	64	55
65	60	47
64	98	48
62	27	27
61	62	23
53	54	48
52	38	24
51	26	42

TABLE 9.2

OLS Output: PROFIT with INCOME and AGE

Source	df	Sum of Squares	Mean Square	F Value	Pr > F
Model	2	460.3044	230.1522	6.01	0.0302
Error	7	268.0957	38.2994		
Corrected total	9	728.4000			
		Root MSE	6.18865	R-square	0.6319
		Dependent mean	62.60000	Adj R-Sq	0.5268
		Coeff var	9.88602		

Variable	df	Parameter Estimate	Standard Error	t Value	Pr > |t|
INTERCEPT	1	52.2778	7.7812	7.78	0.0003
INCOME	1	0.2669	0.2669	0.08	0.0117
AGE	1	-0.1622	-0.1622	0.17	0.3610

"start" value given to all individuals, regardless of their specific values of the predictor variables in the model.

The estimated OLS PROFIT model is defined in Equation (9.2):

$$PROFIT = 52.2778 + 0.2667 * INCOME - 0.1622 * AGE \qquad (9.2)$$

9.2.2 Scoring an OLS Profit Model

The SAS program (Figure 9.1) produces the OLS profit model built with dataset A and scores the external dataset B in Table 9.3. The SAS procedure REG produces the ordinary regression coefficients and puts them in the "ols_coeff" file, as indicated by the code "outest = ols_coeff." The ols_coeff file produced by SAS is in Table 9.4.

The SAS procedure SCORE scores the five individuals in dataset B using the OLS coefficients, as indicated by the code "score = ols_coeff." The procedure appends the predicted Profit variable in Table 9.3 (called pred_Profit as indicated by "pred_Profit" in the second line of code in Figure 9.1) to the output file B_scored, as indicated by the code "out = B_scored."

9.3 Mini Case Study

I present a "big" discussion on ordinary regression modeling with the mini dataset A. I use this extremely small dataset not only to make the discussion of data mining techniques tractable but also to emphasize two aspects

```
/****** Building the OLS Profit Model on dataset A ************/
PROC REG data = A outest = ols_coeff;
pred_Profit: model Profit =
Income Age;
run;

/****** Scoring the OLS Profit Model on dataset B ************/
PROC SCORE data = B predict type = parms score = ols_coeff
out = B_scored;
var Income Age;
run;
```

FIGURE 9.1
SAS program for building and scoring OLS profit model.

TABLE 9.3

Dataset B

Income ($000)	Age (years)	Predicted Profit ($)
148	37	85.78
141	43	82.93
97	70	66.81
90	62	66.24
49	42	58.54

TABLE 9.4

OLS_Coeff File

OBS	_MODEL_	_TYPE_	_DEPVAR_	_RMSE_	Intercept	Income	Age	Profit
1	est_Profit	PARMS	Profit	6.18865	52.52778	0.26688	-0.16217	-1

of data mining. First, data mining techniques of great service should work as well with small data as with big data, as explicitly stated in the definition of data mining in Chapter 1. Second, every fruitful effort of data mining on small data is evidence that big data are not always necessary to uncover structure in the data. This evidence is in keeping with the EDA (exploratory data analysis) philosophy that the data miner should work from simplicity until indicators emerge to go further. If predictions are not acceptable, then increase data size.

The objective of the mini case study is as follows: to build an OLS Profit model based on INCOME and AGE. The ordinary regression model (celebrating over 200 years of popularity since the invention of the method of least squares on March 6, 1805) is the quintessential linear model, which implies the all-important assumption: The underlying relationship between a given predictor variable and the dependent variable is linear. Thus, I use the method of smoothed scatterplots, as described in Chapter 2, to determine whether the linear assumption holds for PROFIT with INCOME and with AGE. For the mini dataset, the smoothed scatterplot is defined by 10 slices, each of size 1. Effectively, the smooth scatterplot is the simple scatterplot of 10 paired (PROFIT, Predictor Variable) points. (Contrasting note: the logit plot as discussed with the logistic regression in Chapter 8 is neither possible nor relevant with OLS methodology. The quantitative dependent variable does not require a transformation, like converting logits into probabilities as found in logistic regression.)

9.3.1 Straight Data for Mini Case Study

Before proceeding with the analysis of the mini case study, I clarify the use of the bulging rule when analysis involves OLS regression. The bulging rule states that the model builder should try reexpressing the predictor variables as well as the dependent variable. As discussed in Chapter 8, it is not possible to reexpress the dependent variable in a logistic regression analysis. However, in performing an ordinary regression analysis, reexpressing the dependent variable is possible, but the bulging rule needs to be supplemented. Consider the illustration discussed next.

A model builder is building a profit model with the quantitative dependent variable Y and three predictor variables X_1, X_2, and X_3. Based on the bulging rule, the model builder determines that the powers of ½ and 2 for Y and X_1, respectively, produce a very adequate straightening of the Y-X_1 relationship. Let us assume that the correlation between the square root Y (sqrt_Y) and square of X_1 (sq_X_1) has a reliable $r_{sqrt\_Y,\ sq\_X1}$ value of 0.85.

Continuing this scenario, the model builder determines that the powers of 0 and ½ and -½ and 1 for Y and X_2 and Y and X_3, respectively, also produce very adequate straightening of the Y-X_2 and Y-X_3 relationships, respectively. Let us assume the correlations between the log of Y (log_Y) and the square root of X_2 (sq_X_2) and between the negative square root of Y (negsqrt_Y) and X_3 have reliable $r_{log\_Y,\ sq\_X1}$ and $r_{negsqrt\_Y,\ X3}$ values of 0.76 and 0.69, respectively. In sum, the model builder has the following results:

1. The best relationship between square root Y (p = ½) and square of X_1 has $r_{sqrt\_Y,\ sq\_X1}$ = 0.85.
2. The best relationship between log of Y (p = 0) and square root of X_2 has $r_{log\_Y,\ sq\_X2}$ = 0.76.
3. The best relationship between the negative square root of Y (p = -½) and X_3 has $r_{neg\_sqrt\_Y,\ X3}$ = 0.69.

In pursuit of a good OLS profit model, the following guidelines have proven valuable when several reexpressions of the quantitative dependent variable are suggested by the bulging rule.

1. If there is a small range of *dependent-variable powers* (powers used in reexpressing the dependent variable), then the best reexpressed dependent variable is the one with the noticeably largest correlation coefficient. In the illustration, the best reexpression of Y is the square root Y: Its correlation has the largest value: r_{sqrt_Y, sq_X1} equals 0.85. Thus, the data analyst builds the model with the square root Y and the square of X_1 and needs to reexpress X_2 and X_3 again with respect to the square root of Y.

2. If there is a small range of dependent-variable powers and the correlation coefficient values are comparable, then the best reexpressed dependent variable is defined by the *average power* among the dependent variable powers. In the illustration, if the model builder were to consider the r values (0.85, 0.76, and 0.69) comparable, then the average power would be 0, which is one of the powers used. Thus, the modeler builds the model with the log of Y and square root of X_2 and needs to reexpress X_1 and X_3 again with respect to the log of Y.

    If the average power were not one of the dependent-variable powers used, then all predictor variables would need to be reexpressed again with the newly assigned reexpressed dependent variable, Y raised to the "average power."

3. When there is a large range of dependent-variable powers, which is likely when there are many predictor variables, the practical and productive approach to the bulging rule for building an OLS profit model consists of initially reexpressing only the predictor variables, leaving the dependent variable unaltered. Choose several handfuls of reexpressed predictor variables, which have the largest correlation coefficients with the unaltered dependent variable. Then, proceed as usual, invoking the bulging rule for exploring the best reexpressions of the dependent variable and the predictor variables. If the dependent variable is reexpressed, then apply steps 1 or 2.

Meanwhile, there is an approach considered the most desirable for picking out the best reexpressed quantitative dependent variable; it is, however, neither practical nor easily assessable. It is outlined in Tukey and Mosteller's *Data Analysis and Regression* ("Graphical Fitting by Stages," pages 271–279). However, this approach is extremely tedious to perform manually as is required because there is no commercially available software for its calculations. Its inaccessibility has no consequence to the model builders' quality of model, as the approach has not provided noticeable improvement over the procedure in step 3 for marketing applications where models are implemented at the decile level.

Now that I have examined all the issues surrounding the quantitative dependent variable, I return to a discussion of reexpressing the predictor variables, starting with INCOME and then AGE.

9.3.1.1 Reexpressing INCOME

I envision an underlying positively sloped straight line running through the 10 points in the PROFIT-INCOME smooth plot in Figure 9.2, even though the smooth trace reveals four severe kinks. Based on the general association test with the test statistic (TS) value of 6, which is *almost* equal to the cutoff score of 7, as presented in Chapter 2, I conclude there is an *almost noticeable* straight-line relationship between PROFIT and INCOME. The correlation coefficient for the relationship is a reliable $r_{PROFIT, INCOME}$ of 0.763. Notwithstanding these indicators of straightness, the relationship could use some straightening, but clearly, the bulging rule does not apply.

An alternative method for straightening data, especially characterized by nonlinearities, is the GenIQ Model, a machine-learning, genetic-based data mining method. As I extensively cover this model in Chapters 29 and 30, suffice it to say that I use GenIQ to reexpress INCOME. The genetic structure, which represents the reexpressed INCOME variable, labeled gINCOME, is defined in Equation (9.3):

$$gINCOME = \sin(\sin(\sin(\sin(INCOME)*INCOME))) + \log(INCOME) \quad (9.3)$$

The structure uses the nonlinear reexpressions of the trigonometric sine function (four times) and the log (to base 10) function to loosen the "kinky"

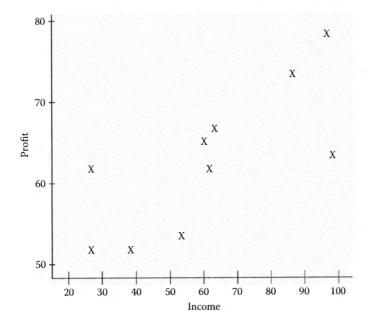

FIGURE 9.2
Plot of PROFIT and INCOME.

PROFIT-INCOME relationship. The relationship between PROFIT and INCOME (via gINCOME) has indeed been smoothed out, as the smooth trace reveals no serious kinks in Figure 9.3. Based on TS equal to 6, which again is almost equal to the cutoff score of 7, I conclude there is an almost noticeable straight-line PROFIT-gINCOME relationship, a nonrandom scatter about an underlying positively sloped straight line. The correlation coefficient for the reexpressed relationship is a reliable $r_{PROFIT, gINCOME}$ of 0.894.

Visually, the effectiveness of the GenIQ procedure in straightening the data is obvious: the sharp peaks and valleys in the original PROFIT smooth plots versus the smooth wave of the reexpressed smooth plot. Quantitatively, the gINCOME-based relationship represents a noticeable improvement of 7.24% (= (0.894 - 0.763)/0.763) increase in correlation coefficient "points" over the INCOME-based relationship.

Two points are noteworthy: Recall that I previously invoked the statistical factoid that states a dollar-unit variable is often reexpressed with the log function. Thus, it is not surprising that the genetically evolved structure gIN-COME uses the log function. With respect to logging the PROFIT variable, I concede that PROFIT could not benefit from a log reexpression, no doubt due to the "mini" in the dataset (i.e., the small size of the data), so I chose to work with PROFIT, not log of PROFIT, for the sake of simplicity (another EDA mandate, even for instructional purposes).

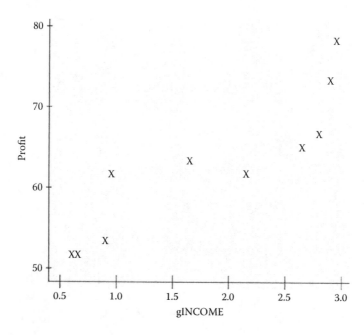

FIGURE 9.3
Plot of PROFIT and gINCOME.

9.3.1.2 Reexpressing AGE

The stormy scatter of the 10-paired (PROFIT, AGE) points in the smooth plot in Figure 9.4 is an exemplary plot of a nonrelationship between two variables. Not surprisingly, the TS value of 3 indicates there is no noticeable PROFIT-AGE relationship. Senselessly, I calculate the correlation coefficient for this nonexistent linear relationship: $r_{PROFIT,\ AGE}$ equals -0.172, which is clearly not meaningful. Clearly, the bulging rule does not apply.

I use GenIQ to reexpress AGE, labeled gAGE. The genetically based structure is defined in Equation (9.4):

$$gAGE = sin(tan(tan(2*AGE) + cos(tan(2*AGE)))) \qquad (9.4)$$

The structure uses the nonlinear reexpressions of the trigonometric sine, cosine, and tangent functions to calm the stormy-nonlinear relationship. The relationship between PROFIT and AGE (via gAGE) has indeed been smoothed out, as the smooth trace reveals in Figure 9.5. There is an almost noticeable PROFIT-gAGE relationship with TS = 6, which favorably compares to the original TS of 3. The reexpressed relationship admittedly does not portray an exemplary straight line, but given its stormy origin, I see a beautiful positively sloped ray, not very straight, but trying to shine through.

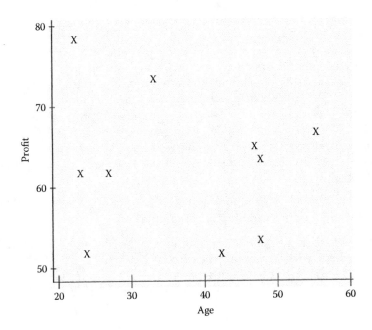

FIGURE 9.4
Plot of PROFIT and AGE.

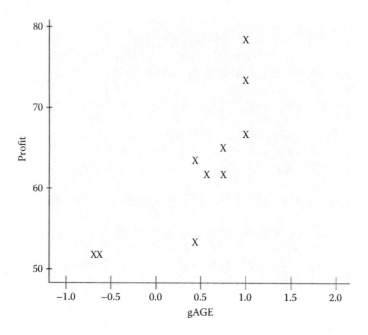

FIGURE 9.5
Plot of PROFIT and gAGE.

I consider the corresponding correlation coefficient $r_{\text{PROFIT, gAGE}}$ value of 0.819 as reliable and remarkable.

Visually, the effectiveness of the GenIQ procedure in straightening the data is obvious: the abrupt spikes in the original smooth plot of PROFIT and AGE versus the rising counterclockwise wave of the second smooth plot of PROFIT and gAGE. With enthusiasm and without quantitative restraint, the gAGE-based relationship represents a noticeable improvement—a whopping 376.2% (= (0.819 - 0.172)/0.172; disregarding the sign) improvement in correlation coefficient points over the AGE-based relationship. Since the original correlation coefficient is meaningless, the improvement percentage is also meaningless.

9.3.2 Plot of Smooth Predicted versus Actual

For a closer look at the detail of the strength (or weakness) of the gINCOME and gAGE structures, I construct the corresponding plots of PROFIT smooth predicted versus actual. The scatter about the 45° lines in the smooth plots for both gINCOME and gAGE in Figures 9.6 and 9.7, respectively, indicate a reasonable level of certainty in the reliability of the structures. In other words, both gINCOME and gAGE should be important variables for predicting PROFIT. The correlations between gINCOME-based predicted and

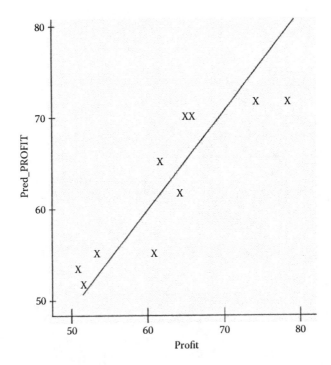

FIGURE 9.6
Smooth PROFIT predicted versus actual based on gINCOME.

actual smooth PROFIT values and between gAGE-based predicted and actual smooth PROFIT values have $r_{sm.PROFIT, sm.gINCOME}$ and $r_{sm.PROFIT, sm.gAGE}$ values equal to 0.894 and 0.819, respectively. (Why are these r values equal to $r_{PROFIT, INCOME}$ and $r_{PROFIT, AGE}$, respectively?)

9.3.3 Assessing the Importance of Variables

As in the correlating section of Chapter 8, the classical approach of assessing the statistical significance of a variable for model inclusion is the well-known null hypothesis-significance testing procedure,* which is based on the reduction in prediction error (actual PROFIT minus predicted PROFIT) associated with the variable in question. The only difference between the discussions of the logistic regression in Chapter 8 is the apparatus used. The statistical apparatus of the formal testing procedure for ordinary regression consists of the sum of squares (total; due to regression; due to error), the F statistic, degrees of freedom (df), and the p value. The procedure uses

---

* "What If There Were No Significance Testing?" (on author's Web site, http://www.geniq.net/res/What-If-There-Were-No-Significance-Testing.html).

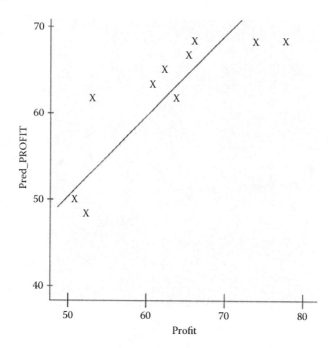

FIGURE 9.7
Smooth PROFIT predicted versus actual based on gAGE.

the apparatus within a theoretical framework with weighty and untenable assumptions, which, from a purist's point of view, can cast doubt on
findings of statistical significance. Even if findings of statistical significance are accepted as correct, it may not be of practical importance or have
noticeable value to the study at hand. For the data miner with a pragmatist
slant, the limitations and lack of scalability of the classical system of variable assessment cannot be overlooked, especially within big data settings.
In contrast, the data mining approach uses the F statistic, R-squared, and
degrees of freedom in an informal data-guided search for variables that
suggest a noticeable reduction in prediction error. Note that the informality
of the data mining approach calls for suitable change in terminology, from
declaring a result as statistically significant to worthy of notice or noticeably important.

9.3.3.1 Defining the F Statistic and R-Squared

In data mining, the assessment of the importance of a subset of variables for
predicting profit involves the notion of a noticeable reduction in prediction
error due to the subset of variables. It is based on the F statistic, R-squared,
and degrees of freedom, which are always reported in the ordinary regression

output. For the sake of reference, I provide their definitions and relationship with each other in Equations (9.5), (9.6), and (9.7).

$$F = \frac{\text{Sum of squares due to regression/df due to regression model}}{\text{Sum of squares due to error/df due to error in regression model}} \quad (9.5)$$

$$R\text{-squared} = \frac{\text{Sum of squares due to regression}}{\text{Total Sum of squares}} \quad (9.6)$$

$$F = \frac{R\text{-squared/number of variables in model}}{(1 - R\text{-squared})/(\text{sample size} - \text{number of variables in model} - 1)} \quad (9.7)$$

For the sake of completion, I provide an additional statistic: the adjusted R-squared. R-squared is affected, among other things, by the ratio of the number of predictor variables in the model to the size of the sample. The larger the ratio, the greater the overestimation of R-squared is. Thus, the adjusted R-squared as defined in Equation (9.8) is not particularly useful in big data settings.

Adjusted R-squared =

$$1 - (1 - R\text{-squared}) \frac{(\text{sample size} - 1)}{(\text{sample size} - \text{number of variables in model} - 1)} \quad (9.8)$$

In the following sections, I detail the decision rules for three scenarios for assessing the importance of variables (i.e., the likelihood the variables have some predictive power). In brief, the larger the F statistic, R-squared, and adjusted R-squared values, the more important the variables are in predicting profit.

9.3.3.2 Importance of a Single Variable

If X is the only variable considered for inclusion into the model, the decision rule for declaring X an important predictor variable in predicting profit is if the F value due to X is greater than the *standard F value* 4, then X is an important predictor variable and should be considered for inclusion in the model. Note that the decision rule only indicates that the variable has some importance, not how much importance. The decision rule implies that a variable with a greater F value has a greater *likelihood of some importance* than a variable with a smaller F value, not that it has greater importance.

9.3.3.3 *Importance of a Subset of Variables*

When subset A consisting of k variables is the only subset considered for model inclusion, the decision rule for declaring subset A important in predicting profit is as follows: If the average F value per number of variables (the degrees of freedom) in the subset A—F/df or F/k—is greater than standard F value 4, then subset A is an important subset of predictor variable and should be considered for inclusion in the model. As before, the decision rule only indicates that the subset has some importance, not how much importance.

9.3.3.4 *Comparing the Importance of Different Subsets of Variables*

Let subsets A and B consist of k and p variables, respectively. The number of variables in each subset does not have to be equal. If the number of variables is equal, then all but one variable can be the same in both subsets. Let $F(k)$ and $F(p)$ be the F values corresponding to the models with subsets A and B, respectively.

The decision rule for declaring which of the two subsets is more important (greater likelihood of some predictive power) in predicting profit is

1. If $F(k)/k$ is greater than $F(p)/p$, then subset $A(k)$ is the more important predictor variable subset; otherwise, $B(p)$ is the more important subset.
2. If $F(k)/k$ and $F(p)/p$ are equal or have comparable values, then both subsets are to be regarded tentatively as of comparable importance. The model builder should consider additional indicators to assist in the decision about which subset is better. It clearly follows from the decision rule that the model defined by the more important subset is the better model. (Of course, the rule assumes that F/k and F/p are greater than the standard F value 4.)

Equivalently, the decision rule either can use R-squared or adjusted R-squared in place of F/df. The R-squared statistic is a friendly concept in that its values serve as an indicator of the percentage of variation explained by the model.

9.4 Important Variables for Mini Case Study

I perform two ordinary regressions, regressing PROFIT on gINCOME and on gAGE; the outputs are in Tables 9.5 and 9.6, respectively. The F values are

TABLE 9.5

OLS Output: PROFIT with gINCOME

Source	df	Sum of Squares	Mean Square	F Value	Pr > F
Model	1	582.1000	582.1000	31.83	0.0005
Error	8	146.3000	18.2875		
Corrected total	9	728.4000			
		Root MSE	4.2764	R-square	0.7991
		Dependent mean	62.6000	Adj R-sq	0.7740
		Coeff var	6.8313		

Variable	df	Parameter Estimate	Standard Error	t Value	Pr > |t|
INTERCEPT	1	47.6432	2.9760	16.01	<.0001
gINCOME	1	8.1972	1.4529	5.64	0.0005

TABLE 9.6

OLS Output: PROFIT with gAGE

Source	df	Sum of Squares	Mean Square	F Value	Pr > F
Model	1	488.4073	488.4073	16.28	0.0038
Error	8	239.9927	29.9991		
Corrected total	9	728.4000			
		Root MSE	5.4771	R-Square	0.6705
		Dependent mean	62.6000	Adj R-sq	0.6293
		Coeff var	8.7494		

Variable	df	Parameter Estimate	Standard Error	t Value	Pr > |t|
INTERCEPT	1	57.2114	2.1871	26.16	< .0001
gAGE	1	11.7116	2.9025	4.03	0.0038

31.83 and 16.28, respectively, which are greater than the standard F value 4. Thus, both gINCOME and gAGE are declared important predictor variables of PROFIT.

9.4.1 Relative Importance of the Variables

Chapter 8 contains the same heading (Section 8.12), with only a minor variation with respect to the statistic used. The *t statistic* as posted in ordinary regression output can serve as an indicator of the relative importance of a variable and for selecting the best subset, which is discussed next.

9.4.2 Selecting the Best Subset

The decision rules for finding the best subset of important variables are nearly the same as those discussed in Chapter 8; refer to Section 8.12.1. Point 1 remains the same for this discussion. However, the second and third points change, as follows:

1. Select an initial subset of important variables.

2. For the variables in the initial subset, generate smooth plots and straighten the variables as required. The most noticeable handfuls of original and reexpressed variables form the starter subset.

3. Perform the preliminary ordinary regression on the starter subset. Delete one or two variables with absolute t-statistic values less than the *t cutoff value 2* from the model. This results in the first incipient subset of important variables. Note the changes to points 4, 5, and 6 with respect to the topic of this chapter.

4. Perform another ordinary regression on the incipient subset. Delete one or two variables with t values less than the t cutoff value 2 from the model. The data analyst can create an illusion of important variables appearing and disappearing with the deletion of different variables. The remainder of the discussion in Chapter 8 remains the same.

5. Repeat step 4 until all retained predictor variables have comparable t values. This step often results in different subsets as the data analyst deletes judicially different pairings of variables.

6. Declare the best subset by comparing the relative importance of the different subsets using the decision rule in Section 9.3.3.4.

9.5 Best Subset of Variables for Case Study

I build a preliminary model by regressing PROFIT on gINCOME and gAGE; the output is in Table 9.7. The two-variable subset has an F/df value of 7.725 (= 15.45/2), which is greater than the standard F value 4. But, the t value for gAGE is 0.78 less than the t cutoff value (see bottom section of Table 9.7). If I follow step 4, then I would have to delete gAGE, yielding a simple regression model with the lowly, albeit straight, predictor variable gINCOME. By the way, the adjusted R-squared is 0.7625 (after all, the entire minisample is not big).

Before I dismiss the two-variable (gINCOME, gAGE) model, I construct the smooth residual plot in Figure 9.8 to determine the quality of the predictions

TABLE 9.7

OLS Output: PROFIT with gINCOME and gAGE

Source	df	Sum of Squares	Mean Square	F Value	Pr > F
Model	2	593.8646	296.9323	15.45	0.0027
Error	7	134.5354	19.2193		
Corrected total	9	728.4000			
		Root MSE	4.3840	R-squared	0.8153
		Dependent mean	62.6000	Adj R-sq	0.7625
		Coeff var	7.0032		

Variable	df	Parameter Estimate	Standard Error	t Value	Pr > |t|
INTERCEPT	1	49.3807	3.7736	13.09	< .0001
gINCOME	1	6.4037	2.7338	2.34	0.0517
gAGE	1	3.3361	4.2640	0.78	0.4596

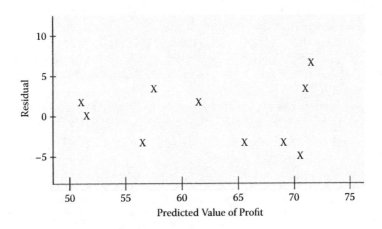

FIGURE 9.8
Smooth residual plot for (gINCOME, gAGE) model.

of the model. The smooth residual plot is declared to be equivalent to the null plot-based general association test (TS = 5). Thus, the overall quality of the predictions is considered good. That is, on average, the predicted PROFIT is equal to the actual PROFIT. Regarding the descriptive statistics for the smooth residual plot, for the smooth residual, the minimum and maximum values and range are −4.567, 6.508, and 11.075, respectively; the standard deviation of the smooth residuals is 3.866.

9.5.1 PROFIT Model with gINCOME and AGE

With a vigilance in explaining the unexpected, I suspect the reason for the relative nonimportance of gAGE (i.e., gAGE is not important in the presence of gINCOME) is the strong correlation of gAGE with gINCOME: $r_{gINCOME,\ gAGE} = 0.839$. This supports my contention but does not confirm it.

In view of the foregoing, I build another two-variable model regressing PROFIT on gINCOME and AGE; the output is in Table 9.8. The (gINCOME, AGE) subset has an F/df value of 12.08 (= 24.15/2), which is greater than the standard F value 4. Statistic happy, I see the t values for both variables are greater than the t cutoff value 2. I cannot overlook the fact that the raw variable AGE, which by itself is not important, now has relative importance in the presence of gINCOME. (More about this "phenomenon" at the end of the chapter.) Thus, the evidence is that the subset of gINCOME and AGE is better than the original (gINCOME, gAGE) subset. By the way, the adjusted R-squared is 0.8373, representing a 9.81% (= (0.8373 - 0.7625)/0.7625) improvement in "adjusted R-squared" points over the original-variable adjusted R-squared.

The smooth residual plot for the (gINCOME, AGE) model in Figure 9.9 is declared to be equivalent to the null plot-based general association test with TS = 4. Thus, there is indication that the overall quality of the predictions is good. Regarding the descriptive statistics for the two-variable smooth residual plot, for the smooth residual, the minimum and maximum values and range are −5.527, 4.915, and 10.442, respectively, and the standard deviation of the smooth residual is 3.200.

To obtain further indication of the quality of the predictions of the model, I construct the plot of the smooth actual versus predicted for the (gINCOME,

TABLE 9.8

OLS Output: PROFIT with gINCOME and AGE

Source	df	Sum of Squares	Mean Square	F Value	Pr > F
Model	2	636.2031	318.1016	24.15	0.0007
Error	7	92.1969	13.1710		
Corrected total	9	728.4000			
		Root MSE	3.6291	R-squared	0.8734
		Dependent mean	62.6000	Adj R-sq	0.8373
		Coeff var	5.79742		

Variable	df	Parameter Estimate	Standard Error	t Value	Pr > |t|
INTERCEPT	1	54.4422	4.1991	12.97	< .0001
gINCOME	1	8.4756	1.2407	6.83	0.0002
AGE	1	-0.1980	0.0977	-2.03	0.0823

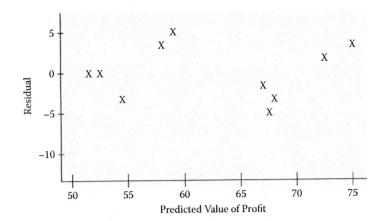

FIGURE 9.9
Smooth residual plot for (gINCOME, AGE) model.

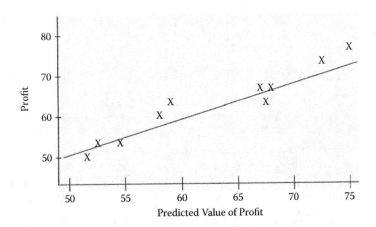

FIGURE 9.10
Smooth actual versus predicted plot for (gINCOME, AGE) model.

AGE) model in Figure 9.10. The smooth plot is acceptable with minimal scatter of the 10 smooth points about the 45 line. The correlation between smooth actual versus predicted PROFIT based on gINCOME and AGE has an $r_{sm.}$ $_{gINCOME, sm.AGE}$ value of 0.93. (This value is the square root of R-squared for the model. Why?)

For a point of comparison, I construct the plot of smooth actual versus predicted for the (gINCOME, gAGE) model in Figure 9.11. The smooth plot is acceptable with minimal scatter of the 10 smooth points about the 45° line, with a noted exception of some wild scatter for PROFIT values greater than $65. The correlation between smooth actual versus predicted PROFIT based

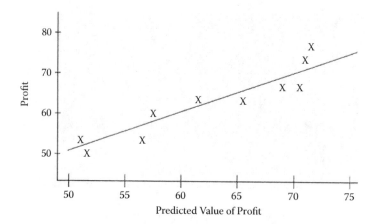

FIGURE 9.11
Smooth actual versus predicted plot for (gINCOME, gAGE) model.

on gINCOME and gAGE has an $r_{sm.gINCOME,\, sm.gAGE}$ value of 0.90. (This value is the square root of R-squared for the model. Why?)

9.5.2 Best PROFIT Model

To decide which of the two PROFIT models is better, I put the vital statistics of the preceding analyses in Table 9.9. Based on the consensus of a committee of one, I prefer the gINCOME-AGE model to the gINCOME-gAGE model because there are noticeable indications that the predictions of the former model are better. The former model offers a 9.8% increase in the adjusted R-squared (less bias), a 17.2% decrease in the smooth residual standard deviation (more stable), and a 5.7% decrease in the smooth residual range (more stable).

9.6 Suppressor Variable AGE

A variable whose behavior is like that of AGE—poorly correlated with the dependent variable Y, but becomes important by its inclusion in a model for predicting Y—is known as a *suppressor variable*. [1, 2] The consequence of a suppressor variable is that it increases the R-squared of the model.

I explain the behavior of the suppressor variable within the context of the mini case study. The presence of AGE in the model removes or suppresses the information (variance) in gINCOME that is not related to the variance in PROFIT; that is, AGE suppresses the unreliable noise in gINCOME. This

TABLE 9.9

Comparison of Vital Statistics for Two PROFIT Models
Smooth Residual

Model	Predictor Variables	F Value	t Value	Range	StdDev	Adjusted R-Square
First	gINCOME, gAGE	Greater than cutoff value	Greater than cutoff value for only gINCOME	11.075	3.866	0.7625
Second	gINCOME, AGE	Greater than cutoff value	Greater than cutoff value for both variablest	10.442	3.200	0.8373
Indication	Improvement of 2nd model over 1st model	NA	Because t value for gAGE is less than t cutoff value, gAGE contributes "noise" in model, as evidenced in range, stdDev and adjusted R-square	-5.7%	-17.2%	9.8%

renders the AGE-adjusted variance in gINCOME more reliable or potent for predicting PROFIT.

I analyze the paired correlations among the three variables to clarify exactly what AGE is doing. Recall that squaring the correlation coefficient represents the "shared variance" between the two variables under consideration. The paired bits of information are in Table 9.10. I know from the prior analysis that PROFIT and AGE have no noticeable relationship; their shared variance of 3% confirms this. I also know that PROFIT and gINCOME do have a noticeable relationship; their shared variance of 80% confirms this as well. In the presence of AGE, the relationship between PROFIT and gINCOME, specifically, the relationship between PROFIT and gINCOME adjusted for AGE has a shared variance of 87%. This represents an improvement of 8.75% (= (0.87 - 0.80/0.80) in shared variance. This "new" variance is now available for predicting PROFIT, increasing the R-squared (from 79.91% to 87.34%).

It is a pleasant surprise in several ways that AGE turns out to be a suppressor variable. First, suppressor variables occur most often in big data settings, not often with small data, and are truly unexpected with minidata. Second, the suppressor variable scenario serves as object lessons for the EDA

TABLE 9.10

Comparison of Pairwise Correlations among PROFIT, AGE, and gINCOME

Correlation Pair	Correlation Coefficient	Shared Variance
PROFIT and AGE	-0.172	3%
PROFIT and gINCOME	0.894	80%
PROFIT and AGE in the presence of gINCOME	0.608	37%
PROFIT and gINCOME in the presence of AGE	0.933	87%

paradigm: Dig, dig, dig into the data, and you will find gold or some reward for your effort. Third, the suppressor variable scenario is a small reminder of a big issue: The model builder must not rely solely on predictor variables that are highly correlated with the dependent variable, but also must consider the poorly correlated predictor variables as they are a great source of latent predictive importance.

9.7 Summary

The ordinary regression model is presented as the workhorse of profit modeling as it has been in steady use for almost 200 years. As such, I illustrated in an orderly and detailed way the essentials of ordinary regression. Moreover, I showed the enduring usefulness of this popular analysis and modeling technique as it works well within the EDA/data mining paradigm of today.

I first illustrated the rudiments of the ordinary regression model by discussing the SAS program for building and scoring an ordinary regression model. The program is a welcome addition to the tool kit of techniques used by model builders working on predicting a quantitative dependent variable.

Then, I discussed ordinary regression modeling with minidata. I used this extremely small dataset not only to make the discussion of the data mining techniques tractable but also to emphasize two aspects of data mining. First, data mining techniques of great service should work as well with big data as with small data. Second, every fruitful effort of data mining on small data is evidence that big data are not always necessary to uncover structure in the data. This evidence is in keeping with the EDA philosophy that the data miner should work from simplicity until indicators emerge to go further: If predictions are not acceptable, then increase data size. The data mining techniques discussed are those introduced in the logistic regression framework of Chapter 8 and carry over with minor modification to ordinary regression.

Before proceeding with the analysis of the mini case study, I supplemented the bulging rule when analysis involves ordinary regression. Unlike in logistic regression, for which the logit dependent variable cannot be reexpressed, in ordinary regression the quantitative dependent variable can be reexpressed. The bulging rule as introduced within the logistic regression framework in Chapter 8 can put several reexpressions of the quantitative dependent variable up for consideration. For such cases, I provided additional guidelines to the bulging rule, which will prove valuable:

1. If there is a small range of dependent-variable powers, then the best reexpressed dependent variable is the one with the noticeably largest correlation coefficient.

2. If there is a small range of dependent-variable powers and the correlation coefficient values are comparable, then the best reexpressed dependent variable is defined by the average power among the dependent-variable powers.

3. When there is a large range of dependent-variable powers, choose several handfuls of reexpressed predictor variables, which have the largest correlation coefficients with the unaltered quantitative dependent variable. Then, proceed as usual, invoking the bulging rule for exploring the best reexpressions of the dependent variable and the predictor variables. If the dependent variable is reexpressed, then apply steps 1 or 2.

With the minidataset selected for regressing PROFIT on INCOME and AGE, I introduced the alternative GenIQ Model, a machine-learning, genetic-based data mining method for straightening data. I illustrated the data mining procedure of smoothing and assessing the smooth with the general association test and determined that both predictor variables need straightening, but the bulging rule does not apply. The GenIQ Model evolved reasonably straight relationships between PROFIT and each reexpressed predictor variable, gINCOME and gAGE, respectively. I generated the plots of PROFIT smooth predictive versus actual, which provided further indication that gINCOME and gAGE should be important variables for predicting PROFIT.

Continuing with the mini case study, I demonstrated an ordinary regression-specific data mining alternative approach to the classical method of assessing. (The data mining techniques discussed are those introduced in the logistic regression framework of Chapter 8 and carry over with minor modification to ordinary regression.) This alternative approach involves the importance of individual predictor variables, as well as the importance of a subset of predictor variables and the relative importance of individual predictor variables, in addition to the goodness of model predictions. Additional methods that set out specific to ordinary regression included selecting the best subset of predictor variables and comparing the importance between two subsets of predictor variables.

Within my illustration of the case study is a pleasant surprise—the existence of a suppressor variable (AGE). A variable whose behavior is poorly correlated with the dependent variable but becomes important by its inclusion in a model for predicting the dependent variable is known as a suppressor variable. The consequence of a suppressor variable is that it increases the R-squared of the model. A suppressor variable occurs most often in big data settings, not often with small data, and is truly unexpected with minidata. The suppressor variable scenario served as an object lesson for the EDA paradigm: Dig deeply into the data, and you will find a reward for your effort. And, the suppressor variable scenario is a small reminder of a bigger issue: The model builder must not rely solely on predictor variables that are highly correlated with the dependent variable but also should consider the poorly correlated predictor variables as they are a great source of latent predictive importance.

References

1. Horst, P., The role of predictor variables which are independent of the criterion, *Social Science Research Bulletin*, 48, 431–436, 1941.
2. Conger, A.J., A revised definition for suppressor variables: A guide to their identification and interpretation, *Educational and Psychological Measurement*, 34, 35–46, 1974.

10

Variable Selection Methods in Regression: Ignorable Problem, Notable Solution

10.1 Introduction

Variable selection in regression—identifying the best subset among many variables to include in a model—is arguably the hardest part of model building. Many variable selection methods exist. Many statisticians know them, but few know they produce poorly performing models. The deficient variable selection methods are a miscarriage of statistics because they are developed by debasing sound statistical theory into a misguided pseudotheoretical foundation. The goal of this chapter is twofold: (1) resurface the scope of literature on the weaknesses of variable selection methods and (2) enliven anew a notable solution for defining a substantial performing regression model. To achieve my goal tactically, I divide the chapter into two objectives. First, I review the five frequently used variable selection methods. Second, I present Tukey's exploratory data analysis (EDA) relevant to the titled topic: the natural seven-step cycle of statistical modeling and analysis (previously discussed in chapter page 2). The "seven-step cycle" serves as a notable solution to variable selection in regression. I feel that newcomers to Tukey's EDA need the seven-step cycle introduced within the narrative of Tukey's analytic philosophy. Accordingly, I enfold the solution with front and back matter: the essence of EDA and the EDA school of thought, respectively. John W. Tukey (1915–2000) was a mega-contributor to the field of statistics and was a humble, unpretentious man, as he always considered himself as a data analyst. Tukey's seminal book, Exploratory Data Analysis [1] is uniquely known by the book's initialed title, EDA.

10.2 Background

Classic statistics dictates that the statistician set about dealing with a given problem with a prespecified procedure designed for that problem. For example, the problem of predicting a continuous dependent variable (e.g., profit) is solved by using the ordinary least squares (OLS) regression model *along with checking* the well-known underlying OLS assumptions [2]. At hand, there are *several* candidate predictor variables, allowing a workable task for the statistician to check assumptions (e.g., predictor variables are linearly independent). Likewise, the dataset has a *practicable* number of observations, making it also a workable task for the statistician to check assumptions (e.g., the errors are uncorrelated). As well, the statistician can perform the *regarded yet often-discarded* exploratory data analysis (also known as EDA), such as examine and apply the appropriate remedies for individual records that contribute to sticky data characteristics (e.g., gaps, clumps, and outliers). It is important that EDA allows the statistician to assess whether a given variable, say, X, needs a transformation/reexpression (e.g., $\log(X)$, $\sin(X)$ or $1/X$). The traditional variable selection methods cannot do such transformations or a priori construction of new variables from the original variables [3]. *Inability* of construction of new variables is a serious weakness of the variable selection methodology [1].

Now, building an OLS regression model or a logistic regression model (LRM; the dependent variable is binary) is problematic because of the size of the dataset. Model builders work on *big data*—consisting of a teeming multitude of variables and an army of observations. The workable tasks are no longer feasible. Model builders cannot sure-footedly use OLS regression and LRM on big data as the two statistical regression models were conceived, tested, and experimented within the small-data setting of the day over 60 and 200 years ago for LRM and OLS regression, respectively. The regression theoretical foundation and the tool of significance testing* employed on big data are without statistical binding force. Thus, fitting big data to a prespecified small-framed model produces a skewed model with doubtful interpretability and questionable results.

Folklore has it that the knowledge and practice of variable selection methods were developed when small data slowly grew into the early size of big data around the late 1960s and early 1970s. With only a single bibliographic citation ascribing variable selection methods to unsupported notions, I believe a reasonable scenario of the genesis of the methods was as follows [4]: College statistics nerds (intelligent thinkers) and computer science geeks (intelligent doers) put together the variable selection methodology using a *trinity of selection components*:

1. Statistical tests (e.g., F, chi-square, and t tests) and significance testing
2. Statistical criteria (e.g., R-squared [R-sq], adjusted R-sq, Mallows's C_p, and MSE [mean squared error]) [5]
3. Statistical stopping rules (e.g., p value flags for variable entry/deletion/staying in a model)

---

* "What If There Were No Significance Testing?" (on the author's Web site, http://www.geniq.net/res/What-If-There-Were-No-Significance-Testing.html).

The newborn-developed variable selection methods were on bearing soil of expertness and adroitness in computer-automated misguided statistics. The trinity distorts its components' original theoretical and inferential meanings when they are framed within the newborn methods. The statistician executing the computer-driven trinity of statistical apparatus in a seemingly intuitive and insightful way gave proof—*face validity*—that the problem of variable selection (also known as subset selection) was solved (at least to the uninitiated statistician).

The newbie subset selection methods initially enjoyed wide acceptance with extensive use and still do presently. Statisticians build *at-risk* accurate and stable models—either *unknowingly* using these unconfirmed methods or *knowingly* exercising these methods because *they know not what else to do*. It was not long before the weaknesses of these methods, some contradictory, generated many commentaries in the literature. I itemize nine ever-present weaknesses for two of the traditional variable selection methods, *all-subset*, and *stepwise* (SW). I concisely describe the five frequently used variable selection methods in the next section.

1. For all-subset selection with more than 40 variables [4]:
   a. The number of possible subsets can be huge.
   b. Often, there are several good models, although some are unstable.
   c. The best X variables may be no better than random variables if sample size is relatively small compared to the number of all variables.
   d. The regression statistics and regression coefficients are biased.
2. All-subset selection regression can yield models that are too small [6].
3. The number of candidate variables and not the number in the final model is the number of degrees of freedom to consider [7].
4. The data analyst knows more than the computer—and failure to use that knowledge produces inadequate data analysis [8].
5. SW selection yields confidence limits that are far too narrow [9].
6. Regarding frequency of obtaining authentic and noise variables: The degree of correlation among the predictor variables affected the frequency with which authentic predictor variables found their way into the final model. The number of candidate predictor variables affected the number of noise variables that gained entry to the model [10].
7. SW selection will not necessarily produce the best model if there are redundant predictors (common problem) [11].
8. There are two distinct questions here: (a) When is SW selection appropriate? (b) Why is it so popular [12]?

9. Regarding question 8b, there are two groups that are inclined to favor its usage. One consists of individuals with little formal training in data analysis; this group confuses knowledge of data analysis with knowledge of the syntax of SAS, SPSS, and the like. They seem to figure that *if its there in a program, it has to be good and better than actually thinking about what the data might look like.* They are fairly easy to spot and to condemn in a right-thinking group of well-trained data analysts. However, there is also a second group who is often well trained: They believe in statistics; given any properly obtained database, a suitable computer program can objectively make substantive inferences without active consideration of the underlying hypotheses. *Stepwise selection* is the parent of this line of *blind data analysis* [13].

Currently, there is burgeoning research that continues the original efforts of subset selection by shoring up its pseudotheoretical foundation. It follows a line of examination that adds assumptions and makes modifications for eliminating the weaknesses. As the traditional methods are being mended, there are innovative approaches with starting points far afield from their traditional counterparts. There are freshly minted methods, like the *enhanced variable selection method* built in the GenIQ Model, constantly being developed [14–17].

10.3 Frequently Used Variable Selection Methods

Variable selection in regression—identifying the best subset among many variables to include in a model—is arguably the hardest part of model building. Many variable selection methods exist because they provide a solution to one of the most important problems in statistics [18, 19]. Many statisticians know them, but few know they produce poorly performing models. The deficient variable selection methods are a miscarriage of statistics because they were developed by debasing sound statistical theory into a misguided pseudotheoretical foundation. They are executed with computer-intensive search heuristics guided by rules of thumb. Each method uses a unique trio of elements, one from each component of the trinity of selection components [20]. Different sets of elements typically produce different subsets. The number of variables in common with the different subsets is small, and the sizes of the subsets can vary considerably.

An alternative view of the problem of variable selection is to examine certain subsets and select the best subset that either maximizes or minimizes an appropriate criterion. Two subsets are obvious: the best single variable

and the complete set of variables. The problem lies in selecting an intermediate subset that is better than both of these extremes. Therefore, the issue
is how to find the *necessary variables* among the complete set of variables by
deleting both *irrelevant variables* (variables not affecting the dependent variable) and *redundant variables* (variables not adding anything to the dependent
variable) [21].

I review five frequently used variable selection methods. These everyday
methods are found in major statistical software packages [22]. The test statistic (TS) for the first three methods uses either the F statistic for a continuous
dependent variable or the G statistic for a binary dependent variable. The TS
for the fourth method is either R-sq for a continuous dependent variable or
the Score statistic for a binary dependent variable. The last method uses one
of the following criteria: R-sq, adjusted R-sq, Mallows's C_p.

1. Forward Selection (FS). This method adds variables to the model
   until no remaining variable (outside the model) can add anything significant to the dependent variable. FS begins with no variable in the
   model. For each variable, the TS, a measure of the contribution of the
   variable to the model, is calculated. The variable with the largest TS
   value that is greater than a preset value C is added to the model. Then,
   the TS is calculated again for the variables still remaining, and the
   evaluation process is repeated. Thus, variables are added to the model
   one by one until no remaining variable produces a TS value that is
   greater than C. Once a variable is in the model, it remains there.

2. Backward Elimination (BE). This method deletes variables one by
   one from the model until all remaining variables contribute something significant to the dependent variable. BE begins with a model
   that includes all variables. Variables are then deleted from the model
   one by one until all the variables remaining in the model have TS
   values greater than C. At each step, the variable showing the smallest contribution to the model (i.e., with the smallest TS value that is
   less than C) is deleted.

3. Stepwise (SW). This method is a modification of the FS approach
   and differs in that variables already in the model do not necessarily stay. As in FS, SW adds variables to the model one at a time.
   Variables that have a TS value greater than C are added to the
   model. After a variable is added, however, SW looks at all the variables already included to delete any variable that does not have a TS
   value greater than C.

4. R-squared (R-sq). This method finds several subsets of different sizes
   that best predict the dependent variable. R-sq finds subsets of variables that best predict the dependent variable based on the appropriate TS. The best subset of size k has the largest TS value. For a

continuous dependent variable, TS is the popular measure R-sq, the coefficient of multiple determination, which measures the proportion of the *explained* variance in the dependent variable by the multiple regression. For a binary dependent variable, TS is the theoretically correct but less-known Score statistic [23]. R-sq finds the best one-variable model, the best two-variable model, and so forth. However, it is unlikely that one subset will stand out as clearly the best as TS values are often bunched together. For example, they are equal in value when rounded at the, say, third place after the decimal point [24]. R-sq generates a number of subsets of each size, which allows the user to select a subset, possibly using nonstatistical conditions.

5. All-possible subsets. This method builds all one-variable models, all two-variable models, and so on, until the last all-variable model is generated. The method requires a powerful computer (because many models are produced) and selection of any one of the criteria: R-sq, adjusted R-sq, Mallows's C_p.

10.4 Weakness in the Stepwise

An ideal variable selection method for regression models would find one or more subsets of variables that produce an *optimal* model [25]. The objective of the ideal method states that the resultant models include the following elements: accuracy, stability, parsimony, interpretability, and lack of bias in drawing inferences. Needless to say, the methods enumerated do not satisfy most of these elements. Each method has at least one drawback specific to its selection criterion. In addition to the nine weaknesses mentioned, I itemize a compiled list of weaknesses of the *most popular SW* method [26].

1. It yields R-sq values that are badly biased high.
2. The F and chi-squared tests quoted next to each variable on the printout do not have the claimed distribution.
3. The method yields confidence intervals for effects and predicted values that are falsely narrow.
4. It yields p values that do not have the proper meaning, and the proper correction for them is a very difficult problem.
5. It gives biased regression coefficients that need shrinkage (the coefficients for remaining variables are too large).
6. It has severe problems in the presence of collinearity.
7. It is based on methods (e.g., F tests) intended to be used to test prespecified hypotheses.

8. Increasing the sample size does not help very much.

9. It allows us not to think about the problem.

10. It uses a lot of paper.

11. The number of candidate predictor variables affected the number of noise variables that gained entry to the model.

I add to the tally of weaknesses by stating common weaknesses in regression models, as well as those specifically related to the OLS regression model and LRM:

> The everyday variable selection methods in the regression model result typically in models having too many variables, an indicator of overfit. The prediction errors, which are inflated by outliers, are not stable. Thus, model implementation results in unsatisfactory performance. For OLS regression, it is well known that in the absence of normality or absence of linearity assumption or outlier presence in the data, variable selection methods perform poorly. For logistic regression, the reproducibility of the computer-automated variable selection models is poor. The variables selected as predictor variables in the models are sensitive to unaccounted sample variation in the data.

Given the litany of weaknesses cited, the lingering question is: Why do statisticians use variable selection methods to build regression models? To paraphrase Mark Twain: "Get your [data] first, and then you can distort them as you please" [27]. My answer is: "Modeler builders use variable selection methods every day because they can." As a counterpoint to the absurdity of "because they can," I enliven anew Tukey's solution of the natural seven-step cycle of statistical modeling and analysis to define a substantial performing regression model. I feel that newcomers to Tukey's EDA need the seven-step cycle introduced within the narrative of Tukey's analytic philosophy. Accordingly, I enfold the solution with front and back matter—the essence of EDA and the EDA school of thought, respectively. I delve into the trinity of Tukey's masterwork. But, first I discuss an enhanced variable selection method for which I might be the only exponent for appending this method to the current baseless arsenal of variable selection.

10.5 Enhanced Variable Selection Method

In lay terms, the variable selection problem in regression can be stated as follows:

> Find the best combination of the original variables to include in a model. The variable selection method neither states nor implies that it has an attribute to concoct new variables stirred up by mixtures of the original variables.

The attribute—data mining—is either overlooked, perhaps because it is reflective of the simple-mindedness of the problem solution at the onset, or currently sidestepped as the problem is too difficult to solve. A variable selection method without a data mining attribute obviously hits a wall, beyond which it would otherwise increase the predictiveness of the technique. In today's terms, the variable selection methods are *without* data mining capability. They cannot dig the data for the mining of potentially important new variables. (This attribute, which has never surfaced during my literature search, is a partial mystery to me.) Accordingly, I put forth a definition of an enhanced variable selection method:

> An enhanced variable selection method is one that identifies a subset that consists of the original variables *and* data-mined variables, whereby *the latter are a result of the data mining attribute of the method itself.*

The following five discussion points clarify the attribute weakness and illustrate the concept of an enhanced variable selection method:

1. Consider the complete set of variables X_1, X_2, ¼ , X_{10}. Any of the current variable selection methods in use finds the best combination of the original variables (say X_1, X_3, X_7, X_{10}), but it can never automatically transform a variable (say transform X_1 to log X_1) if it were needed to increase the *information content* (*predictive power*) of that variable. Furthermore, none of the methods can generate a reexpression of the original variables (perhaps X_3/X_7) if the constructed variable (*structure*) were to offer more predictive power than the original component variables combined. In other words, current variable selection methods cannot find an *enhanced subset*, which needs, say, to include transformed and reexpressed variables (possibly X_1, X_3, X_7, X_{10}, log X_1, X_3/X_7). A subset of variables without the potential of new structure offering more predictive power clearly limits the model builder in building the best model.

2. Specifically, the current variable selection methods fail to identify structure of the types discussed here: *transformed variables* with a *preferred* shape. A variable selection procedure should have the ability to transform an individual variable, if necessary, to induce a symmetric distribution. Symmetry is the preferred shape of an individual variable. For example, the workhorse of statistical measures—the mean and variance—is based on symmetric distribution. A skewed distribution produces inaccurate estimates for means, variances, and related statistics, such as the correlation coefficient. Symmetry facilitates the interpretation of the effect of the variable in an analysis. A skewed distribution is difficult to examine because most of the observations are bunched together at one end of the distribution.

Modeling and analyses based on skewed distributions typically provide a model with doubtful interpretability and questionable results.

3. The current variable selection method also should have the ability to *straighten* nonlinear relationships. A linear or straight-line relationship is the *preferred* shape when considering two variables. A straight-line relationship between independent and dependent variables is an assumption of the popular statistical linear regression models (e.g., OLS regression and LRM). (Remember that a linear model is defined as a sum of weighted variables, such as $Y = b_0 + b_1{}^*X_1 + b_2{}^*X_2 + b_3{}^*X_3$ [28].) Moreover, straight-line relationships *among all* the independent variables constitute a *desirable property* [29]. In brief, straight-line relationships are easy to interpret: A unit of increase in one variable produces an expected constant increase in a second variable.

4. *Constructed variables* are derived from the original variables using simple arithmetic functions. A variable selection method should have the ability to construct simple reexpressions of the the original variables. Sum, difference, ratio, or product variables potentially offer more information than the original variables themselves. For example, when analyzing the efficiency of an automobile engine, two important variables are miles traveled and fuel used (gallons). However, it is known that the ratio variable of miles per gallon is the best variable for assessing the performance of the engine.

5. *Constructed variables* are derived from the original variables using *a set of functions* (e.g., arithmetic, trigonometric, or Boolean functions). A variable selection method should have the ability to construct complex reexpressions with mathematical functions to capture the complex relationships in the data and offer potentially more information than the original variables themselves. In an era of data warehouses and the Internet, big data consisting of hundreds of thousands to millions of individual records and hundreds to thousands of variables are commonplace. Relationships among many variables produced by so many individuals are sure to be complex, beyond the simple straight-line pattern. Discovering the mathematical expressions of these relationships, although difficult with theoretical guidance, should be the hallmark of a high-performance variable selection method. For example, consider the well-known relationship among three variables: the lengths of the three sides of a right triangle. A powerful variable selection procedure would identify the relationship among the sides, even in the presence of measurement error: The longer side (diagonal) is the square root of the sum of squares of the two shorter sides.

In sum, the attribute weakness implies that a variable selection method should have the ability of generating an enhanced subset of candidate predictor variables.

10.6 Exploratory Data Analysis

I present the trinity of Tukey's EDA that is relevant to the titled topic: (I) the essence of EDA; (II) the natural seven-step cycle of statistical modeling and analysis, serving as a notable solution to variable selection in regression; and (III) the EDA school of thought.

I. The essence of EDA is best described in Tukey's own words: "Exploratory data analysis is detective work—numerical detective work—or counting detective work—or graphical detective work. ... [It is] about looking at data to see what it seems to say. It concentrates on simple arithmetic and easy-to-draw pictures. It regards whatever appearances we have recognized as partial descriptions, and tries to look beneath them for new insights." EDA includes the following characteristics:

  a. Flexibility: techniques with greater flexibility to delve into the data
  b. Practicality: advice for procedures of analyzing data
  c. Innovation: techniques for interpreting results
  d. Universality: use all statistics that apply to analyzing data
  e. Simplicity: above all, the belief that simplicity is the golden rule

The statistician has also been empowered by the computational strength of the personal computer (PC), which without the natural seven-step cycle of statistical modeling and analysis would not be possible. The PC and the analytical cycle comprise the perfect pairing as long as the steps are followed in order and the information obtained from one step is used in the next step. Unfortunately, statisticians are human and succumb to taking shortcuts through the seven-step cycle. They ignore the cycle and focus solely on the sixth step, delineated next. However, careful statistical endeavor requires additional procedures, as described in the originally outlined seven-step cycle that follows [30]:

II. The natural seven-step cycle of statistical modeling and analysis

  a. Definition of the problem.

Determining the best way to tackle the problem is not always obvious. Management objectives are often expressed qualitatively, in which case the selection of the outcome or target (dependent) variable is subjectively biased. When the objectives are clearly stated, the appropriate dependent variable is often not available, in which case a surrogate must be used.

b. Determining technique

The technique first selected is often the one with which the data analyst is most comfortable; it is not necessarily the best technique for solving the problem.

c. Use of competing techniques

Applying alternative techniques increases the odds that a thorough analysis is conducted.

d. Rough comparisons of efficacy

Comparing variability of results across techniques can suggest additional techniques or the deletion of alternative techniques.

e. Comparison in terms of a precise (and thereby inadequate) criterion

Explicit criterion is difficult to define; therefore, precise surrogates are often used.

f. Optimization in terms of a precise and similarly inadequate criterion

An explicit criterion is difficult to define; therefore, precise surrogates are often used.

g. Comparison in terms of several optimization criteria

This constitutes the final step in determining the best solution.

The founding fathers of classical statistics—Karl Pearson and Sir Ronald Fisher—would have delighted in the ability of the PC to free them from time-consuming empirical validations of their concepts. Pearson, whose contributions include regression analysis, the correlation coefficient, the standard deviation (a term he coined), and the chi-square test of statistical significance, would have likely developed even more concepts with the free time afforded by the PC. One can speculate further that the functionality of PCs would have allowed Fisher's methods of maximum likelihood estimation, hypothesis testing, and analysis of variance to have immediate, practical applications.

III. The EDA school of thought

Tukey's book is more than a collection of new and creative rules and operations; it defines EDA as a discipline that holds that data analysts fail only if they fail to try many things. It further espouses the belief that data analysts are especially successful if their detective work forces them to notice the unexpected. In other words, the philosophy of EDA is a trinity of *attitude* and *flexibility* to do whatever it takes to refine the analysis and *sharp-sightedness* to observe the unexpected when it does appear. EDA is thus a self-propagating theory; each data analyst adds his or her own contribution, thereby contributing to the discipline.

The sharp-sightedness of EDA warrants more attention as it is a very important feature of the EDA approach. The data analyst should be a keen observer of those indicators that are capable of being dealt with successfully and use them to paint an analytical picture of the data. In addition to the ever-ready visual graphical displays as an indicator of what the data reveal, there are numerical indicators, such as counts, percentages, averages, and the other classical descriptive statistics (e.g., standard deviation, minimum, maximum, and missing values). The data analyst's personal judgment and interpretation of indicators are not considered a bad thing, as the goal is to draw informal inferences rather than those statistically significant inferences that are the hallmark of statistical formality.

In addition to visual and numerical indicators, there are the *indirect messages* in the data that force the data analyst to take notice, prompting responses such as "The data look like … ," or, "It appears to be … " Indirect messages may be vague, but their importance is to help the data analyst draw informal inferences. Thus, indicators do not include any of the hard statistical apparatus, such as confidence limits, significance tests, or standard errors.

With EDA, a new trend in statistics was born. Tukey and Mosteller quickly followed up in 1977 with the second EDA book (commonly referred to as EDA II), *Data Analysis and Regression,* which recasts the basics of classical inferential procedures of data analysis and regression as an assumption-free, nonparametric approach guided by "(a) a sequence of philosophical attitudes ¼ for effective data analysis, and (b) a flow of useful and adaptable techniques that make it possible to put these attitudes to work" [31].

Hoaglin, Mosteller, and Tukey in 1983 succeeded in advancing EDA with *Understanding Robust and Exploratory Data Analysis,* which provides an understanding of how badly the classical methods behave when their restrictive assumptions do not hold and offers alternative robust and exploratory methods to broaden the effectiveness of statistical analysis [32]. It includes a collection of methods to cope with data in an informal way, guiding the identification of data structures relatively quickly and easily and trading off optimization of objectives for stability of results.

Hoaglin, Mosteller, and Tukey in 1991 continued their fruitful EDA efforts with *Fundamentals of Exploratory Analysis of Variance* [33]. They recast the basics of the analysis of variance with the classical statistical apparatus (e.g., degrees of freedom, F ratios, and p values) in a host of numerical and graphical displays, which often give insight into the structure of the data, such as size effects, patterns and interaction, and behavior of residuals.

EDA set off a burst of activity in the visual portrayal of data. Published in 1983, *Graphical Methods for Data Analysis* presents new and old methods—some of which require a computer, while others only paper and a pencil—but all are powerful data analysis tools to learn more about data structure [34]. In 1986, du Toit et al. came out with *Graphical Exploratory Data Analysis*, providing a comprehensive, yet simple presentation of the topic [35]. Jacoby, with *Statistical Graphics for Visualizing Univariate and Bivariate Data* (1997) and *Statistical Graphics for Visualizing Multivariate Data* (1998) carries out his objective to obtain pictorial representations of quantitative information by elucidating histograms, one-dimensional and enhanced scatterplots, and nonparametric smoothing [36, 37]. In addition, he successfully transfers graphical displays of multivariate data on a single sheet of paper, a two-dimensional space.

EDA presents a major paradigm shift, depicted in Figure 10.1, in the ways models are built. With the mantra "Let your data be your guide," EDA offers a view that is a complete reversal of the classical principles that govern the usual steps of model building. The EDA declares the model must always follow the data, not the other way around, as in the classical approach.

In the classical approach, the problem is stated and formulated in terms of an outcome variable Y. It is assumed that the *true* model explaining all the variation in Y is known. Specifically, it is assumed that all the structures (predictor variables, X_i's) affecting Y and their forms are known and present in the model. For example, if Age affects Y, but the log of Age reflects the

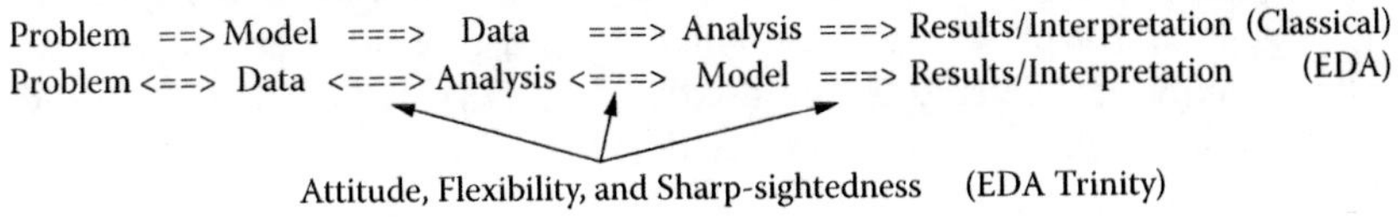

Problem ==> Model ===> Data ===> Analysis ===> Results/Interpretation (Classical)
Problem <==> Data <===> Analysis <===> Model ===> Results/Interpretation (EDA)

Attitude, Flexibility, and Sharp-sightedness (EDA Trinity)

FIGURE 10.1
EDA Paradigm

true relationship with Y, then log of Age must be present in the model. Once the model is specified, the data are taken through the model-specific analysis, which provides the results in terms of numerical values associated with the structures or estimates of the true predictor variables' coefficients. Then, interpretation is made for declaring X_i an important predictor, assessing how X_i affects the prediction of Y, and ranking X_i in order of predictive importance.

Of course, the data analyst never knows the true model. So, familiarity with the content domain of the problem is used to put forth explicitly the true *surrogate* model, from which good predictions of Y can be made. According to Box, "All models are wrong, but some are useful" [38]. In this case, the model selected provides serviceable predictions of Y. Regardless of the model used, the assumption of knowing the truth about Y sets the statistical logic in motion to cause likely bias in the analysis, results, and interpretation.

In the EDA approach, not much is assumed beyond having some prior experience with the content domain of the problem. The right attitude, flexibility, and sharp-sightedness are the forces behind the data analyst, who assesses the problem and lets the data guide the analysis, which then suggests the structures and their forms of the model. If the model passes the validity check, then it is considered final and ready for results and interpretation to be made. If not, with the force still behind the data analyst, the analysis or data are revisited until new structures produce a sound and validated model, after which final results and interpretation are made. Take a second look at Figure 10.1. Without exposure to assumption violations, the EDA paradigm offers a degree of confidence that its prescribed exploratory efforts are not biased, at least in the manner of the classical approach. Of course, no analysis is bias free as all analysts admit their own bias into the equation.

With all its strengths and determination, EDA as originally developed had two minor weaknesses that could have hindered its wide acceptance and great success. One is of a subjective or psychological nature, and the other is a misconceived notion. Data analysts know that failure to look into a multitude of possibilities can result in a flawed analysis; thus, they find themselves in a competitive struggle against the data itself. So, EDA can foster data analysts with insecurity that their work is never done. The PC can assist data analysts in being thorough with their analytical due diligence but bears no responsibility for the arrogance EDA engenders.

The belief that EDA, which was originally developed for the small-data setting, does not work as well with large samples is a misconception. Indeed, some of the graphical methods, such as the stem-and-leaf plots, and some of the numerical and counting methods, such as folding and binning, do break down with large samples. However, the majority of the EDA methodology is unaffected by data size. Neither the manner in which the methods are carried out nor the reliability of the results is changed. In fact, some of the most powerful EDA techniques scale up quite nicely, but do require the PC to do the serious number crunching of the big data [39]. For example, techniques such as ladder of powers, reexpressing, and smoothing are valuable tools for large sample or big data applications.

10.7 Summary

Finding the best possible subset of variables to put in a model has been a frustrating exercise. Many variable selection methods exist. Many statisticians know them, but few know they produce poorly performing models. The deficient variable selection methods are a miscarriage of statistics because they are developed by debasing sound statistical theory into a misguided pseudotheoretical foundation. I reviewed the five widely used variable selection methods, itemized some of their weaknesses, and answered why they are used. Then, I presented the notable solution to variable selection in regression: the natural seven-step cycle of statistical modeling and analysis. I feel that newcomers to Tukey's EDA need the seven-step cycle introduced within the narrative of Tukey's analytic philosophy. Accordingly, I enfolded the solution with front and back matter—the essence of EDA and the EDA school of thought, respectively.

References

1. Tukey, J.W., *The Exploratory Data Analysis*, Addison-Wesley, Reading, MA, 1977.
2. Classical underlying assumptions, http://en.wikipedia.org/wiki/Regression_analysis, 2009.
3. The variable selection methods do not include the new breed of methods that have data mining capability.
4. Miller, A.J., *Subset Selection in Regression*, Chapman and Hall, New York, 1990, iii–x.

5. Statistica-Criteria-Supported-by-SAS.pdf, http://www.geniq.net/res/Statistical-Criteria-Supported-by-SAS.pdf, 2010.
6. Roecker, E.B., Prediction error and its estimation for subset-selected models, *Technometrics, 33, 459–468, 1991.*
7. Copas, J.B., Regression, prediction and shrinkage (with discussion), *Journal of the Royal Statistical Society, B 45, 311–354, 1983.*
8. Henderson, H.V., and Velleman, P.F., Building multiple regression models interactively, *Biometrics, 37, 391–411, 1981.*
9. Altman, D.G., and Andersen, P.K., Bootstrap investigation of the stability of a Cox regression model, *Statistics in Medicine, 8, 771–783, 1989.*
10. Derksen, S., and Keselman, H.J., Backward, forward and stepwise automated subset selection algorithms, *British Journal of Mathematical and Statistical Psychology, 45, 265–282, 1992.*
11. Judd, C.M., and McClelland, G.H., *Data Analysis: A Model Comparison Approach,* Harcourt Brace Jovanovich, New York, 1989.
12. Bernstein, I.H., *Applied Multivariate Analysis,* Springer-Verlag, New York, 1988.
13. Comment without an attributed citation: Frank Harrell, Vanderbilt University School of Medicine, Department of Biostatistics, professor of biostatistics and department chair, 2009.
14. Kashid, D.N., and Kulkarni, S.R., A more general criterion for subset selection in multiple linear regression, *Communication in Statistics–Theory & Method*, 31(5), 795–811, 2002.
15. Tibshirani, R., Regression shrinkage and selection via the Lasso, *Journal of the Royal Statistical Society, B 58(1), 267–288, 1996.*
16. Ratner, B., *Statistical Modeling and Analysis for Database Marketing: Effective Techniques for Mining Big Data,* CRC Press, Boca Raton, FL, 2003, Chapter 15, which presents the GenIQ Model (http://www.GenIQModel.com).
17. Chen, S.-M., and Shie, J.-D., *A New Method for Feature Subset Selection for Handling Classification Problems, Journal Expert Systems with Applications: An International Journal* Volume 37 Issue 4, Pergamon Press, Inc. Tarrytown, NY, April, 2010.
18. SAS Proc Reg Variable Selection Methods.pdf, support.sas.com, 2011.
19. Comment without an attributed citation: In 1996, Tim C. Hesterberg, research scientist at Insightful Corporation, asked Brad Efron for the most important problems in statistics, fully expecting the answer to involve the bootstrap given Efron's status as inventor. Instead, Efron named a single problem, variable selection in regression. This entails selecting variables from among a set of candidate variables, estimating parameters for those variables, and inference—hypotheses tests, standard errors, and confidence intervals.
20. Other criteria are based on information theory and Bayesian rules.
21. Dash, M., and Liu, H, Feature selection for classification, *Intelligent Data Analysis,* 1, 131–156, 1997.
22. SAS/STAT Manual. See PROC REG, and PROC LOGISTIC, support.sas.com, 2011.
23. R-squared theoretically is not the appropriate measure for a binary dependent variable. However, many analysts use it with varying degrees of success.
24. For example, consider two TS values: 1.934056 and 1.934069. These values are equal when rounding occurs at the third place after the decimal point: 1.934.

25. Even if there were a perfect variable selection method, it is unrealistic to believe there is a unique best subset of variables.
26. Comment without an attributed citation: Frank Harrell, Vanderbilt University School of Medicine, Department of Biostatistics, professor of biostatistics and department chair, 2010.
27. Mark Twain quotation: "Get your facts first, then you can distort them as you please." http://thinkexist.com/quotes/mark_twain/, 2011.
28. The weights or coefficients (b_0, b_1, b_2, and b_3) are derived to satisfy some criterion, such as minimize the mean squared error used in ordinary least squares regression or minimize the joint probability function used in logistic regression.
29. Fox, J., *Applied Regression Analysis, Linear Models, and Related Methods*, Sage, Thousand Oaks, CA, 1997.
30. The seven steps are Tukey's. The annotations are mine.
31. Mosteller, F., and Tukey, J.W., *Data Analysis and Regression*, Addison-Wesley, Reading, MA, 1977.
32. Hoaglin, D.C., Mosteller, F., and Tukey, J.W., *Understanding Robust and Exploratory Data Analysis*, Wiley, New York, 1983.
33. Hoaglin, D.C., Mosteller, F., and Tukey, J.W., *Fundamentals of Exploratory Analysis of Variance*, Wiley, New York, 1991.
34. Chambers, M.J., Cleveland, W.S., Kleiner, B., and Tukey, P.A., *Graphical Methods for Data Analysis*, Wadsworth & Brooks/Cole, Pacific Grove, CA, 1983.
35. du Toit, S.H.C., Steyn, A.G.W., and Stumpf, R.H., *Graphical Exploratory Data Analysis*, Springer-Verlag, New York, 1986.
36. Jacoby, W.G., *Statistical Graphics for Visualizing Univariate and Bivariate Data*, Sage, Thousand Oaks, CA, 1997.
37. Jacoby, W.G., *Statistical Graphics for Visualizing Multivariate Data*, Sage, Thousand Oaks, CA, 1998.
38. Box, G.E.P., Science and statistics, *Journal of the American Statistical Association*, 71, 791–799, 1976.
39. Weiss, S.M., and Indurkhya, N., *Predictive Data Mining*, Morgan Kaufman, San Francisco, 1998.

11

CHAID for Interpreting a Logistic Regression Model*

11.1 Introduction

The logistic regression model is the standard technique for building a response model. Its theory is well established, and its estimation algorithm is available in all major statistical software packages. The literature on the theoretical aspects of logistic regression is large and rapidly growing. However, little attention is paid to the interpretation of the logistic regression response model. The purpose of this chapter is to present a *data mining* method based on CHAID (chi-squared automatic interaction detection) for interpreting a logistic regression model, specifically to provide a complete assessment of the effects of the predictor variables on response.

11.2 Logistic Regression Model

I state briefly the definition of the logistic regression model. Let Y be a binary response (dependent) variable, which takes on yes/no values (typically coded 1/0, respectively), and X_1, X_2, X_3, X_n be the predictor (independent) variables. The logistic regression model estimates the *logit of Y*—the log of the odds of an individual responding yes as defined in Equation (11.1), from which an individual's probability of responding yes is obtained in Equation (11.2):

$$\text{Logit } Y = b_0 + b_1{}^*X_1 + b_2{}^*X_2 + X_3 + b_n{}^*X_n \tag{11.1}$$

$$\text{Prob}(Y = 1) = \frac{\exp(\text{Logit } Y)}{1 + \exp(\text{Logit } Y)} \tag{11.2}$$

An individual's predicted probability of responding yes is calculated by "plugging in" the values of the predictor variables for that individual in

---

* This chapter is based on an article with the same title in *Journal of Targeting, Measurement and Analysis for Marketing*, 6, 2, 1997. Used with permission.

Equations (11.1) and (11.2). The b's are the logistic regression coefficients. The coefficient b_0 is referred to as the Intercept and has no predictor variable with which it is multiplied.

The *odds ratio* is the traditional measure of assessing the effect of a predictor variable on the response variable (actually on the odds of response = 1) given that the other predictor variables are "held constant." The phrase "given that … " implies that the odds ratio is the average effect of a predictor variable on response when the effects of the other predictor variables are "partialled out" of the relationship between response and the predictor variable. Thus, the odds ratio does not explicitly reflect the variation of the other predictor variables. The odds ratio for a predictor variable is obtained by exponentiating the coefficient of the predictor variable. That is, the odds ratio for X_i equals $exp(b_i)$, where exp is the exponential function, and b_i is the coefficient of X_i.

11.3 Database Marketing Response Model Case Study

A woodworker's tool supplier, who wants to increase response to her catalog in an upcoming campaign, needs a model for generating a list of her most responsive customers. The response model, which is built on a sample drawn from a recent catalog mailing with a 2.35% response rate, is defined by the following variables:

1. The response variable is RESPONSE, which indicates whether a customer made a purchase (yes = 1, no = 0) to the recent catalog mailing.
2. The predictor variables are (a) CUST_AGE, the customer age in years; (b) LOG_LIFE, the log of total purchases in dollars since the customer's first purchase, that is, the log of lifetime dollars; and (c) PRIOR_BY, the dummy variable indicating whether a purchase was made in the 3 months prior to the recent catalog mailing (yes = 1, no = 0).

The logistic regression analysis on RESPONSE with the three predictor variables produces the output in Table 11.1. The "Parameter Estimate" column contains the logistic regression coefficients, from which the RESPONSE model is defined in Equation (11.3).

Logit RESPONSE

$$= -8.43 + 0.02*CUST\_AGE + 0.74*LOG\_LIFE + 0.82*PRIOR\_BY \quad (11.3)$$

11.3.1 Odds Ratio

The following discussion illustrates two weaknesses in the odds ratio. First, the odds ratio is in terms of odds-of-responding-yes units, with

TABLE 11.1

Logistic Regression Output

Variable		Parameter Estimate	Standard Error	Wald Chi-square	Pr > Chi-square	Odds Ratio
INTERCEPT	1	-8.4349	0.0854	9760.7175	1.E + 00	
CUST_AGE	1	0.0223	0.0004	2967.8450	1.E + 00	1.023
LOG_LIFE	1	0.7431	0.0191	1512.4483	1.E + 00	2.102
PRIOR_BY	1	0.8237	0.0186	1962.4750	1.E + 00	2.279

which all but the mathematically adept feel comfortable. Second, the odds ratio provides a "static" assessment of the effect of a predictor variable, as its value is constant regardless of the relationship among the other predictor variables. The odds ratio is part of the standard output of the logistic regression analysis. For the case study, the odds ratio is in the rightmost column in Table 11.1.

1. For PRIOR_BY with coefficient value 0.8237, the odds ratio is 2.279 (= exp(0.8237)). This means that for a unit increase in PRIOR_BY (i.e., going from 0 to 1), the odds (of responding yes) for an individual who has made a purchase within the prior 3 months is 2.279 times the odds of an individual who has *not* made a purchase within the prior 3 months—given CUST_AGE and LOG_LIFE are held constant.*

2. CUST_AGE has an odds ratio of 1.023. This indicates that for every 1 year increase in a customer's age, the odds increase by 2.3%—given PRIOR_BY and LOG_LIFE are held constant.

3. LOG_LIFE has an odds ratio of 2.102, which indicates that for every one log-lifetime-dollar-unit increase, the odds increase by 110.2%—given PRIOR_BY and CUST_AGE are held constant.

The proposed CHAID-based data mining method supplements the odds ratio for interpreting the effects of a predictor on response. It provides treelike displays in terms of nonthreatening probability units, everyday values ranging from 0% to 100%. Moreover, the CHAID-based graphics provide a complete assessment of the effect of a predictor variable on response. It brings forth the simple, unconditional relationship between a given predictor variable and response, as well as the conditional relationship between a given predictor variable and response shaped by the

---

* For given values of CUST_AGE and LOG_LIFE, say, a and b, respectively, the odds ratio for PRIOR_BY is defined as:

$$= \frac{\text{odds (PRIOR\_BY} = 1 \text{ given CUST\_AGE} = a \text{ and LOG\_LIFE} = b)}{\text{odds (PRIOR\_BY} = 0 \text{ given CUST\_AGE} = a \text{ and LOG\_LIFE} = b)}$$

relationships between the other X's and a given predictor variable and the other X's and response.

11.4 CHAID

Briefly stated, CHAID is a technique that recursively partitions a population into separate and distinct subpopulations or segments such that the variation of the dependent variable is minimized within the segments and maximized among the segments. A CHAID analysis results in a treelike diagram, commonly called a CHAID tree. CHAID was originally developed as a method of finding "combination" or interaction variables. In database marketing today, CHAID primarily serves as a market segmentation technique. The use of CHAID in the proposed application for interpreting a logistic regression model is possible because of the salient data mining features of CHAID. CHAID is eminently good in uncovering structure, in this application, within the conditional and unconditional relationships among response and predictor variables. Moreover, CHAID is excellent in graphically displaying multivariable relationships; its tree output is easy to read and interpret.

It is worth emphasizing that in this application CHAID is not used as an alternative method for analyzing or modeling the data at hand. CHAID is used as a visual aid for depicting the "statistical" mechanics of a logistic regression model, such as how the variables of the model work together in contributing to the predicted probability of response. It is assumed that the response model is already built by any method, not necessarily logistic regression. As such, the proposed method can enhance the interpretation of the predictor variables of any model built by statistical or machine-learning techniques.

11.4.1 Proposed CHAID-Based Method

To perform an ordinary CHAID analysis, the model builder is required to select both the response variable and a set of predictor variables. For the proposed CHAID-based data mining method, the already-built response model's *estimated probability of response* is selected as the CHAID response variable. The CHAID set of predictor variables consists of the predictor variables that defined the response model in their original units, not in reexpressed units (if reexpressing was necessary). Reexpressed variables are invariably in units that hinder the interpretation of the CHAID-based analysis and, consequently, the logistic regression model. Moreover, to facilitate the analysis, the continuous predictor variables are categorized into meaningful intervals based on content domain of the problem under study.

For the case study, the CHAID response variable is the estimated probability of RESPONSE, called Prob_est, which is obtained from Equations (11.2) and (11.3) and defined in Equation (11.4):

$$\text{Prob\_est} = \frac{\exp(-8.43 + 0.02 * \text{CUST\_AGE} + 0.74 * \text{LOG\_LIFE} + 0.82 * \text{PRIOR\_BY})}{1 + \exp(-8.43 + 0.02 * \text{CUST\_AGE} + 0.74 * \text{LOG\_LIFE} + 0.82 * \text{PRIOR\_BY})} \quad (11.4)$$

The CHAID set of predictor variables is CUST_AGE, categorized into two classes, PRIOR_BY, and the original variable LIFETIME DOLLARS (log-lifetime-dollar units are hard to understand), categorized into three classes. The woodworker views her customers in terms of the following classes:

1. The two CUST_AGE classes are "less than 35 years" and "35 years and up." CHAID uses bracket and parenthesis symbols in its display of intervals, denoting two customer age intervals: [18, 35) and [35, 93], respectively. CHAID defines intervals as closed interval and left-closed/right-open interval. The former is denoted by [a, b], indicating all values between and including a and b. The latter is denoted by [a, b), indicating all values greater than/equal to a and less than b. Minimum and maximum ages in the sample are 18 and 93, respectively.

2. The three LIFETIME DOLLARS classes are less than $15,000; $15,001 to $29,999; and equal to or greater than $30,000. CHAID denotes the three lifetime dollar intervals: [12, 1500), [1500, 30000), and [30000, 675014], respectively. Minimum and maximum lifetime dollars in the sample are $12 and $675,014, respectively.

The CHAID trees in Figures 11.1 to 11.3 are based on the Prob_est variable with three predictor variables and are read as follows:

1. All CHAID trees have a top box (root node), which represents the sample under study: sample size and response rate. For the proposed CHAID application, the top box reports the sample size and the average estimated probability (AEP) of response. For the case study, the sample size is 858,963, and the AEP of response is 0.0235.*

2. The CHAID tree for PRIOR_BY is in Figure 11.1: The left leaf node represents a segment (size 333,408) defined by PRIOR_BY = no. These customers have *not* made a purchase in the prior 3 months; their AEP of response is 0.0112. The right leaf node represents a segment (size

---

* The average estimated probability or response rate is always equal to the true response rate.

FIGURE 11.1
CHAID tree for PRIOR_BY.

FIGURE 11.2
CHAID tree for CUST_AGE.

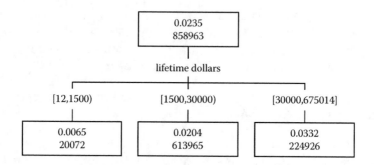

FIGURE 11.3
CHAID tree for LIFETIME DOLLARS.

525,555) defined by PRIOR_BY = yes. These customers have made a purchase in the prior three months; their AEP of response is 0.0312.

3. The CHAID tree for CUST_AGE is in Figure 11.2: The left leaf node represents a segment (size 420,312) defined by customers whose ages are in the interval [18, 35); their AEP of response is 0.0146. The right leaf node represents a segment (size 438,651) defined by customers whose ages are in the interval [35, 93]; their AEP of response is 0.0320.

4. The CHAID tree for LIFETIME DOLLARS is in Figure 11.3: The left leaf node represents a segment (size 20,072) defined by customers whose lifetime dollars are in the interval [12, 1500); their AEP of response is 0.0065. The middle leaf node represents a segment (size 613,965) defined by customers whose lifetime dollars are in the interval [1500, 30000); their AEP of response is 0.0204. The left leaf node represents a segment (size 224,926) defined by customers whose lifetime dollars are in the interval [30000, 675014]; their AEP of response is 0.0332.

At this point, the single predictor variable CHAID tree shows the effect of the predictor variable on response. Reading the leaf nodes from left to right, it is revealed clearly how response changes as the values of the predictor variable increase. Although the single-predictor-variable CHAID tree is easy to interpret with probability units, it is still like the odds ratio in that it does not reveal the effects of a predictor variable with respect to the variation of the other predictor variables in the model. For a complete visual display of the effect of a predictor variable on response accounting for the presence of other predictor variables, a *multivariable CHAID tree* is required.

11.5 Multivariable CHAID Trees

The multivariable CHAID tree in Figure 11.4 shows the effects of LIFETIME DOLLARS on response with respect to the variation of CUST_AGE and PRIOR_BY = *no*. The LIFETIME DOLLARS-PRIOR_BY = no CHAID tree is read and interpreted as follows:

1. The root node represents the sample (size 858,963); the AEP of response is 0.0235.
2. The tree has six *branches*, which are defined by the intersection of the CUST_AGE and LIFETIME DOLLARS intervals/nodes. Branches are read from an end leaf node (bottom box) upward to and through intermediate leaf nodes, stopping at the first-level leaf node below the root node.
3. Reading the tree starting at the bottom of the multivariable CHAID tree in Figure 11.4 from left to right: The leftmost branch 1 is defined by LIFETIME DOLLARS = [12, 1500), CUST_AGE = [18, 35), and PRIOR_BY = no.

The second leftmost branch 2 is defined by LIFETIME DOLLARS = [1500, 30000), CUST_AGE = [18, 35), and PRIOR_BY = no.

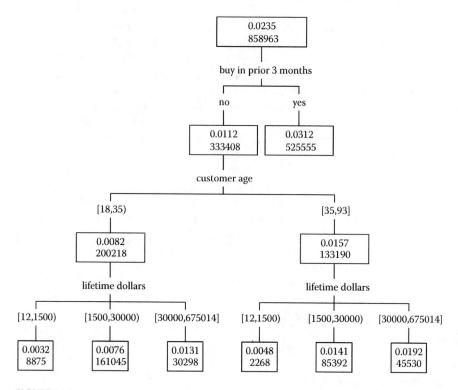

FIGURE 11.4
Multivariable CHAID tree for effects of LIFETIME DOLLARS, accounting for CUST_AGE and PRIOR_BY = no.

Branches 3 to 5 are similarly defined.

The rightmost branch 6 is defined by LIFETIME DOLLARS = [30000, 675014], CUST_AGE = [35, 93], and PRIOR_BY = no.

4. Focusing on the three leftmost branches 1 to 3, as LIFETIME DOLLARS increase, its effect on RESPONSE is gleaned from the multivariable CHAID tree: The AEP of response goes from 0.0032 to 0.0076 to 0.0131 for customers whose ages are in the interval [18, 35) *and* have not purchased in the prior 3 months.

5. Focusing on the three rightmost branches 4 to 6, as LIFETIME DOLLARS increase, its effect on RESPONSE is gleaned from the multivariable CHAID tree in Figure 11.4: The AEP of response ranges from 0.0048 to 0.0141 to 0.0192 for customers whose ages are in the interval [35, 93] *and* have not purchased in the prior 3 months.

The multivariable CHAID tree in Figure 11.5 shows the effect of LIFETIME DOLLARS on response with respect to the variation of CUST_AGE and

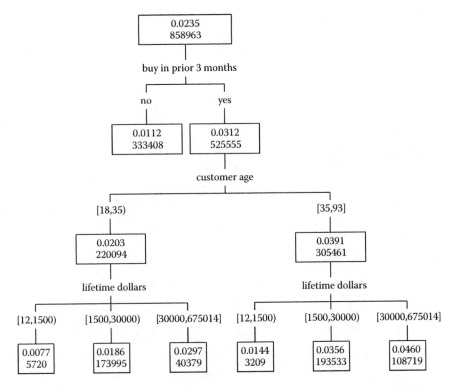

FIGURE 11.5
Multivariable CHAID tree for effects of LIFETIME DOLLARS, accounting for CUST_AGE and
PRIOR_BY = yes.

PRIOR_BY = yes. Briefly, the LIFETIME DOLLARS-PRIOR_BY = yes CHAID
tree is interpreted as follows:

1. Focusing on the three leftmost branches 1 to 3, as LIFETIME
   DOLLARS increase, its effect on RESPONSE is gleaned from the
   multivariable CHAID tree: The AEP of response goes from 0.0077 to
   0.0186 to 0.0297 for customers whose ages are in the interval [35, 93]
   and have purchased in the prior 3 months.
2. Focusing on the three rightmost branches 4 to 6, as LIFETIME
   DOLLARS increase, its effect on RESPONSE is gleaned from the
   multivariable CHAID tree: The AEP of response goes from 0.0144 to
   0.0356 to 0.0460 for customers whose ages are in the interval [35, 93]
   and have purchased in the prior 3 months.

The multivariable CHAID trees for the effects of PRIOR_BY on RESPONSE
with respect to the variation of CUST_AGE = [18, 35] and LIFETIME DOLLARS
and with respect to the variation of CUST_AGE = [35, 93] and LIFETIME
DOLLARS are in Figures 11.6 and 11.7, respectively. They are similarly read

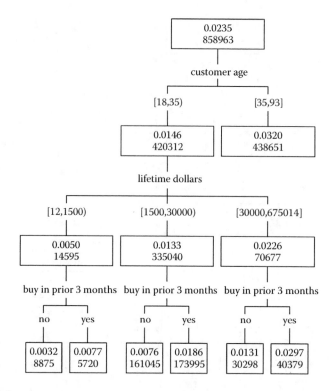

FIGURE 11.6
Multivariable CHAID tree for effect of PRIOR_BY, accounting for LIFETIME DOLLARS and
CUST_AGE = [18, 35).

and interpreted in the same manner as the LIFETIME DOLLARS multivariable CHAID trees in Figures 11.4 and 11.5.

The multivariable CHAID trees for the effects of CUST_AGE on
RESPONSE with respect to the variation of PRIOR_BY = no and LIFETIME
DOLLARS, and with respect to the variation of PRIOR_BY = yes and
LIFETIME DOLLARS are in Figures 11.8 and 11.9, respectively. There are
similarly read and interpreted as the LIFETIME DOLLARS multivariable
CHAID trees.

11.6 CHAID Market Segmentation

I take this opportunity to use the analysis (so far) to illustrate CHAID as
a market segmentation technique. A closer look at the full CHAID trees
in Figures 11.4 and 11.5 identifies three market segment pairs, which
show three levels of response performance, 0.76%/0.77%, 4.8%/4.6%, and

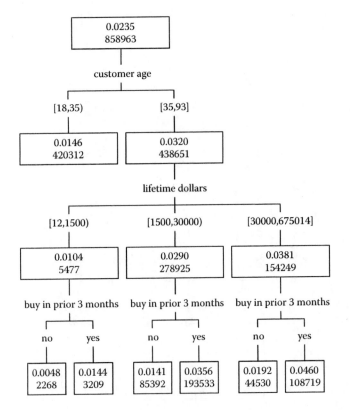

FIGURE 11.7
Multivariable CHAID tree for effect of PRIOR_BY, accounting for LIFETIME DOLLARS and
CUST_AGE = [35, 93].

1.41%/1.46%. Respectively, CHAID provides the cataloguer with marketing intelligence for high-, medium-, and low-performing segments. Marketing strategy can be developed to stimulate the high performers with techniques such as cross selling, or pique interest in the medium performers with new products, as well as prod the low performers with incentives and discounts.

The descriptive profiles of the three market segments are as follows:

> *Market segment 1:* There are customers whose ages are in the interval [18, 35), who have *not* purchased in the prior 3 months, and who have lifetime dollars in the interval [1500, 30000). The AEP of response is 0.0076. See branch 2 in Figure 11.4. Also, there are customers whose ages are in the interval [18, 35), who have purchased in the prior 3 months, and who have lifetime dollars in the interval [12, 1500). The AEP of response is 0.0077. See branch 1 in Figure 11.5.

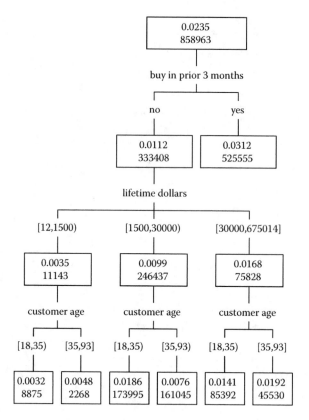

FIGURE 11.8
Multivariable CHAID tree for effect of CUST_AGE, accounting for LIFETIME DOLLARS and
PRIOR_BY = no.

Market segment 2: There are customers whose ages are in the interval
[35, 93], who have *not* purchased in the prior 3 months, and who
have lifetime dollars in the interval [12, 1500). The AEP of response
is 0.0048. See branch 4 in Figure 11.4. Also, there are customers
whose ages are in the interval [35, 93], who have purchased in the
prior 3 months, and who have lifetime dollars in the interval [30000,
675014]. The AEP of response is 0.0460. See branch 6 in Figure 11.5.

Market segment 3: There are customers whose ages are in the interval [35, 93], who have *not* purchased in the prior 3 months, and
who have lifetime dollars in the interval [1500, 30000). The AEP
of response is 0.0141. See branch 5 in Figure 11.4. Also, there are
customers whose ages are in the interval [35, 93], who have purchased in the prior 3 months, and who have lifetime dollars in the
interval [12, 1500). The AEP of response is 0.0144. See branch 4 in
Figure 11.5.

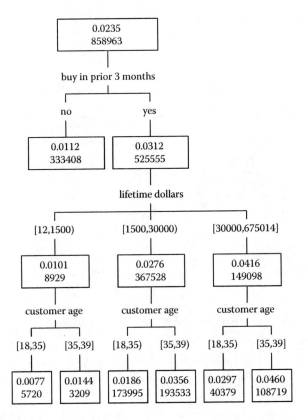

FIGURE 11.9
Multivariable CHAID tree for effect of CUST_AGE, accounting for LIFETIME DOLLARS and PRIOR_BY = yes.

11.7 CHAID Tree Graphs

Displaying the multivariable CHAID trees in a *single* graph provides the desired displays of a *complete* assessment of the effects of the predictor variables on response. Construction and interpretation of the *CHAID tree graph* for a given predictor is as follows:

1. Collect the set of multivariable CHAID trees for a given predictor variable. For example, for PRIOR_BY there are two trees, which correspond to the two values of PRIOR_BY, yes and no.
2. For each branch, plot the AEP of response values (Y-axis) and the minimum values of the end leaf nodes of the given predictor variable (X-axis).*

---

* The minimum value is one of several values that can be used; alternatives are the mean or median of each predefined interval.

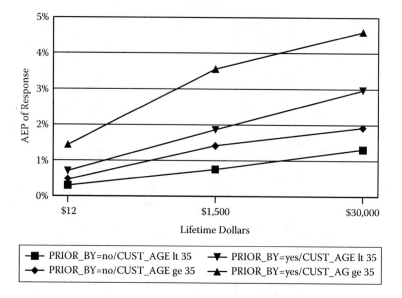

FIGURE 11.10
CHAID tree graph for effects of LIFETIME on RESPONSE by CUST_AGE and PRIOR_BY.

3. For each branch, connect the *nominal points* (AEP response value, minimum value). The resultant *trace line* segment represents a market or customer segment defined by the branch's intermediate leaf intervals/ nodes.

4. The shape of the trace line indicates the effect of the predictor variable on response for that segment. A comparison of the trace lines provides a total view of how the predictor variable affects response, accounting for the presence of the other predictor variables.

The LIFETIME DOLLARS CHAID tree graph in Figure 11.10 is based on the multivariable LIFETIME DOLLARS CHAID trees in Figures 11.4 and 11.5. The top trace line with a noticeable bend corresponds to older customers (age 35 years and older) who have made purchases in the prior 3 months. The implication is LIFETIME DOLLARS has a nonlinear effect on response for this customer segment. As LIFETIME DOLLARS goes from a nominal $12 to a nominal $1,500 to a nominal $30,000, RESPONSE increases at a nonconstant rate as depicted in the tree graph.

The other trace lines are viewed as straight lines* with various slopes. The implication is LIFETIME DOLLARS has various constant effects on response across the corresponding customer segments. As LIFETIME DOLLARS goes

---

* The segment that is defined by PRIOR_BY = no and CUST_AGE greater than or equal to 35 years appears to have a very slight bend. However, I treat its trace line as straight because the bend is very slight.

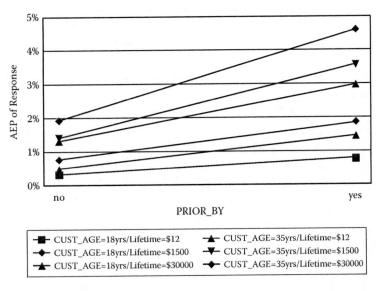

FIGURE 11.11
CHAID tree graph for effects of PRIOR_BY on RESPONSE by CUST_AGE and LIFETIME DOLLARS.

from a nominal $12 to a nominal $1,500 to a nominal $30,000, RESPONSE increases at various constant rates as depicted in the tree graph.

The PRIOR_BY CHAID tree graph in Figure 11.11 is based on the multivariable PRIOR_BY CHAID trees in Figures 11.6 and 11.7. I focus on the slopes of the trace lines.* The evaluation rule is as follows: The steeper the slope is, the greater the constant effect on response will be. Among the six trace lines, the top trace line for the "top" customer segment of older customers with lifetime dollars equal to or greater than $30,000 has the steepest slope. The implications are as follows: (1) PRIOR_BY has a rather noticeable constant effect on response for the top customer segment, and (2) the size of the PRIOR_BY effect for the top customer segment is greater than the PRIOR_BY effect for the remaining five customer segments. As PRIOR_BY goes from no to yes, RESPONSE increases at a rather noticeable constant rate for the top customer segment as depicted in the tree graph.

The remaining five trace lines have slopes of varying steepness. The implication is PRIOR_BY has various constant effects on RESPONSE across the corresponding five customer segments. As PRIOR_BY goes from no to yes, RESPONSE increases at various constant rates as depicted in the tree graph.

The CUST_AGE CHAID tree graph in Figure 11.12 is based on the multivariable CUST_AGE CHAID trees in Figures 11.8 and 11.9. There are two sets

---

* The trace lines are necessarily straight because two points (PRIOR_BY points no and yes) always determine a straight line.

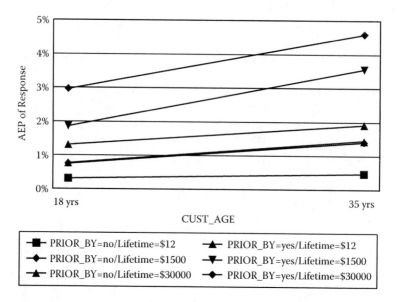

FIGURE 11.12
CHAID Tree Graph for Effects of CUST_AGE on RESPONSE by PRIOR_BY and LIFE_DOLLARS.

of parallel trace lines with different slopes. The first set of the top two parallel trace lines correspond to two customer segments defined by

1. PRIOR_BY = no and LIFETIME DOLLARS in the interval [1500, 30000)
2. PRIOR_BY = yes and LIFETIME DOLLARS in the interval [30000, 675014]

The implication is CUST_AGE has the same constant effect on RESPONSE for the two customer segments. As CUST_AGE goes from a nominal age of 18 years to a nominal age of 35 years, RESPONSE increases at a constant rate as depicted in the tree graph.

The second set of the next three parallel trace lines (two trace lines virtually overlap each other) corresponds to three customer segments defined by

1. PRIOR_BY = no and LIFETIME DOLLARS in the interval [30000, 675014]
2. PRIOR_BY = yes and LIFETIME DOLLARS in the interval [1500, 30000)
3. PRIOR_BY = yes and LIFETIME DOLLARS in the interval [12, 1500]

The implication is CUST_AGE has the same constant effect on RESPONSE for the three customer segments. As CUST_AGE goes from a nominal age of

18 years to a nominal age of 35 years, RESPONSE increases at a constant rate as depicted in the tree graph. Note that the constant CUST_AGE effect for the three customer segments is less than the CUST_AGE effect for the former two customer segments, as the slope of the former segments is less steep than that of the latter segments.

Last, the bottom trace line, which corresponds to the customer segment defined by PRIOR_BY = no and LIFETIME DOLLARS in the interval [12, 1500), has virtually no slope as it is nearly horizontal. The implication is CUST_AGE has no effect on RESPONSE for the corresponding customer segment.

11.8 Summary

After a brief introduction of the logistic regression model as the standard technique for building a response model, I focused on its interpretation, an area in the literature that has not received a lot of attention. I discussed the traditional approach to interpreting a logistic regression model using the odds ratio statistic, which measures the effect of a predictor variable on the odds of response. Then, I introduced two weaknesses of the odds ratio. First, the odds ratio is reported in terms of unwieldy odds-of-responding-yes units. Second, the odds ratio only provides an incomplete assessment of the effect of a predictor variable on response since it does not explicitly reflect the relationship of the other predictor variables.

I proposed a CHAID-based data mining method to supplement the odds ratio as it supplements its two weaknesses. The CHAID-based method adds a visual touch to the original concept of the odds ratio. I illustrated the new method, which exports the information of the odds ratio into CHAID trees, visual displays in terms of simple probability values. More important, the CHAID-based method makes possible the desired complete assessment of the effect of a predictor variable on response explicitly reflecting the relationship of the other predictor variables. I illustrated the new method, which combines individual predictor variable CHAID trees into a multivariable CHAID tree graph as a visual complete assessment of a predictor variable.

Moreover, I pointed out that the CHAID-based method can be used to enhance the interpretation of any model—response or profit—built by either statistical or machine-learning techniques. The only assumption of the method—that the predicted dependent variable values along with the predictor variables used in the modeling process are available—allows the new method to be applied to any model.

12

The Importance of the Regression Coefficient

12.1 Introduction

Interpretation of the ordinary regression model—the most popular technique for making predictions of a single continuous variable Y—focuses on the model's coefficients with the aid of three concepts: the statistical p value, variables "held constant," and the standardized regression coefficient. The purpose of this chapter is to take a closer look at these widely used, yet often misinterpreted, concepts. In this chapter, I demonstrate the statistical p value, the sole measure for declaring X_i an important predictor variable, is sometimes problematic; the concept of variables held constant is critical for reliable assessment of how X_i affects the prediction of Y; and the standardized regression coefficient provides the correct ranking of variables in order of predictive importance—only under special circumstances.

12.2 The Ordinary Regression Model

The ordinary regression model, formally known as the ordinary least squares multiple linear regression model, is the most popular technique for making predictions of a single continuous variable. Its theory is well established, and the estimation algorithm is available in all major statistical computer software packages. The model is relatively easy to build and usually produces usable results.

Let Y be a continuous dependent variable (e.g., sales) and $X_1, X_2, X_3, X_i, \ldots, X_n$ be the predictor variables. The regression model (prediction equation) is defined in Equation (12.1):

$$Y = b_0 + b_1X_1 + b_2X_2 + \ldots + b_iX_i + \ldots + b_nX_n \qquad (12.1)$$

The b's are the regression coefficients,* which are estimated by the method of ordinary least squares. Once the coefficients are estimated, an individual's predicted Y value (estimated sales) is calculated by "plugging in" the values of the predictor variables for that individual in the equation.

12.3 Four Questions

Interpretation of the model focuses on the regression coefficient with the aid of three concepts: the statistical p value, the average change in Y associated with a unit change in X_i when the other X's[†] are held constant, and the standardized regression coefficient. The following four questions apply universally to any discussion of the regression coefficient; they are discussed in detail to provide a better understanding of why the regression coefficient is important.

1. Is X_i important for making good predictions? The usual answer is that X_i is an important predictor variable if its associated p value is less than 5%. This answer is correct for experimental studies, but may not be correct for big data[‡] applications.

2. How does X_i affect the prediction of Y? The usual answer is that Y experiences an average change of b_i with a unit increase in X_i when the other X's are held constant. This answer is an "honest" one, which often is not accompanied by a caveat.

3. Which variables in the model have the *greatest effect* on the prediction of Y, in ranked order? The usual answer is that the variable with the largest regression coefficient has the greatest effect; the variable with the next-largest regression coefficient has the next-greatest effect, and so on. This answer is usually incorrect.

4. Which variables in the model are the *most important* predictors, in ranked order? The usual answer is that the variable with the largest standardized regression coefficient is the most important variable; the variable with the next-largest standardized regression coefficient is the next important variable, and so on. This answer is usually incorrect.

---

* b_0 is called the intercept, which serves as a mathematical necessity for the regression model. However, b_0 can be considered the coefficient of $X_0 = 1$.
† The other X's consist of n - 1 variables without X_i, namely, $X_1, X_2, ..., X_{i-1}, X_{i+1}, ..., X_n$.
‡ Big data are defined in Section 1.6 of Chapter 1.

12.4 Important Predictor Variables

X_i is declared an important predictor variable if it significantly reduces the prediction error of (actual Y - predicted Y) the regression model. The size of reduction in prediction error due to X_i can be tested for significance with the null hypothesis (NH)-significance testing procedure. Briefly, I outline the procedure[*] as follows:

1. The NH and alternative hypothesis (AH) are defined as follows:

   NH: The change in mean squared prediction error due to X_i (cMSE_X_i) is equal to zero.

   AH: cMSE_X_i is not equal to zero.

2. Significance testing for cMSE_X_i is equivalent to significance testing for the regression coefficient for Xi, b_i. Therefore, NH and AH can be alternatively stated:

   NH: b_i is equal to zero.

   AH: b_i is not equal to zero.

3. The working assumptions[†] of the testing procedure are that the sample size is correct, and the sample accurately reflects the relevant population. (Well-established procedures for determining the correct sample size for experimental studies are readily available in intermediate-level statistics textbooks.)

4. The decision to reject or fail to reject NH is based on the statistical p value.[‡] The statistical p value is the probability of observing a value of the sample statistic (cMSE or b_i) as extreme or more extreme than the observed value (sample evidence) given NH is true.[§]

5. The decision rule:

   a. If the p value is *not* "very small," typically greater than 5%, then the sample evidence supports the decision to fail to reject NH.[¶] It is concluded that b_i is zero, and X_i does not significantly contribute to the reduction in prediction error. Thus, X_i is not an important predictor variable.

---

[*] See any good textbook on the null hypothesis-significance testing procedure, such as Chow, S.L., *Statistical Significance*, Sage, Thousand Oaks, CA, 1996.

[†] A suite of classical assumptions is required for the proper testing of the least squares estimate of b_i. See any good mathematical statistics textbook, such as Ryan, T.P., *Modern Regression Methods*, Wiley, New York, 1997.

[‡] Failure to reject NH is not equivalent to accepting NH.

[§] The p value is a conditional probability.

[¶] The choice of "very small" is arbitrary, but convention sets it at 5% or less.

b. If the p value is very small, typically less than 5%, then the sample
   evidence supports the decision to reject NH in favor of accepting
   AH. It is concluded that b_i (or cMSE_X_i) has some nonzero value, and
   X_i contributes a significant reduction in prediction error. Thus, X_i is
   an important predictor variable.

The decision rule makes it clear that the p value is an indicator of the *likelihood* that the variable has *some* predictive importance, *not* an indicator of how
much importance (AH does not specify a value of b_i). Thus, *a smaller p value
implies a greater likelihood of some predictive importance, not a greater predictive
importance.* This is contrary to the common misinterpretation of the p value:
The smaller the p value is, the greater the predictive importance of the associated variable will be.

12.5 P Values and Big Data

Relying solely on the p value for declaring important predictor variables
is problematic in big data applications, which are characterized by "large"
samples drawn from populations with unknown spread of the X's. The p
value is affected by the sample size (as sample size increases, the p value
decreases) and by the spread of X_i (as spread of X_i increases, the p value
decreases)* [1]. Accordingly, a small p value may be due to a large sample or
a large spread of the X's. Thus, *in big data applications, a small p value is only an
indicator of a potentially important predictor variable.*

The issue of how the p value is affected by moderating sample size
is currently unresolved. Big data are nonexperimental data for which
there are no procedures for determining the correct large sample size.
A large sample produces many small p values—a spurious result. The
associated variables are often declared important when in fact they are
not important; this reduces the stability of a model[t,‡] [2]. A procedure
that adjusts the p values when working with big data is needed.§ Until
such procedures are established, the recommended ad hoc approach is
as follows: *In big data applications, variables with small p values must undergo
a final assessment of importance based on their actual reduction in prediction*

---

* The size of b_i also affects the p value: As b_i increases, the p value decreases. This factor cannot
  be controlled by the analyst.
† Falsely reject NH.
‡ Falsely fail to reject NH. The effect of a "small" sample is that a variable can be declared
  unimportant when, in fact, it is important.
§ This is a procedure similar to the Bonferroni method, which adjusts p values downward
  because repeated testing increases the chance of incorrectly declaring a relationship or coefficient significant.

error. Variables associated with the greatest reduction in prediction error can be declared important predictors. When problematic cases are eliminated, the effects of spread of the X's can be moderated. For example, if the relevant population consists of 35- to 65-year-olds and the big data includes 18- to 65-year-olds, then simply excluding the 18- to 34-year-olds eliminates the spurious effects of spread.

12.6 Returning to Question 1

Is X_i important for making good predictions? The usual answer is that X_i is an important predictor variable if it has an associated p value less than 5%.

This answer is correct for experimental studies, in which sample sizes are correctly determined and samples have presumably known spread. For big data applications, in which large samples with unknown spread adversely affect the p value, a small p value is only an indicator of a potentially important predictor variable. The associated variables must go through an ad hoc evaluation of their actual reduction in the prediction errors before ultimately being declared important predictors.

12.7 Effect of Predictor Variable on Prediction

An assessment of the effect of predictor variable X_i on the prediction of Y focuses on the regression coefficient b_i. The common interpretation is that b_i is the average change in the predicted Y value associated with a unit change in X_i when the other X's are held constant. A detailed discussion of what b_i actually measures and how it is calculated shows that this is an honest interpretation, which must be accompanied by a caveat.

The regression coefficient b_i, also known as a partial regression coefficient, is a measure of the linear relationship between Y and X_i when the influences of the other X's are partialled out or held constant. The expression *held constant* implies that the calculation of b_i involves the removal of the effects of the other X's. While the details of the calculation are beyond the scope of this chapter, it suffices to outline the steps involved, as delineated next.*

The calculation of b_i uses a method of statistical control in a three-step process:

1. The removal of the linear effects of the other Xs *from* Y produces a new variable Y-adj (= Y adjusted linearly for the other X's).

---

* The procedure for statistical control can be found in most basic statistics textbooks.

2. The removal of the linear effects of the other Xs *from* X_i produces a new variable X_i-adj (= X_i adjusted linearly for the other Xs).

3. The regression of Y-adj on X_i-adj produces the desired partial regression coefficient b_i.

The partial regression coefficient b_i is an honest estimate of the relationship between Y and X_i (with the effects of the other X's partialled out) because it is based on statistical control, not experimental control. The statistical control method estimates b_i without data for the relationship between Y and X_i when the other X's are actually held constant. The rigor of this method ensures the estimate is an honest one.

In contrast, the experimental control method involves collecting data for which the relationship between X_i and Y is observed when the other X's are actually held constant. The resultant partial regression coefficient is directly measured and therefore yields a "true" estimate of b_i. Regrettably, experimental control data are difficult and expensive to collect.

12.8 The Caveat

There is one caveat to ensure the proper interpretation of the partial regression coefficient. *It is not enough to know the variable being multiplied by the regression coefficient; the other X's must also be known* [3]. Recognition of the other X's in terms of their values in the sample ensures that interpretation is valid. Specifically, the average change b_i in Y is valid for each and every unit change in X_i, within the range of X_i values in the sample when the other X's are held constant within the ranges or region of the values of the other X's in the sample.* This point is clarified in the following illustration:

I regress SALES (in dollar units) on EDUCATION (in year units); AGE (in year units); GENDER (in female gender units; 1 = female and 0 = male); and INCOME (in thousand-dollar units). The regression equation is defined in Equation (12.2):

$$SALES = 68.5 + 0.75*AGE + 1.03*EDUC + 0.25*INCOME + 6.49*GENDER \quad (12.2)$$

Interpretation of the regression coefficients:

1. The individual variable ranges in Table 12.1 suffice to mark the boundary of the region of the values of the other X's.

2. For AGE, the average change in SALES is 0.75 dollars for each 1-year increase in AGE within the range of AGE values in the sample when

---

* The region of the values of the other X's is defined as the values in the sample common to the entire individual variable ranges of the other X's.

EDUC, INCOME, and GENDER (E-I-G) are held constant within the E-I-G region in the sample.

3. For EDUC, the average change in SALES is 1.03 dollars for each 1-year increase in EDUC within the range of EDUC values in the sample when AGE, INCOME, and Gender (A-I-G) are held constant within the A-I-G region in the sample.

4. For INCOME, the average change in SALES is 0.25 dollars for each $1,000 increase in INCOME within the range of Income values in the sample when AGE, EDUC, and GENDER (A-E-G) are held constant within the A-E-G region in the sample.

5. For GENDER, the average change in SALES is 6.49 dollars for a "female-gender" unit increase (a change from male to female) when AGE, INCOME and EDUCATION (A-I-E) are held constant within the A-I-E region in the sample.

To further the discussion on the proper interpretation of the regression coefficient, I consider a composite variable (e.g., $X_1 + X_2$, X_1*X_2, or X_1/X_2), which often is included in regression models. I show that the regression coefficients of the composite variable and the variables defining the composite variable are not interpretable.

I add a composite product variable to the original regression model: EDUC_INC (= EDUC*INCOME, in years*thousand-dollar units). The resultant regression model is in Equation (12.3):

$$SALES = 72.3 + 0.77*AGE + 1.25*EDUC + 0.17*INCOME +$$
$$6.24*GENDER + 0.006*EDUC\_INC \qquad (12.3)$$

Interpretation of the new regression model and its coefficients is as follows:

1. The coefficients of the original variables have changed. This is expected because the value of the regression coefficient for X_i depends not only on the relationship between Y and X_i but also on the relationships between the other X's and X_i and the other X's and Y.

2. Coefficients for AGE and GENDER changed from 0.75 to 0.77 and from 6.49 to 6.24, respectively.

3. For AGE, the average change in SALES is 0.77 dollars for each 1-year increase in AGE within the range of AGE values in the sample when EDUC, INCOME, GENDER, and EDUC_INC (E-I-G-E_I) are held constant within the E-I-G-E_I region in the sample.

4. For GENDER, the average change in SALES is 6.24 dollars for a "female-gender" unit increase (a change from male to female) when AGE, EDUC, INCOME, and EDUC_INC (A-E-I-E_I) are held constant within the A-E-I-E_I region in the sample.

5. Unfortunately, the inclusion of EDUC_INC in the model compromises the interpretation of the regression coefficients for EDUC and INCOME—two variables that cannot be interpreted. Consider the following:

   a. For EDUC, the usual interpretation is that the average change in SALES is 1.25 dollars for each 1-year increase in EDUC within the range of EDUC values in the sample when Age, Income, Gender, and EDUC_INC (A-I-G-E_I) are held constant within the A-I-G-E_I region in the sample. *This statement is meaningless.* It is not possible to hold constant EDUC_INC for any 1-year increase in EDUC; as EDUC would vary, so must EDUC_INC. Thus, no meaningful interpretation can be given to the regression coefficient for EDUC.

   b. Similarly, no meaningful interpretations can be given to the regression coefficients for INCOME and EDUC_INC. It is not possible to hold constant EDUC_INC for INCOME. And, for EDUC_INC, it is not possible to hold constant EDUC and INCOME.

12.9 Returning to Question 2

How does the X_i affect the prediction of Y? The usual answer is that Y experiences an average change of b_i with a unit increase in X_i when the other X's are held constant.

This answer is an honest one (because of the statistical control method that estimates b_i) that must be accompanied by the values of the X_i range and the region of the other X's. Unfortunately, the effects of a composite variable and the variables defining the composite variable cannot be assessed because their regression coefficients are not interpretable.

12.10 Ranking Predictor Variables by Effect on Prediction

I return to the first regression model in Equation (12.2).

$$SALES = 68.5 + 0.75*AGE + 1.03*EDUC + 0.25*INCOME + 6.49*GENDER$$

(12.2)

A common misinterpretation of the regression coefficient is that GENDER has the greatest effect on SALES, followed by EDUC, AGE, and INCOME because the coefficients can be ranked in that order. The problem with this interpretation is discussed in the following paragraphs. I also present the correct rule for ranking predictor variables in terms of their effects on the prediction of dependent variable Y.

This regression model illustrates the difficulty in relying on the regression coefficient for ranking predictor variables. The regression coefficients are incomparable because different units are involved. No meaningful comparison can be made between AGE and INCOME because the variables have different units (years and thousand-dollar units, respectively). Comparing GENDER and EDUC is another mixed-unit comparison between female gender and years, respectively. Even a comparison between AGE and EDUC, whose units are the same (i.e., years), is problematic because the variables have unequal spreads (e.g., standard deviation, StdDev) in Table 12.1.

The correct ranking of predictor variables in terms of their effects on the prediction of Y is the ranking of the variables by the magnitude of the standardized regression coefficient. The sign of the standardized coefficient is disregarded as it only indicates direction. (Exception to this rule is discussed further in the chapter.) The standardized regression coefficient (also known as the beta regression coefficient) is produced by multiplying the original regression coefficient (also called the raw regression coefficient) by a conversion factor (CF). The standardized regression coefficient is unitless, just a plain number allowing meaningful comparisons among the variables. The transformation equation that converts a unit-specific raw regression coefficient into a unitless standardized regression coefficient is in Equation (12.4):

$$\text{Standardized regression coefficient for } X_i = CF*\text{Raw} \text{ regression coefficient for } X_i \tag{12.4}$$

The CF is defined as the ratio of a unit measure of Y variation to a unit measure of X_i variation. The StdDev is the usual measure used. However, if the distribution of variable is not bell shaped, then the StdDev is not reliable,

TABLE 12.1

Descriptive Statistics of Sample

Variable	Mean	Ranges (min, max)	StdDev	H spread
SALES	30.1	(8, 110)	23.5	22
AGE	55.8	(44, 76)	7.2	8
EDUC	11.3	(7, 15)	1.9	2
INCOME	46.3	(35.5, 334.4)	56.3	28
GENDER	0.58	(0, 1)	0.5	1

and the resultant standardized regression coefficient is questionable. An alternative measure, one that is not affected by the shape of the variable, is the H-spread. The H-spread is defined as the difference between the 75th percentile and 25th percentile of the variable distribution. Thus, there are two popular CFs and two corresponding transformations in Equations (12.5) and (12.6).

$$\text{Standardized regression coefficient for } X_i = [\text{StdDev of } X_i / \text{StdDev of } Y]*\text{Raw regression coefficient for } X_i \tag{12.5}$$

$$\text{Standardized regression coefficient for } X_i = [\text{H spread of } X_i / \text{H-spread of } Y]*\text{Raw regression coefficient for } X_i \tag{12.6}$$

Returning to the illustration of the first regression model, I note the descriptive statistics in Table 12.1. AGE has equivalent values for both StdDev and H-spread (23.5 and 22, respectively); so does EDUC (7.2 and 8, respectively). INCOME, which is typically not bell shaped, has quite different values for the two measures; the unreliable StdDev is 56.3, and the reliable H-spread is 28.

Dummy variables have no meaningful measure of variation. StdDev and H-spread are often reported for dummy variables as a matter of course, but they have "no value" at all.

The correct ranking of the predictor variables in terms of their effects on SALES—based on the magnitude of the standardized regression coefficients in Table 12.2—puts INCOME first, with the greatest effect, followed by AGE and EDUC. This ordering is obtained with either the StdDev or the H-spread. However, INCOME's standardized coefficient should be based on the H-spread, as INCOME is typically skewed. Because GENDER is a dummy variable with no meaningful CF, its effect on prediction of Y cannot be ranked.

TABLE 12.2

Raw and Standardized Regression Coefficients

Variable (unit)	Raw Coefficients (unit)	Standardized Coefficient (unitless) Based on	
		StdDev	H-spread
AGE (years)	0.75 (dollars/years)	0.23	0.26
EDUC (years)	1.03 (dollars/years)	0.09	0.09
INCOME (000 dollars)	0.25 (dollars/000-dollars)	0.59	0.30
GENDER (female gender)	6.49 (dollars/ female-gender)	0.14	0.27

12.11 Returning to Question 3

Which variables in the model have the greatest effect on the prediction of Y, in ranked order? The usual answer is that the variable with the largest regression coefficient has the greatest effect; the variable with the next-largest regression coefficient has the next-greatest effect, and so on. This answer is usually incorrect.

The correct ranking of predictor variables in terms of their effects on the prediction of Y is the ranking of the variables by the magnitude of the standardized regression coefficient. Predictor variables that cannot be ranked are (1) dummy variables, which have no meaningful measure of variation; and (2) composite variables and the variables defining them, which have both raw regression coefficients and standardized regression coefficients that cannot be interpreted.

12.12 Returning to Question 4

Which variables in the model are the most important predictors, in ranked order? The usual answer is that the variable with the largest standardized regression coefficient is the most important variable; the variable with the next-largest standardized regression coefficient is the next-important variable, and so on.

This answer is correct only when the predictor variables are uncorrelated, a rare occurrence in marketing models. With uncorrelated predictor variables in a regression model, there is a rank order correspondence between the magnitude of the standardized coefficient and reduction in prediction error. Thus, the magnitude of the standardized coefficient can rank the predictor variables in order of most to least important. Unfortunately, there is no rank order correspondence with correlated predictor variables. Thus, the magnitude of standardized coefficient of correlated predictor variables cannot rank the variables in terms of predictive importance. The proof of these facts is beyond the scope of this chapter [4].

12.13 Summary

It should now be clear that the common misconceptions of the regression coefficient lead to an incorrect interpretation of the ordinary regression model. This exposition puts to rest these misconceptions, promoting an appropriate and useful presentation of the regression model.

Common misinterpretations of the regression coefficient are problematic. The use of the statistical p value as a sole measure for declaring X_i an important predictor is problematic because of its sensitivity to sample size and spread of the X values. In experimental studies, the model builder must ensure that the study design takes into account these sensitivities to allow valid inferences to be drawn. In big data applications, variables with small p values must undergo a final assessment of importance based on their actual reduction in prediction error. Variables associated with the greatest reduction in prediction error can be declared important predictors.

When assessing how X_i affects the prediction of Y, the model buider must report the other X's in terms of their values. Moreover, the analyst must not attempt to assess the effects of a composite variable and the variables defining the composite variable because their regression coefficients are not interpretable.

By identifying the variables in the model that have the greatest effect on the prediction of Y—in rank order—the superiority of the standardized regression coefficient for providing correct ranking of predictor variables, rather than the raw regression coefficient, should be apparent. Furthermore, it is important to recognize that the standardized regression coefficient ranks variables in order of predictive importance (most to least) for only uncorrelated predictor variables. This is not true for correlated predictor variables. It is imperative that the model builder does not use the coefficient to rank correlated predictor variables despite the allure of its popular misuse.

References

1. Kraemer, H.C., and Thiemann, S., *How Many Subjects?* Sage, Thousand Oaks, CA, 1987.
2. Dash, M., and Liu, H., *Feature Selection for Classification,* Intelligent Data Analysis, Elsevier Science, New York, 1997.
3. Mosteller, F., and Tukey, J., *Data Analysis and Regression*, Addison-Wesley, Reading, MA, 1977.
4. Hayes, W.L., *Statistics for the Social Sciences*, Holt, Rinehart and Winston, Austin, TX, 1972.

13

The Average Correlation: A Statistical Data Mining Measure for Assessment of Competing Predictive Models and the Importance of the Predictor Variables

13.1 Introduction

The purpose of this chapter is to introduce the number one statistic, the mean, and the runner-up, the correlation coefficient, which when used together as discussed here yield the *average correlation*, providing a fruitful statistical data mining measure. The average correlation, along with the correlation coefficient, provides a quantitative criterion for assessing (1) competing predictive models and (2) the importance of the predictor variables.

13.2 Background

Two essential characteristics of a predictive model are its *reliability* and *validity*, terms that are often misunderstood and misused. *Reliability* refers to a model[*] yielding consistent results.[†] For a predictive model, the proverbial question is how dependable (reliable) the model is for predictive purposes. A predictive model is reliable to the extent the model is producing repeatable predictions. It is normal for an individual to vary in performance, as chance influences are always in operation, but performance is expected to

---

[*] A "model" can be predictive (statistical regression), explanatory (principal components analysis, PCA), or predictive and explanatory (structural equation).

[†] A model should be monitored for consistent results after each implementation. If the results of the model show signs of degradation, then the model should either be recalibrated (update the regression coefficients: keep the same variables in the model but use fresh data) or retrained (update the model: add new variables to the original variables of the model and use fresh data).

fall within narrow limits. Thus, for a predictive model, predictions for the same individual, obtained from repeated implementations of the (reliable) model, are expected to vary closely.

Model *validity* refers to the extent to which the model measures what it is intended to measure with respect to a given criterion (e.g., a predictive model criterion is small prediction errors). One indispensable element of a valid model is that is has high reliability. If the reliability of the model is low, so goes the validity of the model. Reliability is a necessary, but not sufficient, condition for a valid model. Thus, a predictive model is valid to the extent the model is *efficient* (precise and accurate) in predicting at a given time. I give a clarifying explanation of efficiency in the next section.

The common question—Is the model valid?—is not directly answerable. A model does not possess a standard level of validity as it may be highly valid at one point in time, but not at another, as the environment about the initial modeling process is likely to change. The inference is that models have to be kept in a condition of good efficiency. See Footnote 2 for a brief discussion of maintaining a model in good standing regarding validity.

There are two other aspects of model validity: face validity and content validity. *Face validity* is a term used to characterize the notion of a model "looking like it is going to work." It is a subjective criterion valuable to model users. Face validity gives users, especially those who may not have the specialized background to build a model but do have practical knowledge of how models work, what they want of a model. Thus, if the model does not look right for the required objective, the confidence level of the model utility drops, as does the acceptance and implementation of the model.

Content validity refers to the variables in a model in that the content of the individual variables and the ensemble of the variables are relevant to the purpose of the model. As a corollary, the model should not have irrelevant, unnecessary variables. Such variables can be eliminated by assessing preliminarily the correlation of each variable with the dependent variable. The correlation coefficient for this elimination process is not foolproof as it at times allows variables with nonrelevant content to "sneak in" the model. More important, taking out the indicated variables is best achieved by determining the *quality* of the variable selection used for defining the best subset of the predictor variables. In other words, the model builder must assess *subjectively* the content of the variables. An *objective* discussion of content validity is beyond the scope of this chapter.[*]

The literature[†] does not address the "left side" of the model equation, the dependent variable. The dependent variable is often expressed in management terms (i.e., nonoperational terms). The model builder needs the explicit backstory about the management objective to create a valid definition of the

---

[*] An excellent reference is Carmines, E.G., and Zeller, R.A., *Reliability and Viability Assessment*, Sage, Thousand Oaks, CA, 1991.
[†] The literature—to my knowledge—does not address the left side of the equation.

dependent variable. Consider the management objective to build a lifetime value (LTV) model to estimate the expected LTV over the next 5 years (LTV5). Defining the dependent variable LTV5 is not always straightaway. For most modeling projects, there are not enough historical data to build an LTV5 model with a literal definition of dependent variable LTV5 using, say, the current 5 years of data. That is, if using a full 5-year window of data for the definition of LTV5, there are not enough remaining years of data for a sufficient pool of candidate predictor variables. The model builder has to shorten the LTV5 window, say, using LTV2, defined by a window of 2 years, and then define LTV5 = LTV2*2.5. Such a legitimate arithmetic adjustment yields a sizable, promising set of candidate predictor variables along with an acceptable content-valid dependent variable. As a result, the building of a LTV5 model is in a favorable situation for excellent valid and reliable performance.

13.3 Illustration of the *Difference* between Reliability and Validity

Reliability does not imply validity. For example, a reliable regression model is predicting consistently (precisely), but may not be predicting what it is intended to predict. In terms of accuracy and precision, reliability is analogous to precision, while validity is analogous to accuracy.

An example often used to illustrate the difference between reliability and validity involves the bathroom scale. If someone weighing 118 pounds steps on the same scale, say, five consecutive times, yielding varied readings of, say, 115, 125, 195, 140, and 136, then the scale is not reliable or precise. If the scale yields consistent readings of, say, 130, then the scale is reliable but not valid or accurate. If the scale readings are 118 all five times, then the scale is both reliable and valid.

13.4 Illustration of the *Relationship* between Reliability and Validity*

I may have given the impression that reliability and validity are separate concepts. In fact, they complement each other. I discuss a common example of the relationship between reliability and validity. Think of the center of the bull's-eye target as the concept that you seek to measure. You take a shot

---

* Consult http://www.socialresearchmethods.net/kb/relandval.php, 2010.

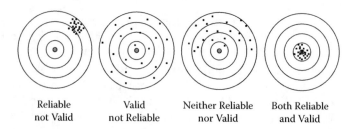

Reliable	Valid	Neither Reliable	Both Reliable
not Valid	not Reliable	nor Valid	and Valid

FIGURE 13.1
Relationship between reliability and validity.

at the target. If you measure the concept perfectly, you hit the center of the target. If you do not, you miss the center. The more you are off the target, the further you are from the center.

There are four situations involving reliability and validity, depicted in Figure 13.1:

1. In the first bull's-eye target, you hit the target consistently, but you miss the center of the target. You are producing consistently poor performances among all hits. Your target practice is reliable but not valid.

2. The second display shows that your hits are spread randomly about the target. You seldom hit the center of the target, but you get good performance among all hits. Your target practice is not reliable but valid. Here, you can see clearly that reliability is directly related to the *variability* of your target practice performance.

3. The third display shows that your hits are spread across the target, and you are consistently missing the center. Your target practice in this case is neither reliable nor valid.

4. Finally, there is the "Wild Bill Hickok" fanfare*: You hit the center of the target consistently. You (and Wild Bill) are both reliable and valid as a sharpshooter.

5. Note: Reliability and validity concepts themselves are independent of each other. But in the context of a predictive model, reliability is necessary but not sufficient for validity.

---

* James Butler Hickok (May 27, 1837–August 2, 1876), better known as Wild Bill Hickok, was a figure in the American Old West. He fought in the Union Army during the American Civil War and gained publicity after the war as a sharpshooter.

13.5 The Average Correlation

The average correlation, along with the correlation coefficient, provides a quantitative criterion for assessing (1) competing predictive models* and (2) the importance of the predictor variables defining a model. With regard to item 2, numerous statistics textbooks are amiss in explaining which variables in the model have the greatest effect on the prediction of Y. The usual answer is that the variable with the largest regression coefficient has the greatest effect; the variable with the next-largest regression coefficient has the next-greatest effect, and so on. This answer is usually incorrect. It is correct only when the predictor variables are uncorrelated, quite a rare occurrence in practice.[†] An illustration is the best route to make the average correlation understandable and, it is hoped, part of the model builders' tool kit for addressing items 1 and 2.

13.5.1 Illustration of the Average Correlation with an LTV5 Model

Let us consider the following objective: to build a LTV5 model. The previous discussion of defining the LTV5 dependent variable serves as background to this illustration. The LTV5 model consists of 11 predictor variables (VAR1–VAR11), whose content domain is the nature of sales performance, sales incentives, sales program enrollment, number of transactions, and the like. The correlation matrix of VAR1–VAR11 is in Table 13.1.

For ease of presentation, I replace the "backslash" diagonal of correlation coefficients of value 1 with "dots" (see Table 13.2). (The correlation coefficient of a variable with itself is obviously equal to a value of 1.) The purpose of replacing the 1's is to show where within the matrix the calculations are performed. The correlation coefficients are now all positive, which is *not always* found in practice.

The dots divide the matrix into upper-triangular and lower-triangular correlation matrices, which are identical. The average correlation is the mean of the absolute values of the pairwise correlation coefficients of either the upper- or lower-triangular matrix, not both. As the correlation coefficient is a measure of the reliability or "closeness" of the relationship between a pair of variables, it follows that the mean of all pairs is pleasing to the mind as an *honest* measure of closeness among the predictor variables in a model.

---

* See "A Dozen Statisticians, a Dozen Outcomes" on my Web site: http://www.geniq.net/res/A-Dozen-Statisticians-with-a-Dozen-Outcomes.html.
[†] I am mindful that if the model builder uses principal component (PC) variables, the PCs are uncorrelated.

TABLE 13.1

LTV Model: Pearson Correlation Coefficients/Prob > |r| under HO: Rho = 0

	VAR1	VAR2	VAR3	VAR4	VAR5	VAR6	VAR7	VAR8	VAR9	VAR10	VAR11
VAR1	1.00000	-0.69609	0.26903	0.30443	0.35499	0.35166	0.37812	0.31297	0.35020	0.29410	0.25545
		<.0001	<.0001	<.0001	<.0001	<.0001	<.0001	<.0001	<.0001	<.0001	<.0001
VAR2	-0.69689	1.00000	-0.18834	0.19207	-0.22438	-0.21200	-0.23549	-0.20007	-0.20309	-0.17179	-0.14809
	<.0001		<.0001	<.0001	<.0001	<.0001	<.0001	<.0001	<.0001	<.0001	<.0001
VAR3	0.26903	-0.10034	1.00000	0.22467	0.37111	0.31629	0.39288	0.38376	0.42024	0.29891	0.26452
	<.0001	<.0001		<.0001	<.0001	<.0001	<.0001	<.0001	<.0001	<.0001	<.0001
VAR4	0.30443	-0.19207	0.22467	1.00000	0.34542	0.29531	0.33502	0.26600	0.24341	0.23202	0.26305
	<.0001	<.0001	<.0001		<.0001	<.0001	<.0001	<.0001	<.0001	<.0001	<.0001
VAR5	0.35499	-0.22438	0.37111	0.34542	1.00000	0.50653	0.46107	0.39413	0.41061	0.44856	0.33444
	<.0001	<.0001	<.0001	<.0001		<.0001	<.0001	<.0001	<.0001	<.0001	<.0001
VAR6	0.35166	-0.21200	0.31629	0.29531	0.50653	1.00000	0.46999	0.35111	0.44730	0.48984	0.34348
	<.0001	<.0001	<.0001	<.0001	<.0001		<.0001	<.0001	<.0001	<.0001	<.0001
VAR7	0.37812	-0.23549	0.39288	0.33582	0.46107	0.46999	1.00000	0.40503	0.44634	0.41615	0.33916
	<.0001	<.0001	<.0001	<.0001	<.0001	<.0001		<.0001	<.0001	<.0001	<.0001
VAR8	0.31297	-0.20007	0.38376	0.26688	0.39413	0.35111	0.40503	1.00000	0.39891	0.29346	0.31058
	<.0001	<.0001	<.0001	<.0001	<.0001	<.0001	<.0001		<.0001	<.0001	<.0001
VAR9	0.35020	-0.20389	0.42024	0.24341	0.41061	0.44730	0.44634	0.39891	1.00000	0.44746	0.35423
	<.0001	<.0001	<.0001	<.0001	<.0001	<.0001	<.0001	<.0001		<.0001	<.0001
VAR10	0.29410	-0.17179	0.29891	0.23282	0.44856	0.48984	0.41615	0.29346	0.44746	1.00000	0.35845
	<.0001	<.0001	<.0001	<.0001	<.0001	<.0001	<.0001	<.0001	<.0001		<.0001
VAR11	0.25545	-0.14809	0.26452	0.26305	0.33444	0.34348	0.33916	0.31058	0.35423	0.35845	1.00000
	<.0001	<.0001	<.0001	<.0001	<.0001	<.0001	<.0001	<.0001	<.0001	<.0001	

TABLE 13.2

The Correlation Matrix of VAR1–VAR11 Divided by Replacing the 1's with Dots (.).

	VAR1	VAR2	VAR3	VAR4	VAR5	VAR6	VAR7	VAR8	VAR9	VAR10	VAR11
VAR1	.	0.69689	0.26903	0.30443	0.35499	0.35166	0.37812	0.31297	0.35020	0.29410	0.25545
VAR2	0.69689	.	0.18834	0.19207	0.22438	0.21200	0.23549	0.20007	0.20389	0.17179	0.14809
VAR3	0.26903	0.18834	.	0.22467	0.37111	0.31629	0.39288	0.38376	0.42024	0.29891	0.26452
VAR4	0.30443	0.19207	0.22467	.	0.34542	0.29531	0.33582	0.26688	0.24341	0.23282	0.26305
VAR5	0.35499	0.22438	0.37111	0.34542	.	0.50653	0.46107	0.39413	0.41061	0.44856	0.33444
VAR6	0.35166	0.21200	0.31629	0.29531	0.50653	.	0.46999	0.35111	0.44730	0.48984	0.34348
VAR7	0.37812	0.23549	0.39288	0.33582	0.46107	0.46999	.	0.40503	0.44634	0.41615	0.33916
VAR8	0.31297	0.20007	0.38376	0.26688	0.39413	0.35111	0.40503	.	0.39891	0.29346	0.31058
VAR9	0.35020	0.20389	0.42024	0.24341	0.41061	0.44730	0.44634	0.39891	.	0.44746	0.35423
VAR10	0.29410	0.17179	0.29891	0.23282	0.44856	0.48984	0.41615	0.29346	0.44746	.	0.35845
VAR11	0.25545	0.14809	0.26452	0.26305	0.33444	0.34348	0.33916	0.31058	0.35423	0.35845	.

I digress to justify my freshly minted measure, the average correlation. One familiar with Cronbach's alpha may think the average correlation is a variant of alpha: It is. I did not realize the relationship between the two measures until the writing of this chapter. My background in the field of psychometrics was in action without my knowing it. Cronbach developed alpha for a measure of reliability within the field of psychometrics. Cronbach started with a variable X, actually a test item score defined as X = t + e, where as X is an observed score, t is the true score, and e is the random error [1]. Alpha does not take absolute values of the correlation coefficients; the correlations of alpha are always positive. I developed the average correlation for a measure of closeness among predictor variables *in a model*. "In a model" implies that the average correlation is a conditional measure. That is, the average correlation is a *relative* measure for the reason that it is anchored with respect to a dependent variable. Alpha measures reliability among variables belonging to a test; there is no "anchor" variable. In practice, the psychometrician seeks large alpha values, whereas the model builder seeks small average correlation values.

A rule of thumb for a desirable value of the average correlation is "small" positive values.

1. A small value indicates the predictor variables are not highly inter-correlated; that is, they do not suffer from the condition of multicollinearity. Multicollinearity makes it virtually impossible to assess the true contribution or importance the predictor variables have on the dependent variable.

   a. Worthy of note is that statistics textbooks refer to multicollinearity as a "data problem, but not a weakness in the model." Students are taught that model performance is not affected by the condition of multicollinearity. Multicollinearity is a data problem because its only affects a clearly assigned contribution of each predictor variable to the dependent: The assigned contribution of each predictor variable is muddied. Unfortunately, students are not taught that model performance is not affected by the condition of multicollinearity *as long as* the condition of multicollinearity remains the same as when the model was initially built. If the condition is the same as when the model was first built, then implementation of the model should yield good performance. However, for every reimplementation of the model after the first, the condition of multicollinearity has *showed in practice* not to remain the same. Hence, I uphold and defend that multicollinearity is a data problem, and multicollinearity *does* affect model performance.

2. Average correlation values in the range of 0.35 or less are desirable. In this situation, a *soundly* honest assessment of the contributions of the predictor variables to the performance of the model can be made.

3. Average correlation values that are greater than 0.35 and less than 0.55 are moderately desirable. In this situation, a *somewhat* honest assessment of the contributions of the predictor variables to the performance of the model can be made.

4. Average correlation values that are greater than 0.55 are not desirable as they indicate the predictor variables are excessively redundant. In this situation, a *questionably* honest assessment of the contributions of the predictor variables to the performance of the model can be made.

As long as the average correlation value is acceptable (less than 0.40), the second proposed item of assessing competing models (every modeler builds several models and must choose the best one) is in play. If a project session brings forth models within the acceptable range of average correlation values, the model builder uses both the average correlation value *and the set* of the individual correlations of predictor variable with dependent variable. The individual correlations indicate the content *validity* of the model. Rules of thumb for the values of the individual correlation coefficients are as follows:

1. Values between 0.0 and 0.3 (0.0 and -0.3) indicate poor validity.
2. Values between 0.3 and 0.7 (-0.3 and -0.7) indicate moderate validity.
3. Values between 0.7 and 1.0 (-0.7 and -1.0) indicate a strong validity.

In sum, the model builder uses the average correlation and the individual correlations to assess competing predictive models and the importance of the predictor variables. I continue with the illustration of the LTV5 model to make sense of these discussions and rules of thumb in the next section.

13.5.2 Continuing with the Illustration of the Average Correlation with an LTV5 Model

The average correlation of the LTV5 model is 0.33502. The individual correlations of the predictor variables with LTV5 (Table 13.3) indicate the variables have moderate to strong validity, except for VAR2. The combination of 0.33502 and values of Table 13.3 is compelling for any modeler to be pleased with the reliability and validity of the LTV5 model.

13.5.3 Continuing with the Illustration with a Competing LTV5 Model

Assessing competing models is a task of the model builder who can compare two sets of measures (the average correlation and the set of individual

TABLE 13.3

Correlations of Predictors (VAR) with LTV5

PREDICTORS	CORR_COEFF_with_ LTV5
VAR8	0.71472
VAR7	0.70277
VAR6	0.68443
VAR5	0.67982
VAR9	0.65602
VAR3	0.61076
VAR10	0.59583
VAR11	0.59087
VAR4	0.52794
VAR1	0.49827
VAR2	-0.27945

TABLE 13.4

Correlations of Predictors (VAR_) with LTV5

PREDICTORS	CORR_COEFF_with_LTV5
VAR_2	0.80377
VAR_3	0.70945
VAR_1	0.62148

correlations). The model builder must have a keen skill to balance the metrics. The continuing illustration makes the "balancing of the metrics" clear.

I build a second model, the competing model, whose average correlation equals 0.37087. The individual correlations of the predictor variables with LTV5 (Table 13.4) indicate the predictor variables have strong validity, except with VAR_1, which has a moderate degree. Clearly, the difference between the average correlations of 0.33502 and 0.37087, for the first and second models, respectively, offers a quick thought for the modeler to contemplate as the difference is minimal. From there, I examine the individual correlations in Tables 13.3 and 13.4. The competing model has large-valued correlation coefficients and has few (three) variables, an indicator of a reliable model. The first model has individual correlations with mostly strong validity, except for VAR2, but there are too many predictor variables (for my taste). My determination of the better model is the competing model because not only are the metrics pointing to that decision, but also I always prefer a model with no more than a handful of predictor variables.

13.5.3.1 *The Importance of the Predictor Variables*

As for the importance of the predictor variables, if the average correlation value is acceptable, the importance of the predictor variables is readily read from the individual correlations table. Thus, for the better model discussed, the most important predictor variable is VAR_2, followed by VAR_3 and VAR_1, successively.

13.6 Summary

I introduced the fruitful statistical data mining measure, the average correlation, that provides, along with the correlation coefficient, a quantitative criterion for assessing both competing predictive models and the importance of the predictor variables. After providing necessary background, which included the essential characteristics of a model, reliability and validity, I chose an illustration with an LTV5 model as the best approach to make the average correlation understandable and, it is hoped, part of the model builders' tool kit for assessing both competing predictive models and the importance of the predictor variables.

Reference

1. Cronbach, L.J., Coefficient alpha and the internal structure of tests, *Psychometrika*, 16(3), 297–334, 1951.

14

CHAID for Specifying a Model with Interaction Variables

14.1 Introduction

To increase the predictive power of a model beyond that provided by its components, data analysts create an interaction variable, which is the product of two or more component variables. However, a compelling case can be made for utilizing CHAID (chi-squared automatic interaction detection) as an alternative data mining method for specifying a model, thereby justifying the omission of the component variables under certain circumstances. Database marketing provides an excellent example for this alternative method. I illustrate the alternative method with a response model case study.

14.2 Interaction Variables

Consider variables X_1 and X_2. The product of these variables, denoted by X_1X_2, is called a *two-way* or *first-order interaction variable*. An obvious property of this interaction variable is that its information or variance is shared with both X_1 and X_2. In other words, X_1X_2 has inherent high correlation with both X_1 and X_2.

If a third variable (X_3) is introduced, then the product of the three variables $(X_1X_2X_3)$ is called a three-way or second-order interaction variable. It is also highly correlated with each of its component variables. Simply multiplying the component variables can create higher-order variables. However, interaction variables of an order greater than three are rarely justified by theory or empirical evidence.

When data have highly correlated variables, they have a condition known as *multicollinearity*. When high correlation exists because of the relationship among these variables, the multicollinearity is referred to as essential ill conditioning. A good example of this is in the correlation between gender and income in the current workforce. The fact that males earn more than their female counterparts creates this "ill conditioning." However, when high correlation is due to an interaction variable, the multicollinearity is referred to as nonessential ill-conditioning [1].

When multicollinearity exists, it is difficult to assess reliably the statistical significance, as well as the informal noticeable importance, of the highly correlated variables. Accordingly, multicollinearity makes it difficult to define a strategy for excluding variables. A sizable literature has developed for essential ill conditioning, which has produced several approaches for specifying models [2]. In contrast, there is a modest collection of articles for nonessential ill conditioning [3, 4].

14.3 Strategy for Modeling with Interaction Variables

The popular strategy for modeling with interaction variables is the *principle of marginality*, which states that a model including an interaction variable should also include the component variables that define the interaction [5, 6]. A cautionary note accompanies this principle: *Neither testing the statistical significance (or noticeable importance) nor interpreting the coefficients of the component variables should be performed* [7]. A significance test, which requires a unique partitioning of the dependent variable variance in terms of the interaction variable and its components, is not possible due to the multicollinearity.

An unfortunate by-product of this principle is that models with *unnecessary* component variables go undetected. Such models are prone to overfit, which results in either unreliable predictions or deterioration of performance, as compared to a well-fit model with the necessary component variables.

An alternative strategy is based on Nelder's *notion of a special point* [8]. Nelder's strategy relies on understanding the functional relationship between the component variables and the dependent variable. When theory or prior knowledge about the relationship is limited or unavailable, Nelder suggests using exploration data analysis to determine the relationship. However, Nelder provides no general procedure or guidelines for uncovering the relationship among the variables.

I propose using CHAID as the data mining method for uncovering the functional relationship among the variables. Higher-order interactions are seldom found to be significant, at least in database marketing. Therefore, I limit this discussion only to first-order interactions. If higher-order interactions are required, the proposed method can be extended.

14.4 Strategy Based on the Notion of a Special Point

For simplicity, consider the full model in Equation (14.1):

$$Y = b_0 + b_1X_1 + b_2X_2 + b_3X_1X_2 \qquad (14.1)$$

$X_1 = 0$ is called a *special point* on the scale if when $X_1 = 0$ there is no relationship between Y and X_2. If $X_1 = 0$ is a special point, then omit X_2 from the full model; otherwise, X_2 should not be omitted.

Similarly for X_2, $X_2 = 0$ is called a special point on the scale if when $X_2 = 0$ there is no relationship between Y and X_1. If $X_2 = 0$ is a special point, then omit X_1; otherwise, X_1 should not be omitted.

If both X_1 and X_2 have special points, then X_1 and X_2 are omitted from the full model, and the model is reduced to $Y = b_0 + b_3X_1X_2$.

If a zero value is not assumed by the component variable, then no special point exists, and the procedure is not applicable.

14.5 Example of a Response Model with an Interaction Variable

Database marketing provides an excellent example for this alternative method. I illustrate the alternative with a response model case study, but it applies as well to a profit model. A music continuity club requires a model to increase response to its solicitations. Based on a random sample (size 299,214) of a recent solicitation with a 1% response, I conduct a logistic regression analysis of RESPONSE on two available predictor variables, X_1 and X_2. The variables are defined as:

1. RESPONSE is the indicator of a response to the solicitation: 0 indicates nonresponse, 1 indicates response.
2. X_1 is the number of months since the last inquiry. A zero month value indicates an inquiry was made within the month of the solicitation.
3. X_2 is a measure of an individual's affinity to the club based on the number of previous purchases and the listening interest categories of the purchases.

The output of the logistic analysis is presented in Table 14.1. X_1 and X_2 have Wald chi-square values of 10.5556 and 2.9985, respectively. Using a Wald cutoff value of 4, as outlined in Chapter 8, Section 8.12.1, there is indication

TABLE 14.1

Logistic Regression of Response on X_1 and X_2

Variable	df	Parameter Estimate	Standard Error	Wald Chi-Square	Pr > Chi-Square
INTERCEPT	1	−4.5414	0.0389	1,3613.7923	0.E + 00
X1	1	−0.0338	0.0104	10.5556	0.0012
X2	1	0.0145	0.0084	2.9985	0.0833

TABLE 14.2

Classification Table of Model with X_1 and X_2

		Classified 0	1	Total
Actual	0	143,012	153,108	296,120
	1	1,390	1,704	3,094
Total		144,402	154,812	299,214
			TCCR	48.37%

that X_1 is an important predictor variable, whereas X_2 is not quite as important. The classification accuracy of the base response model is displayed in classification Table 14.2. The "Total" column represents the actual number of nonresponders and responders: There are 296,120 nonresponders and 3,094 responders. The "Total" row represents the predicted or classified number of nonresponders and responders: There are 144,402 individuals classified as nonresponders and 154,812 classified as responders. The diagonal cells indicate the correct classifications of the model. The upper-left cell (actual = 0 and classified = 0) indicates the model correctly classified 143,012 nonresponders. The lower-right cell (actual = 1 and classified = 1) indicates the model correctly classified 1,704 responders. The total correct classification rate (TCCR) is equal to 48.37% (= [143,012 + 1,704]/299,214).

After creating the interaction variable X_1X_2 (= $X_1{}^*X_2$), I conduct another logistic regression analysis of RESPONSE with X_1, X_2, and X_1X_2; the output is in Table 14.3. I look at the Wald chi-square values to see that no direct statistical assessment is possible as per the cautionary note.

The classification accuracy of this full response model is displayed in classification Table 14.4. TCCR(X_1, X_2, X_1X_2) equals 55.64%, which represents a 15.0% improvement over the TCCR(X_1, X_2) of 48.37% for the model without the interaction variable. These two TCCR values provide a benchmark for assessing the effects of omitting—if possible under the notion of a special point—X_1 or X_2.

Can component variable X_1 or X_2 be omitted from the full RESPONSE model in Table 14.3? To omit a component variable, say X_2, it must be established

TABLE 14.3

Logistic Regression of Response on X_1, X_2, and X_1X_2

Variable	df	Parameter Estimate	Standard Error	Wald Chi-Square	Pr > Chi-Square
INTERCEPT	1	-4.5900	0.0502	8,374.8095	0.E + 00
X1	1	-0.0092	0.0186	0.2468	0.6193
X2	1	0.0292	0.0126	5.3715	0.0205
X1X2	1	-0.0074	0.0047	2.4945	0.1142

TABLE 14.4

Classification Table of Model with X1, X2, and X1X2

		Classified 0	Classified 1	Total
Actual	0	164,997	131,123	296,120
	1	1,616	1,478	3,094
Total		166,613	132,601	299,214
			TCCR	55.64%

that there is no relationship between RESPONSE and X_2 when X_1 is a special point. CHAID can be used to determine whether the relationship exists. Following a brief review of the CHAID technique, I illustrate how to use CHAID for this new approach.

14.6 CHAID for Uncovering Relationships

CHAID is a technique that recursively partitions (or splits) a population into separate and distinct segments. These segments, called *nodes*, are split in such a way that the variation of the dependent variable (categorical or continuous) is minimized within the segments and maximized among the segments. After the initial splitting of the population into two or more nodes (defined by values of an independent or predictor variable), the splitting process is repeated on each of the nodes. Each node is treated like a new subpopulation. It is then split into two or more nodes (defined by the values of another predictor variable) such that the variation of the dependent variable is minimized within the nodes and maximized among the nodes. The splitting process is repeated until stopping rules are met.

The output of CHAID is a *tree* display, where the root is the population and the branches are the connecting segments such that the variation of the dependent variable is minimized within all the segments and maximized among all the segments.

CHAID was originally developed as a method of finding interaction variables. In database marketing, CHAID is primarily used today as a market segmentation technique. Here, I utilize CHAID as an alternative data mining method for uncovering the relationship among component variables and the dependent variable to provide the information needed to test for a special point.

For this application of CHAID, the RESPONSE variable is the dependent variable, and the component variables are predictor variables X_1 and X_2. The CHAID analysis is forced to produce a tree, in which the initial splitting of the population is based on the component variable to be tested for a special point. One of the nodes must be defined by a zero value. Then, the "zero" node is split by the other component variable, producing response rates for testing for a special point.

14.7 Illustration of CHAID for Specifying a Model

Can X_1 be omitted from the full RESPONSE model? If when $X_2 = 0$ there is no relationship between RESPONSE and X_1, then $X_2 = 0$ is a special point, and X_1 can be omitted from the model. The relationship between RESPONSE and X_1 can be assessed easily by the RESPONSE CHAID tree (Figure 14.1), which is read and interpreted as follows:

1. The top box (root node) indicates that for the sample of 299,214 individuals, there are 3,094 responders and 296,120 nonresponders. The response rate is 1% and nonresponse rate is 99%.

2. The left leaf node (of the first level) of the tree represents 208,277 individuals whose X_2 values are not equal to zero. The response rate among these individuals is 1.1%.

3. The right leaf node represents 90,937 individuals whose X_2 values equal zero. The response rate among these individuals is 1.0%.

4. Tree notation: Trees for a continuous predictor variable denote the continuous values in intervals: a closed interval or a left-closed/right-open interval. The former is denoted by [a, b], indicating all values between and including a and b. The latter is denoted by [a, b), indicating all values greater than or equal to a and less than b.

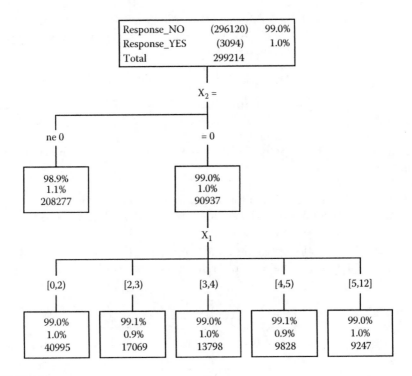

FIGURE 14.1
CHAID tree for testing X_2 for a special point.

5. I reference the bottom row of five branches (defined by the intersection of X_2 and X_1 intervals/nodes), from left to right: 1 through 5.

6. Branch 1 represents 40,995 individuals whose X_2 values equal zero *and* X_1 values lie in [0, 2). The response rate among these individuals is 1.0%.

7. Branch 2 represents 17,069 individuals whose X_2 values equal zero *and* X_1 values lie in [2, 3). The response rate among these individuals is 0.9%

8. Branch 3 represents 13,798 individuals whose X_2 values equal zero *and* X_1 values lie in [3, 4). The response rate among these individuals is 1.0%.

9. Branch 4 represents 9,828 individuals whose X_2 values equal zero *and* X_1 values lie in [4, 5). The response rate among these individual is 0.9%.

10. Node 5 represents 9,247 individuals whose X_2 values equal zero *and* X_1 values lie in [5, 12]. The response rate among these individuals is 1.0%.

11. The pattern of response rates (1.0%, 0.9%, 1.0%, 0.9%, 1.0%) across the five branches reveals there is no relationship between response and X_1 when $X_2 = 0$. Thus, $X_2 = 0$ is a special point.

12. The implication is X_1 can be omitted from the response model.

Can X_2 be omitted from the full RESPONSE model? If when $X_1 = 0$ there is no relationship between RESPONSE and X_2, then $X_1 = 0$ is a special point, and X_2 can be omitted from the model. The relationship between RESPONSE and X_2 can be assessed by the CHAID tree in Figure 14.2, which is read and interpreted as follows.

1. The top box indicates that for the sample of 299,214 individuals there are 3,094 responders and 296,120 nonresponders. The response rate is 1% and nonresponse rate is 99%.

2. The left leaf node represents 229,645 individuals whose X_1 values are not equal to zero. The response rate among these individuals is 1.0%.

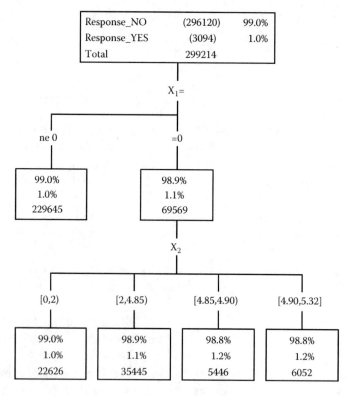

FIGURE 14.2
CHAID tree for testing X_1 for a special point.

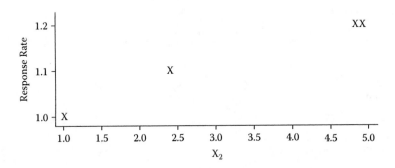

FIGURE 14.3
Smooth plot of response and X_2.

3. The right leaf node represents 69,569 individuals whose X_1 values equal zero. The response rate among these individuals is 1.1%.

4. I reference the bottom row of four branches (defined by the intersection of X_1 and X_2 intervals/nodes), from left to right: 1 through 4.

5. Branch 1 represents 22,626 individuals whose X_1 values equal zero *and* X_2 values lie in [0, 2). The response rate among these individuals is 1.0%.

6. Branch 2 represents 35,445 individuals whose X_1 values equal zero *and* X_2 values lie in [2, 4.85). The response rate among these individuals is 1.1%.

7. Branch 3 represents 5,446 individuals whose X_1 values equal zero *and* X_2 values lie in [4.85, 4.90). The response rate among these individuals is 1.2%.

8. Branch 4 represents 6,052 individuals whose X_1 values equal zero *and* X_2 values lie in [4.90, 5.32]. The response rate among these individuals is 1.2%.

9. The pattern of response rates across the four nodes is best observed in the smooth plot of response rates by the minimum values* of the intervals for the X_2 branches in Figure 14.3. There appears to be a positive straight-line relationship between Response and X_2 when X_1 = 0. Thus, X_1 = 0 is *not* a special point.

10. The implication is X_2 *cannot* be omitted from the response model.

Because I choose to omit X_1, I perform a logistic regression analysis on RESPONSE with X_2 and X_1X_2; the output is in Table 14.5. The classification accuracy of this response model is displayed in the classification Table 14.6:

---

* The minimum value is one of several values that can be used; alternatives are the means or medians of the predefined ranges.

TABLE 14.5

Logistic Regression of Response on X_2 and X_1X_2

Variable	df	Parameter Estimate	Standard Error	Wald Chi-Square	Pr > Chi-Square
INTERCEPT	1	−4.6087	0.0335	18,978.8487	0.E + 00
X1X2	1	−0.0094	0.0026	12.6840	0.0004
X2	1	0.0332	0.0099	11.3378	0.0008

TABLE 14.6

Classification Table of Model with X_2 and X_1X_2

		Classified		
		0	1	Total
Actual	0	165,007	131,113	296,120
	1	1,617	1,477	3,094
Total		166,624	132,590	299,214
			TCCR	55.64%

TCCR(X_2, X_1X_2) equals 55.64%. TCCR for the full model, TCCR(X_1, X_2, X_1X_2), is also equal to 55.64%. The implication is that X_1 is not needed in the model.

In sum, the parsimonious (best-so-far) RESPONSE model is defined by X_2 and X_1X_2. As an alternative method for specifying a model with interaction variables, using CHAID justifies the omission of the component variables in situations like the one illustrated.

14.8 An Exploratory Look

A closer look at the plot in Figure 14.3 seems to indicate that the relationship between Response and X_2 bends slightly in the upper-right portion, which would imply there is a quadratic component in the relationship. Because trees are always exploratory in nature, I choose to test the X_2 squared term (X_2*X_2), denoted by X_2_SQ, in the model. (Note: The bulging rule discussed in Chapter 8, which seeks to straighten unconditional data, does not apply here because the relationship between RESPONSE and X_2 is a conditional one as it is based on individuals with the "condition $X_1 = 0$.")

The logistic regression analysis on RESPONSE with X_2, X_1X_2, and X_2_SQ is in Table 14.7. The classification accuracy of this model is shown in classification Table 14.8. TCCR(X_2, X_1X_2, X_2_SQ) equals 64.59%, which represents a 16.1% improvement over the best-so-far model with TCCR(X_2, X_1X_2) equals 55.64%.

TABLE 14.7

Logistic Regression of Response on X_2, X_1X_2, and X_2_SQ

Variable	df	Parameter Estimate	Standard Error	Wald Chi-Square	Pr > Chi-Square
INTERCEPT	1	-4.6247	0.0338	18,734.1156	0.E + 00
X1X2	1	-0.0075	0.0027	8.1118	0.0040
X2	1	0.7550	0.1151	43.0144	5.E-11
X2_SQ	1	-0.1516	0.0241	39.5087	3.E-10

TABLE 14.8

Classification Table of Model with X_2, X_1X_2, and X_2_SQ

		Classified		
		0	1	Total
Actual	0	191,998	104,122	296,120
	1	1,838	1,256	3,094
Total		193,836	105,378	299,214
			TCCR	64.59%

I conclude that the relationship is quadratic, and the corresponding model is a good fit of the data. Thus, the best RESPONSE model is defined by X_2, X_1X_2, and X_2_SQ.

14.9 Database Implication

Database marketers are among those who use response models to identify individuals most likely to respond to their solicitations and thus place more value in the information in cell (1, 1)—the number of responders correctly classified—than in the TCCR. Table 14.9 indicates the number of responders correctly classified for the models tested. The model that *appears* to be the best is actually not the best for a database marketer because it identifies the least number of responders (1,256).

I summarize the modeling process as follows: The base RESPONSE model with the two original variables, X_1 and X_2, produces TCCR(X_1, X_2) = 48.37%. The interaction variable X_1X_2, which is added to the base model, produces the full model with TCCR(X_1, X_2, X_1X_2) = 55.64%, for a 15.0% classification improvement over the base model.

Using the new CHAID-based data mining approach to determine whether a component variable can be omitted, I observe that X_2 (but not X_1) can be

TABLE 14.9

Summary of Model Performance

Model			Number of Responder	
Type	Defined by	TCCR	Correct Classification	RCCR
Base	X_1, X_2	48.37%	1,704	1.10%
Full	X_1, X_2, X_1X_2	55.64%	1,478	1.11%
Best so far	X_2, X_1X_2	55.64%	1,477	1.11%
Best	X_2, X_1X_2, X_2_SQ	64.59%	1,256	1.19%

dropped from the full model. Thus, the best-so-far model—with X_2 and X_1X_2—has no loss of performance over the full model: $TCCR(X_2, X_1X_2) = TCCR(X_1, X_2, X_1X_2) = 55.64\%$.

A closer look at the smooth plot in RESPONSE and X_2 suggests that X_2_SQ be added to the best-so-far model, producing the best model with a $TCCR(X_2, X_1X_2, X_2\_SQ)$ of 64.59%, which indicates a 16.1% classification improvement over the best-so-far model.

Database marketers assess the performance of a response model by how well the model correctly classifies responders among the total number of individuals classified as responders. That is, the percentage of responders correctly classified, or the responder correct classification rate (RCCR), is the pertinent measure. For the base model, the "Total" row in Table 14.2 indicates the model classifies 154,812 individuals as responders, among whom there are 1,704 correctly classified: RCCR is 1.10% in Table 14.5. RCCR values for the best, best-so-far, and full models are 1.19%, 1.11%, and 1.11%, respectively, in Table 14.5. Accordingly, the best RCCR-based model is still the best model, which was originally based on TCCR.

It is interesting to note that the performance improvement based on RCCR is not as large as the improvement based on TCCR. The best model compared to the best-so-far model has a 7.2% (= 1.19%/1.11%) RCCR improvement versus 16.1% (= 64.59%/55.64%) TCCR improvement.

14.10 Summary

After briefly reviewing the concepts of interaction variables and multicollinearity and the relationship between the two, I restated the popular strategy for modeling with interaction variables: The principle of marginality states that a model including an interaction variable should also include the component variables that define the interaction. I reinforced the cautionary note that accompanies this principle: Neither testing the statistical significance (or noticeable importance) nor interpreting the coefficients of the component variables should be performed. Moreover, I pointed out that an unfortunate by-product of this principle is that models with unnecessary component variables go undetected, resulting in either unreliable predictions or deterioration of performance.

Then, I presented an alternative strategy, which is based on Nelder's notion of a special point. I defined the strategy for first-order interaction, X_1*X_2, as higher-order interactions are seldom found in database marketing applications: Predictor variable $X_1 = 0$ is called a special point on the scale if when $X_1 = 0$ there is no relationship between the dependent variable and a second predictor variable X_2. If $X_1 = 0$ is a special point, then omit X_2 from the model; otherwise, X_2 should not be omitted. I proposed using CHAID as an alternative data mining method for determining whether there is a relationship between the dependent variable and X_2.

I presented a case study involving the building of a database response model to illustrate the special point CHAID alternative method. The resultant response model, which omitted one component variable, clearly demonstrated the utility of the new method. Then, I took advantage of the full-bodied case study, as well as the mantra of data mining (never stop digging into the data), to improve the model. I determined that the additional term, the square of the included component variable, added 16.2% improvement over the original response model in terms of the traditional measure of model performance, the TCCR.

Digging a little deeper, I emphasized the difference between the traditional and database measures of model performance. Database marketers are more concerned about models with a larger RCCR than a larger TCCR. As such, the improved response model, which initially appeared not to have the largest RCCR, was the best model in terms of RCCR as well as TCCR.

References

1. Marquardt, D.W., You should standardize the predictor variables in your regression model, *Journal of the American Statistical Association*, 75, 87–91, 1980.
2. Aiken, L.S., and West, S.G., *Multiple Regression: Testing and Interpreting Interactions*, Sage, Thousand Oaks, CA, 1991.
3. Chipman, H., Bayesian variable selection with related predictors, *Canadian Journal of Statistics*, 24, 17–36, 1996.
4. Peixoto, J.L., Hierarchical variable selection in polynomial regression models, *The American Statistician*, 41, 311–313, 1987.
5. Nelder, J.A., Functional marginality is important (letter to editor), *Applied Statistics*, 46, 281–282, 1997.

6. McCullagh, P.M., and Nelder, J.A., *Generalized Linear Models*, Chapman & Hall, London, 1989.
7. Fox, J., *Applied Regression Analysis, Linear Models, and Related Methods*, Sage, Thousand Oaks, CA, 1997.
8. Nelder, J.A., The selection of terms in response-surface models—how strong is the weak-heredity principle? *The American Statistician*, 52, 4, 1998.

15

Market Segmentation Classification Modeling with Logistic Regression*

15.1 Introduction

Logistic regression analysis is a recognized technique for classifying individuals into two groups. Perhaps less known but equally important, polychotomous logistic regression (PLR) analysis is another method for performing classification. The purpose of this chapter is to present PLR analysis as a multigroup classification technique. I illustrate the technique using a cellular phone market segmentation study to build a market segmentation classification model as part of a customer relationship management, better known as CRM, strategy.

I start the discussion by defining the typical two-group (binary) logistic regression model. After introducing necessary notation for expanding the binary logistic regression (BLR) model, I define the PLR model. For readers uncomfortable with such notation, the PLR model provides several equations for classifying individuals into one of many groups. The number of equations is one less than the number of groups. Each equation looks like the BLR model.

After a brief review of the estimation and modeling processes used in PLR, I illustrate PLR analysis as a multigroup classification technique with a case study based on a survey of cellular phone users. The survey data were used initially to segment the cellular phone market into four groups. I use PLR analysis to build a model for classifying cellular users into one of the four groups.

15.2 Binary Logistic Regression

Let Y be a binary dependent variable that assumes two outcomes or classes, typically labeled 0 and 1. The BLR model classifies an individual into one

---

* This chapter is based on an article with the same title in *Journal of Targeting Measurement and Analysis for Marketing*, 8, 1, 1999. Used with permission.

of the classes based on the values of the predictor (independent) variables $X_1, X_2, ..., X_n$ for that individual. BLR estimates the *logit of Y*—the log of the odds of an individual belonging to class 1; the logit is defined in Equation (15.1). The logit can easily be converted into the probability of an individual belonging to class 1, $Prob(Y = 1)$, which is defined in Equation (15.2).

$$logit\ Y = b_0 + b_1{}^*X_1 + b_2{}^*X_2 + ... + b_n{}^*X_n \tag{15.1}$$

$$Prob(Y = 1) = \frac{exp(logit\ Y)}{1 + exp(logit\ Y)} \tag{15.2}$$

An individual's predicted probability of belonging to class 1 is calculated by "plugging in" the values of the predictor variables for that individual in Equations (15.1) and (15.2). The b's are the logistic regression coefficients, which are determined by the calculus-based method of maximum likelihood. Note that, unlike the other coefficients, b_0 (referred to as the Intercept) has no predictor variable with which it is multiplied. Needless to say, the probability of an individual belonging to class 0 is $1 - Prob(Y = 1)$.*

15.2.1 Necessary Notation

I introduce notation that is needed for the next section. There are several explicit restatements in Equations (15.3), (15.4), (15.5), and (15.6) of the logit of Y of Equation (15.1). They are superfluous when Y takes on only two values, 0 and 1:

$$logit\ Y = b_0 + b_1{}^*X_1 + b_2{}^*X_2 + ¼ + b_n{}^*X_n \tag{15.3}$$

$$logit(Y = 1) = b_0 + b_1{}^*X_1 + b_2{}^*X_2 + ¼ + b_n{}^*X_n \tag{15.4}$$

$$logit(Y = 1\ vs.\ Y = 0) = b_0 + b_1{}^*X_1 + b_2{}^*X_2 + ¼ + b_n{}^*X_n \tag{15.5}$$

$$logit(Y = 0\ vs.\ Y = 1) = -\,[b_0 + b_1{}^*X_1 + b_2{}^*X_2 + ¼ + b_n{}^*X_n] \tag{15.6}$$

Equation (15.3) is the standard notation for the BLR model; it is assumed that Y takes on two values 1 and 0, and class 1 is the outcome being modeled. Equation (15.4) indicates that class 1 is being modeled and assumes class 0 is the other category. Equation (15.5) formally states that Y has two classes, and class 1 is being modeled. Equation (15.6) indicates that class 0 is the outcome being modeled; this expression is the negative of the other expressions, as indicated by the negative sign on the right-hand side of the equation.

---

* Because $Prob(Y = 0) + Prob(Y = 1) = 1$.

15.3 Polychotomous Logistic Regression Model

When the class-dependent variable takes on more than two outcomes or classes, the PLR model, an extension of the BLR model, can be used to predict class membership. For ease of presentation, I discuss Y with three categories, coded 0, 1, and 2.

Three binary logits can be constructed as in Equations (15.7), (15.8), and (15.9).*

$$\text{logit\_10} = \text{logit}(Y = 1 \text{ vs. } Y = 0) \tag{15.7}$$

$$\text{logit\_20} = \text{logit}(Y = 2 \text{ vs. } Y = 0) \tag{15.8}$$

$$\text{logit\_21} = \text{logit}(Y = 2 \text{ vs. } Y = 1) \tag{15.9}$$

I use the first two logits (because of the similarity with the standard expression of the BLR) to define the PLR model in Equations (15.10), (15.11), and (15.12):

$$\text{Prob}(Y = 0) = \frac{1}{1 + \exp(\text{logit\_10}) + \exp(\text{logit\_20})} \tag{15.10}$$

$$\text{Prob}(Y = 1) = \frac{\exp(\text{logit\_10})}{1 + \exp(\text{logit\_10}) + \exp(\text{logit\_20})} \tag{15.11}$$

$$\text{Prob}(Y = 2) = \frac{\exp(\text{logit\_20})}{1 + \exp(\text{logit\_10}) + \exp(\text{logit\_20})} \tag{15.12}$$

The PLR model is easily extended when there are more than three classes. When Y = 0, 1, ¼ , k (i.e., k + 1 outcomes), the model is defined in Equations (15.13), (15.14), and (15.15):

$$\text{Prob}(Y = 0) = \frac{1}{1 + \exp(\text{logit\_10}) + \exp(\text{logit\_20}) + ... + \exp(\text{logit\_k0})} \tag{15.13}$$

$$\text{Prob}(Y = 1) = \frac{\exp(\text{logit\_10})}{1 + \exp(\text{logit\_10}) + \exp(\text{logit\_20}) + ... + \exp(\text{logit\_k0})} \tag{15.14}$$

$$...,$$

---

* It can be shown that from any pair of logits, the remaining logit can be obtained.

$$\text{Prob}(Y = k) = \frac{\exp(\text{logit\_k0})}{1 + \exp(\text{logit\_10}) + \exp(\text{logit\_20}) + \ldots + \exp(\text{logit\_k0})} \quad (15.15)$$

where

$$\text{logit\_10} = \text{logit}(Y = 1 \text{ vs. } Y = 0),$$

$$\text{logit\_20} = \text{logit}(Y = 2 \text{ vs. } Y = 0),$$

$$\text{logit\_30} = \text{logit}(Y = 3 \text{ vs. } Y = 0),$$

$$\ldots,$$

$$\text{logit\_k0} = \text{logit}(Y = k \text{ vs. } Y = 0).$$

Note: There are k logits for a PLR with k + 1 classes.

15.4 Model Building with PLR

PLR is estimated by the same method used to estimate BLR, namely, maximum likelihood estimation. The theory for stepwise variable selection, model assessment, and validation has been worked out for PLR. Some theoretical problems remain. For example, a variable can be declared significant for all but, say, one logit. Because there is no theory for estimating a PLR model with the constraint of setting a coefficient equal to zero for a given logit, the PLR model may produce unreliable classifications.

Choosing the best set of predictor variables is the toughest part of modeling and is perhaps more difficult with PLR because there are k logit equations to consider. The traditional stepwise procedure is the popular variable selection process for PLR. (See Chapter 30 for a discussion of the traditional stepwise procedure and recall Chapter 10.) Without arguing the pros and cons of the stepwise procedure, its use as the determinant of the final model is questionable.* The stepwise approach is best as a rough-cut method for boiling down many variables—about 50 or more—to a manageable set of about 10. I prefer a methodology based on CHAID (chi-squared automatic

---

* Briefly, a stepwise approach is misleading because all possible subsets are not considered; the final selection is too data dependent and sensitive to influential observations. Also, it does not automatically check for model assumptions and does not automatically test for interaction terms. Moreover, the stepwise approach does not guarantee finding the globally best subset of the variables.

interaction detection) as the variable selection process for the PLR as it fits well into the data mining paradigm of digging into the data to find unexpected structure. In the next section, I illustrate CHAID as the variable selection procedure for building a market segmentation classification model for a cellular phone.

15.5 Market Segmentation Classification Model

In this section, I describe the cellular phone user study. Using the four user groups derived from a cluster analysis of the survey data, I build a four-group classification model with PLR. I use CHAID to identify the final set of candidate predictor variables, interaction terms, and variable structures (i.e., reexpressions of original variables defined with functions such as log or square root) for inclusion in the model. After a detailed discussion of the CHAID analysis, I define the market segmentation classification model and assess the total classification accuracy of the resultant model.

15.5.1 Survey of Cellular Phone Users

A survey of 2,005 past and current users of cellular phones from a wireless carrier was conducted to gain an understanding of customer needs, as well as the variables that affect churn (cancellation of cellular service) and long-term value. The survey data were used to segment this market of consumers into homogeneous groups so that group-specific marketing programs, namely, CRM strategies, can then be developed to maximize the individual customer relationship.

A cluster analysis was performed that produced four segments. Segment names and sizes are in Table 15.1. The hassle-free segment is concerned with the ability of the customer to design the contract and rate plan. The service segment is focused on quality of the call, such as no dropped calls and call clarity. The price segment values discounts, such as offering 10% off the base monthly charges and 30 free minutes of use. Last, the features segment represents the latest technology, such as long-lasting batteries and free phone upgrades. A model is needed to divide the entire database of the wireless carrier into these four actionable segments, after which marketing programs can be used in addressing the specific needs of these predefined groups.

The survey was appended with information from the billing records of the carrier. For all respondents, who are now classified into one of four segments, there are 10 usage variables, such as number of mobile phones, minutes of use, peak and off-peak calls, airtime revenue, base charges, roaming charges, and free minutes of use (yes/no).

TABLE 15.1

Cluster Analysis Results

Name	Size
Hassle-free	13.2% (265)
Service	24.7% (495)
Price	38.3% (768)
Features	23.8% (477)
Total	100% (2,005)

15.5.2 CHAID Analysis

Briefly, CHAID is a technique that recursively partitions a population into separate and distinct subpopulations or nodes such that the variation of the dependent variable is minimized within the nodes and maximized among the nodes. The dependent variable can be binary (dichotomous), polychotomous, or continuous. The nodes are defined by independent variables, which pass through an algorithm for partitioning. The independent variables can be categorical or continuous.

To perform a CHAID analysis, I must define the dependent variable and the set of independent variables. For this application of CHAID, the set of independent variables is the set of usage variables appended to the survey data. The dependent variable is the class variable Y identifying the four segments from the cluster analysis. Specifically, I define the dependent variable as follows:

Y = 0 if segment is Hassle-free

= 1 if segment is Service

= 2 if segment is Price

= 3 if segment is Features

The CHAID analysis identifies* four important predictor variables[†]:

1. NUMBER OF MOBILE PHONES: the number of mobile phones a customer has
2. MONTHLY OFF-PEAK Calls: the number of off-peak calls averaged over a 3-month period
3. FREE MINUTES, yes/no: free first 30 minutes of use per month
4. MONTHLY AIRTIME REVENUE: total revenue excluding monthly charges averaged over a 3-month period

---

* Based on the average chi-square value per degrees of freedom (number of nodes).
[†] I do not consider interaction variables identified by CHAID because the sample is too small.

The CHAID tree for NUMBER OF MOBILE PHONES, shown in Figure 15.1, is read as follows:

1. The top box indicates that, for the sample of 2,005 customers, the sizes (and incidences) of the segments are 264 (13.2%), 495 (24.75%), 767 (38.3%), and 479 (23.9%) for hassle-free, service, price, and features segments, respectively.

2. The left node represents 834 customers who have one mobile phone. Within this subsegment, the incidence rates of the four segments are 8.5%, 21.8%, 44.7%, and 24.9% for hassle-free, service, price, and features segments, respectively.

3. The middle node represents 630 customers who have two mobile phones, with incidence rates of 14.8%, 35.7%, 31.3%, and 18.3% for hassle-free, service, price, and features segments, respectively.

4. The right node represents 541 customers who have three mobile phones, with incidence rates of 18.5%, 16.3%, 36.4%, and 28.8% for hassle-free, service, price, and features segments, respectively.

The CHAID tree for MONTHLY OFF-PEAK Calls, shown in Figure 15.2, is read as follows:

1. The top box is the sample breakdown of the four segments; it is identical to the top box for NUMBER OF MOBILE Phones.

2. There are three nodes: (a) The left node is defined by the number of calls in the left-closed and right-open interval [0,1); this means zero

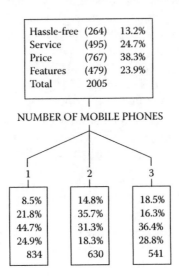

FIGURE 15.1
CHAID tree for NUMBER OF MOBILE PHONES.

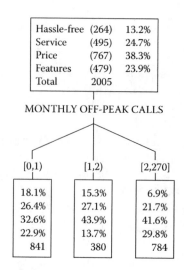

FIGURE 15.2
CHAID tree for MONTHLY OFF-PEAK CALLS.

calls. (b) The middle node is defined by the number of calls in the left-closed and right-open interval [1,2); this means one call. (c) The right node is defined by the number of calls in the closed interval [2, 270]; this means calls between greater than or equal to 2 and less than or equal to 270.

3. The left node represents 841 customers who have zero off-peak calls. Within this subsegment, the incidence rates of the four segments are 18.1%, 26.4%, 32.6%, and 22.9% for hassle-free, service, price, and features, respectively.

4. The middle node represents 380 customers who have one off-peak call, with incidence rates of 15.3%, 27.1%, 43.9%, and 13.7% for hassle-free, service, price, and features, respectively.

5. The right node represents 784 customers who have off-peak calls inclusively between 2 and 270, with incidence rates of 6.9%, 21.7%, 41.6%, and 29.8% for hassle-free, service, price, and features, respectively.

Similar readings can be made for the other predictor variables, FREE MINUTES and MONTHLY AIRTIME REVENUE, identified by CHAID. Their CHAID trees are in Figures 15.3 and 15.4, respectively.

Analytically, CHAID declaring a variable significant means the segment incidence rates (as a column array of rates) differ significantly across the nodes. For example, MONTHLY OFF-PEAK CALLS has three column arrays of segment incidence rates, corresponding to the three nodes: {18.1%, 26.4%, 32.6%, 22.9%}, {15.3%, 27.1%, 43.9%, 13.7%}, and {6.9%, 21.7%, 41.6%, 29.8%}.

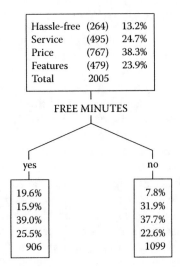

FIGURE 15.3
CHAID tree for FREE MINUTES.

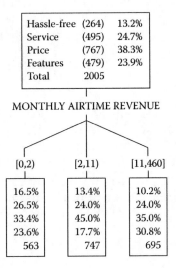

FIGURE 15.4
CHAID tree for MONTHLY AIRTIME REVENUE.

These column arrays are significantly different from each other. This is a complex concept that really has no interpretive value, at least in the context of identifying variables with classification power. However, the CHAID tree can help evaluate the potential predictive power of a variable when it is transformed into a tree graph.

15.5.3 CHAID Tree Graphs

Displaying a CHAID tree in a graph facilitates the evaluation of the potential predictive power of a variable. I plot the incidence rates by the minimum values* of the intervals for the nodes and connect the *smooth* points to form a *trace line*, one for each segment. The shape of the trace line indicates the effect of the predictor variable on identifying individuals in a segment. The baseline plot, which indicates a predictor variable with no classification power, consists of all the segment trace lines being horizontal or "flat." The extent to which the segment trace lines are not flat indicates the potential predictive power of the variable for identifying an individual belonging to the segments. A comparison of all trace lines (one for each segment) provides a total view of how the variable affects classification *across* the segments.

The following discussion relies on an understanding of the basics of reexpressing data as discussed in Chapter 8. Suffice it to say that sometimes the predictive power offered by a variable can be increased by reexpressing or transforming the original form of the variable. The final forms of the four predictor variables identified by CHAID variables are given in the next section.

The PLR is a linear model,[†] which requires a linear or straight-line relationship between predictor variable and each implicit binary segment dependent variable.[‡] The tree graph suggests the appropriate reexpression when the empirical relationships between predictor and binary segment variables are not linear.[§]

The CHAID tree graph for NUMBER OF MOBILE PHONES in Figure 15.5 indicates the following:

1. There is a positive and nearly linear relationship[¶] between NUMBER OF MOBILE PHONES and the identification of customers in the hassle-free segment. This implies that only the variable NUMBER OF MOBILE PHONES in its raw form, no reexpression, may be needed.

2. The relationship for the Features segment also has a positive relationship but with a bend from below.[**] This implies that the variable NUMBER OF MOBILE PHONES in its raw form and its square may be required.

---

* The minimum value is one of several values that can be used; alternatives are the mean or median of each predefined interval.
† That is, each individual logit is a sum of weighted predictor variables.
‡ For example, for segment Hassle-free (Y = 0), binary Hassle-free segment variable = 1 if Y = 0; otherwise, Hassle-free segment variable = 0.
§ The suggestions are determined from the ladder of powers and the bulging rule discussed in Chapter 8.
¶ A relationship is assessed by determining the slope between the left and right node smooth points.
** The position of a bend is determined by the middle-node smooth point.

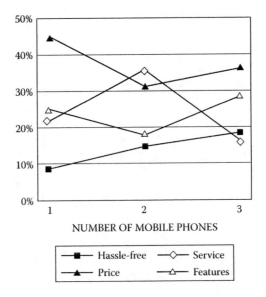

FIGURE 15.5
CHAID tree graph for NUMBER OF MOBILE PHONES.

3. Price segment has a negative effect with a bend from below. This implies that the variable NUMBER OF MOBILE PHONES itself in raw form and its square root may be required.

4. The service segment has a negative relationship but with a bend from above. This implies that the variable NUMBER OF MOBILE PHONES itself in raw form and its square may be required.

The CHAID tree graphs for the other predictor variables, MONTHLY OFF-PEAK CALLs, MONTHLY AIRTIME REVENUE, and FREE MINUTES are in Figures 15.6, 15.7, and 15.8, respectively. Interpreting the graphs, I diagnostically determine the following:

1. MONTHLY OFF-PEAK Calls: This variable in its raw, square, and square root forms may be required.

2. MONTHLY AIRTIME REVENUE: This variable in its raw, square, and square root forms may be required.

3. FREE MINUTES: This variable as is, in its raw form, may be needed.

The preceding CHAID tree graph analysis serves as the initial variable selection process for PLR. The final selection criterion for including a variable in the PLR model is that the variable must be significantly/noticeably important on no less than three of the four logit equations based on the techniques discussed in Chapter 8.

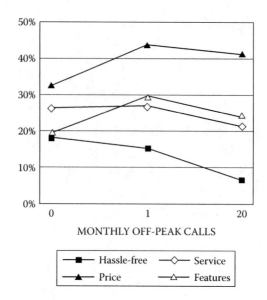

FIGURE 15.6
CHAID tree graph for MONTHLY OFF-PEAK CALLS.

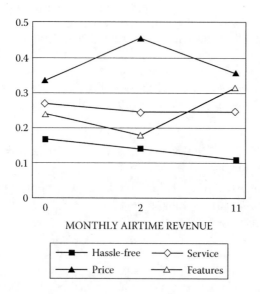

FIGURE 15.7
CHAID tree graph for MONTHLY AIRTIME REVENUE.

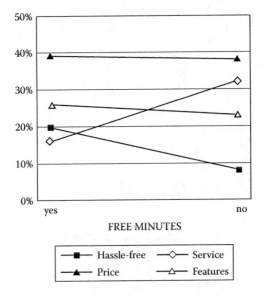

FIGURE 15.8
CHAID tree graph for FREE MINUTES.

15.5.4 Market Segmentation Classification Model

The final PLR model for classifying customers into one of the four cellular phone behavioral market segments has the following variables:

1. Number of Mobile Phones (NMP)
2. Square of NMP
3. Square root of NMP
4. Monthly Off-Peak Calls
5. Monthly Airtime Revenue (MAR)
6. Square of MAR
7. Square root of MAR
8. Free Minutes

Without arguing the pros and cons of validation procedures, I draw a fresh holdout sample of size 5,000 to assess the total classification accuracy of the model.* The classification results of the model in Table 15.2 are read as follows:

---

* Here is a great opportunity to debate and test various ways of calibrating and validating a "difficult" model under the best of conditions.

TABLE 15.2

Market Segment Model: Classification Table of Cell Counts

		PREDICTED				
		Hassle-free	Service	Price	Features	Total
	Hassle-free	326 (50.0%)	68	158	98	650 (13.0%)
ACTUAL	Service	79	460 (36.3%)	410	275	1,224 (24.5%)
	Price	147	431	922 (49.0%)	416	1,916 (38.3%)
	Features	103	309	380	418 (34.6%)	1,210 (24.2%)
	Total	655 (13.1%)	1,268 (25.4%)	1,870 (37.4%)	1,207 (24.1%)	5,000 (100%)

1. The row totals are the actual counts in the sample. The sample consists of 650 Hassle-free customers, 1,224 Service customers, 1,916 Price customers, and 1,210 Features customers. The percentage figures are the percentage compositions of the segments with respect to the total sample. For example, 13.0% of the sample consists of actual Hassle-free customers.

2. The column totals are *predicted* counts. The model predicts 655 Hassle-free customers, 1,268 Service customers, 1,870 Price customers, and 1,207 Features customers. The percentage figures are the percentage compositions with respect to the predicted counts. For example, the model predicts 13.1% of the sample as Hassle-free customers.

3. Given that the sample consists of 13.0% Hassle-free customers and the model predicts 13.1% Hassle-free customers, the model has no *bias* with respect to classifying Hassle-free customers. Similarly, the model shows no bias in classifying the other groups: For Service, the actual incidence is 24.5% versus the predicted 25.4%; for Price, the actual incidence is 38.3% versus the predicted 37.4%; and for Features, the actual incidence is 24.2% versus the predicted 24.1%.

4. Although the model shows no bias, the big question is how accurate the predictions are. Among those customers predicted to be in the Hassle-free segment, how many actually are? Among those customers predicted to be in the Service segment, how many are actually service in the segment? Similarly, for Price and Features segments, among those customers predicted to be in the segments, how many actually are? The percentages in the table cells provide the answer. For hassle-free customers, the model correctly classifies 50.0% (= 326/655) of the time. Without a model, I would expect 13.0% correct

classifications of Hassle-free customers. Thus, the model has a lift of 385 (50.0%/13.0%); that is, the model provides 3.85 times the number of correct classifications of Hassle-free customers obtained by chance.

5. The model has a lift of 148 (36.3%/24.5%) for Service, a lift of 128 (49.0%/38.3%) for Price, and a lift of 143 (34.6%/24.2%) for Features.

6. As a summary measure of how well the model makes correct classifications, I look at the total correct classification rate (TCCR). Simply put, the TCCR is the total number of correct classifications across all groups divided by total sample size. Accordingly, I have 326 + 460 + 922 + 418 = 2,126 divided by 5,000, which yields TCCR = 42.52%.

7. To assess the improvement of total correct classification provided by the model, I must compare it to the total correct classification provided by the model chance. TCCR(chance model) is defined as the sum of the squared actual group incidence. For the data at hand, TCCR(chance model) is 28.22% (= (13.0%*13.0%) + (24.5%*24.5%) + (38.3%*38.3%) + (24.2%*24.2%)).

8. Thus, the model lift is 151 (= 42.52%/28.22%). That is, the model provides 51% more total correct classifications across all groups than obtained by chance.

15.6 Summary

I cast the multigroup classification technique of PLR as an extension of the more familiar two-group (binary) logistic regression model. I derived the PLR model for a dependent variable assuming k + 1 groups by "piecing together" k individual binary logistic models. I proposed a CHAID-based data mining methodology as the variable selection process for the PLR as it fits well into the data mining paradigm to dig into the data to find important predictor variables and unexpected structure.

To bring into practice the PLR, I illustrated the building of a classification model based on a four-group market segmentation of cellular phone users. For the variable selection procedure, I demonstrated how CHAID could be used. For this application of CHAID, the dependent variable is the variable identifying the four market segments. CHAID trees were used to identify the starter subset of predictor variables. Then, the CHAID trees were transformed into CHAID tree graphs, which offer potential reexpressions—or identification of structures—of the starter variables. The final market segmentation classification model had a few reexpressed variables (involving square roots and squares).

Last, I assessed the performance of the final four-group/market segment classification model in terms of TCCR. TCCR for the final model was 42.52%, which represented a 51% improvement over the TCCR (28.22%) when no model was used to classify customers into one of the four market segments.

16

CHAID as a Method for Filling in Missing Values

16.1 Introduction

The problem of analyzing data with missing values is well known to data analysts. Data analysts know that almost all standard statistical analyses require complete data for reliable results. These analyses performed with incomplete data assuredly produce biased results. Thus, data analysts make every effort to fill in the missing data values in their datasets. The popular solutions to the problem of handling missing data belong to the collection of imputation, or fill-in techniques. This chapter presents CHAID (chi-squared automatic interaction detection) as an alternative data mining method for filling in missing data.

16.2 Introduction to the Problem of Missing Data

Missing data are a pervasive problem in data analysis. It is the rare exception when the data at hand have no missing data values. The objective of filling in missing data is to recover or minimize the loss of information due to the incomplete data. I introduce briefly the problem of handling missing data.

Consider a random sample of 10 individuals in Table 16.1; these individuals are described by three variables: AGE, GENDER, and INCOME. There are missing values, which are denoted by a dot (.). Eight of 10 individuals provide their age; seven of 10 individuals provide their gender and income.

Two common solutions for handling missing data are available-case analysis and complete-case analysis.* The available-case analysis uses only the cases for which the variable of interest is available. Consider the calculation

---

* Available-case analysis is also known as pairwise deletion. Complete-case analysis is also known as listwise deletion or casewise deletion.

TABLE 16.1

Random Sample of 10 Individuals

Individual	Age (years)	Gender (0 = male, 1 = female)	Income
1	35	0	$50,000
2	.	.	$55,000
3	32	0	$75,000
4	25	1	$100,000
5	41	.	.
6	37	1	$135,000
7	45	.	.
8	.	1	$125,000
9	50	1	.
10	52	0	$65,000
Total	317	4	$605,000
Number of nonmissing values	8	7	7
Mean	39.6	57%	$86,429

. = missing value.

of the mean AGE: The available sample size (number of nonmissing values) is 8, not the original sample size of 10. The calculation for the means of INCOME and GENDER* uses two different available samples of size 7. The calculation on different samples points to a weakness in available-case analysis. Unequal sample sizes create practical problems. Comparative analysis across variables is difficult because different subsamples of the original sample are used. Also, estimates of multivariable statistics are prone to illogical values.†

The popular complete-case analysis uses examples for which all variables are present. A complete-case analysis of the original sample in Table 16.1 includes only five cases, as reported in Table 16.2. The advantage of this type of analysis is simplicity, because standard statistical analysis can be applied without modification for incomplete data. Comparative analysis across variables is not complicated because only one common subsample of the original sample is used. The disadvantage of discarding incomplete cases is the resulting loss of information.

---

* The mean of GENDER is the incidence of females in the sample.
† Consider the correlation coefficient of X1 and X2. If the available sample sizes for X1 and X2 are unequal, it is possible to obtain a correlation coefficient value that lies outside the theoretical [−1, 1] range.

TABLE 16.2

Complete-Case Version

Individual	Age (years)	Gender (0 = male, 1 = female)	Income
1	35	0	$50,000
3	32	0	$75,000
4	25	1	$100,000
6	37	1	$135,000
10	52	0	$65,000
Total	181	2	$425,000
Number of nonmissing values	5	5	5
Mean	36.2	40%	$85,000

Another solution is dummy variable adjustment [1]. For a variable X with missing data, two new variables are used in its place. X_filled and X_dum are defined as follows:

1. X_filled = X if X is not missing; X_filled = 0 if X is missing.
2. X_dum = 0 if X is not missing; X_dum = 1 if X is missing.

The advantage of this solution is its simplicity of use without having to discard cases. The disadvantage is that the analysis can become unwieldy when there are many variables with missing data. In addition, filling in the missing value with a zero is arbitrary, which is unsettling for some data analysts.

Among the missing data solutions is the imputation method, which is defined as any process that fills in missing data to produce a complete dataset. The simplest and most popular imputation method is *mean-value imputation*. The mean of the nonmissing values of the variable of interest is used to fill in the missing data. Consider individuals 2 and 8 in Table 16.1. Their missing ages are replaced with the mean AGE of the file, namely, 40 years (rounded from 39.6). The advantage of this method is undoubtedly its ease of use. The means may be calculated within classes, which are predefined by other variables related to the study at hand.

Another popular method is *regression-based imputation*. Missing values are replaced by the predicted values from a regression analysis. The dependent variable Y is the variable whose missing values need to be imputed. The predictor variables, the X's, are the *matching variables*. Y is regressed on the X's using a complete-case analysis dataset. If Y is continuous, then ordinary least squares (OLS) regression is appropriate. If Y is categorical, then the logistic regression model (LRM) is used. For example, I wish to impute AGE for individual 8 in Table 16.1. I regress AGE on GENDER and INCOME (the matching variables) based on the complete-case dataset consisting of five individuals (IDs 1, 3, 4, 6, and 10). The OLS regression imputation model is defined in Equation (16.1):

$$\text{AGE\_IMPUTED} = 25.8 - 20.5*\text{GENDER} + 0.0002*\text{INCOME} \qquad (16.1)$$

Plugging in the values of GENDER (= 1) and INCOME (= $125,000) for individual 8, the imputed AGE is 53 years.

16.3 Missing Data Assumption

Missing data methods presuppose that the missing data are "missing at random" (MAR). Rubin formalized this condition into two separate assumptions [2]:

1. Missing at random (MAR) means that what is missing does not depend on the missing values but may depend on the observed values.

2. Missing completely at random (MCAR) means that what is missing does not depend on either the observed values or the missing values. When this assumption is satisfied for all variables, the reduced sample of individuals with only complete data can be regarded as a simple random subsample from the original data. Note that the second assumption MCAR represents a stronger condition than the MAR assumption.

The missing data assumptions are problematic. The MCAR assumption can be tested to some extent by comparing the information from complete cases to the information from incomplete cases. A procedure often used is to compare the distribution of the variable of interest, say, Y, based on nonmissing data with the distribution of Y based on missing data. If there are significant differences, then the assumption is considered not met. If no significant differences are indicated, then the test offers no direct evidence of assumption violation. In this case, the assumption is considered cautiously to be satisfied.* The MAR assumption is impossible to test for validity. (Why?)

It is accepted wisdom that missing data solutions at best perform satisfactorily, even when the amount of missing data is moderate and the missing data assumption has been met. The potential of the two new imputation methods, maximum likelihood and multiple imputation, which offer substantial improvement over the complete-case analysis, is questionable as

---

* This test proves the necessary condition for MCAR. It remains to be shown that there is no relationship between missingness on a given variable and the values of that variable.

their assumptions are easily violated. Moreover, their utility in big data applications has not been established.

Nothing can take the place of the missing data. Allison noted, "The best solution to the missing data problem is not to have any missing data" [3]. Dempster and Rubin warned that "imputation is both seductive and dangerous," seductive because it gives a false sense of confidence that the data are complete and dangerous because it can produce misguided analyses and untrue models [4].

The above admonitions are without reference to the impact of big data on filling in missing data. In big data applications, the problem of missing data is severe as it is common for at least one variable to have 30%–90% of its values missing. Thus, I strongly argue that the imputation of big data applications must be used with restraint, and their findings must be used judiciously.

In the spirit of the EDA (exploratory data analysis) tenet—that failure is when you fail to try—I advance the proposed data mining/EDA CHAID imputation method as a hybrid mean-value/regression-based imputation method that explicitly accommodates missing data without imposing additional assumptions. The salient features of the new method are the EDA characteristics:

1. Flexibility: Assumption-free CHAID work especially well with big data containing large amounts of missing data.
2. Practicality: A descriptive CHAID tree provides analysis of data.
3. Innovation: The CHAID algorithm defines imputation classes.
4. Universality: Blending of two diverse traditions occurs: traditional imputation methods and machine-learning algorithm for data structure identification.
5. Simplicity: CHAID tree imputation estimates are easy to use.

16.4 CHAID Imputation

I introduce a couple of terms required for the discussion of CHAID imputation. Imputation methods require the sample to be divided into groups or classes, called *imputation classes*, which are defined by variables called *matching variables*. The formation of imputation classes is an important step to ensure the reliability of the imputation estimates. As the homogeneity of the classes increases, so does the accuracy and stability of the estimates. It is assumed that the variance (with respect to the variable whose missing values are to be imputed) within each class is small.

CHAID is a technique that recursively partitions a population into separate and distinct groups, which are defined by the predictor variables, such that the variance of the dependent variable is minimized within the groups and maximized across the groups. CHAID was originally developed as a method of detecting "combination" or interaction variables. In database marketing today, CHAID primarily serves as a market segmentation technique. Here, I propose CHAID as an alternative data mining method for mean-value/regression-based imputation.

The justification for CHAID as a method of imputation is as follows: By definition, CHAID creates optimal homogeneous groups, which can be used effectively as trusty imputation classes.* Accordingly, CHAID provides a reliable method of mean-value/regression-based imputation.

The CHAID methodology provides the following:

CHAID is a tree-structured, assumption-free modeling alternative to OLS regression. It provides reliable estimates without the assumption of specifying the true structural form of the model (i.e., knowing the correct independent variables and their correct reexpressed forms) and without regard for the weighty classical assumptions of the underlying OLS model. Thus, CHAID with its trusty imputation classes provides reliable *regression tree* imputation for a continuous variable.

CHAID can be used as a nonparametric tree-based alternative to the binary and polychotomous LRM without the assumption of specifying the true structural form of the model. Thus, CHAID with its trusty imputation classes provides reliable *classification tree* imputation for a categorical variable.

CHAID potentially offers more reliable imputation estimates due to its ability to use most of the analysis sample. The analysis sample for CHAID is not as severely reduced by the pattern of missing values in the matching variables, as is the case for regression-based models, because CHAID can accommodate missing values for the matching variables in its analysis.† The regression-based imputation methods cannot make such an accommodation.

16.5 Illustration

Consider a sample of 29,132 customers from a cataloguer's database. The following is known about the customers: their ages (AGE_CUST), GENDER, total lifetime dollars (LIFE_DOL), and whether a purchase was made within the past 3 months (PRIOR_3). Missing values for each variable are denoted by "???."

---

* Some analysts may argue about the optimality of the homogeneous groups but not the trustworthiness of the imputation classes.
† Missing values are allowed to "float" within the range of the matching variable and rest at the position that optimizes the homogeneity of the groups.

The counts and percentages of missing and nonmissing values for the variables are in Table 16.3. For example, there are 691 missing values for LIFE_DOL, resulting in a 2.4% missing rate. It is interesting to note that the complete-case sample size for an analysis with all four variables is 27,025. This represents a 7.2% (= 2,107/29,132) loss of information from discarding incomplete cases from the original sample.

16.5.1 CHAID Mean-Value Imputation for a Continuous Variable

I wish to impute the missing values for LIFE_DOL. I perform a mean-value imputation with CHAID using LIFE_DOL as the dependent variable and AGE_CUST as the predictor (matching) variable. The AGE_CUST CHAID tree is in Figure 16.1.

I set some conventions to simplify the discussions of the CHAID analyses:

1. The left-closed and right-opened interval [x, y) indicates values between x and y, including x and excluding y.
2. The closed interval [x, y] indicates values between x and y, including both x and y.
3. The distinction is made between nodes and imputation classes. Nodes are the visual displays of the CHAID groups. Imputation classes are defined by the nodes.
4. Nodes are referenced by numbers (1, 2, ...) from left to right as they appear in the CHAID tree.

The AGE_CUST CHAID tree, in Figure 16.1, is read as follows:

1. The top box indicates a mean LIFE_DOL of $27,288.47 for the available sample of 28,441 (nonmissing) observations for LIFE_DOL.
2. The CHAID creates four nodes with respect to AGE_CUST. Node 1 consists of 6,499 individuals whose ages are in the interval [18, 30) with a mean LIFE_DOL of $14,876.75. Node 2 consists of 7,160

TABLE 16.3

Counts and Percentages of Missing and Nonmissing Values

Variable	Missing		Nonmissing	
	Count	%	Count	%
AGE_CUST	1,142	3.9	27,990	96.1
GENDER	2,744	9.4	26,388	90.6
LIFE_DOL	691	2.4	28,441	97.6
PRIOR_3	965	3.3	28,167	96.7
All Variables	2,096	7.2	27,025	92.8

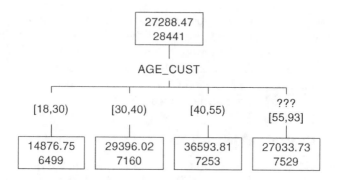

FIGURE 16.1
AGE_CUST CHAID tree for LIFE_DOL.

TABLE 16.4

CHAID Imputation Estimates for Missing Values of LIFE_DOL

AGE_CUST Class	Class Size	Imputation Estimate
[18, 30)	0	$14,876.75
[30,40)	0	$29,396.02
[40, 55)	0	$36,593.81
[55, 93] or ???	691	$27,033.73

individuals whose ages are in the interval [30, 40) with a mean LIFE_ DOL of $29,396.02. Node 3 consists of 7,253 individuals whose ages are in the interval [40, 55) with a mean LIFE_DOL of $36,593.81. Node 4 consists of 7,529 individuals whose ages are either in the interval [55, 93] or missing. The mean LIFE_DOL is $27,033.73.

3. Note: CHAID positions the missing values of AGE_CUST in the oldest-age missing node.

The set of AGE_CUST CHAID mean-value imputation estimates for LIFE_ DOL is the mean values of nodes 1 to 4: $14,876.75, $29,396.02, $36,593.81, and $27,033.73, respectively. The AGE_CUST distribution of the 691 individuals with missing LIFE_DOL is in Table 16.4. All the 691 individuals belong to the last class, and their imputed LIFE_DOL value is $27,033.73. Of course, if any of the 691 individuals were in the other AGE_CUST classes, the corresponding mean values would be used.

16.5.2 Many Mean-Value CHAID Imputations for a Continuous Variable

CHAID provides many mean-value imputations—as many as there are matching variables—as well as a measure to determine which imputation estimates to use. The "goodness" of a CHAID tree is assessed by the measure percentage variance explained (PVE). The imputation estimates based on the matching variable with the largest PVE value are often selected as the preferred estimates. Note, however, a large PVE value does not necessarily guarantee reliable imputation estimates, and the largest PVE value does not necessarily guarantee the best imputation estimates. The data analyst may have to perform the analysis at hand with imputations based on several of the large PVE value-matching variables.

Continuing with imputation for LIFE_DOL, I perform two additional CHAID mean-value imputations. The first CHAID imputation, in Figure 16.2, uses LIFE_DOL as the dependent variable and GENDER as the matching variable. The second imputation, in Figure 16.3, uses LIFE_DOL as the dependent variable and PRIOR_3 as the matching variable. The PVE values for the matching variables AGE_CUST, GENDER, and PRIOR_3 are 10.20%, 1.45%, and 1.52%, respectively. Thus, the preferred imputation estimates for LIFE_DOL are based on AGE_CUST because the AGE_CUST-PVE value is noticeably the largest.

Comparative note: Unlike CHAID mean-value imputation, traditional mean-value imputation provides no guideline for selecting an all-case

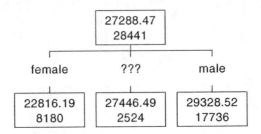

FIGURE 16.2
GENDER CHAID tree for LIFE_DOL.

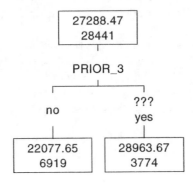

FIGURE 16.3
PRIOR_3 CHAID tree for LIFE_DOL.

continuous matching variable (e.g., AGE_CUST), whose imputation estimates are preferred.

16.5.3 Regression Tree Imputation for LIFE_DOL

I can selectively add matching variables* to the preferred single-variable CHAID tree—generating a regression tree—to increase the reliability of the imputation estimates (i.e., increase the PVE value). Adding GENDER and PRIOR_3 to the AGE_CUST tree, I obtain a PVE value of 12.32%, which represents an increase of 20.8% (= 2.12%/10.20%) over the AGE_CUST-PVE value. The AGE_CUST-GENDER-PRIOR_3 regression tree is displayed in Figure 16.4.

The AGE_CUST-GENDER-PRIOR_3 regression tree is read as follows:

1. Extending the AGE_CUST CHAID tree, I obtain a regression tree with 13 end nodes.
2. Node 1 (two levels deep) consists of 2,725 individuals whose ages are in the interval [18, 30) *and* have not made a purchase in the past 3 months (PRIOR_3 = no). The mean LIFE_DOL is $13,352.11.
3. Node 2 (three levels deep) consists of 1,641 individuals whose ages are in the interval [18, 30) *and* PRIOR_3 = ??? or yes *and* whose GENDER = ??? or female. The mean LIFE_DOL is $15,186.52.
4. The remaining nodes are interpreted similarly.

The AGE_CUST-GENDER-PRIOR_3 regression tree imputation estimates for LIFE_DOL are the mean values of 13 end nodes. An individual's missing LIFE_DOL value is replaced with the mean value of the imputation class to which the individual matches. The distribution of the 691 individuals with missing LIFE_DOL values in terms of the three matching variables is in Table 16.5. All the missing LIFE_DOL values come from the five rightmost end nodes (nodes 9 to 14). As before, if any of the 691 individuals were in the other 8 nodes, the corresponding mean values would be used.

Comparative note: Traditional OLS regression-based imputation for LIFE_DOL based on four matching variables—AGE_CUST, two dummy variables for GENDER ("missing" GENDER is considered a category), and one dummy variable for PRIOR_3—results in a complete-case sample size of 27,245, which represents a 4.2% (= 1,196/28,441) loss of information from the CHAID analysis sample.

---

* Here is a great place to discuss the relationship between increasing the number of variables in a (tree) model and its effects on bias and stability of the estimates provided by the model.

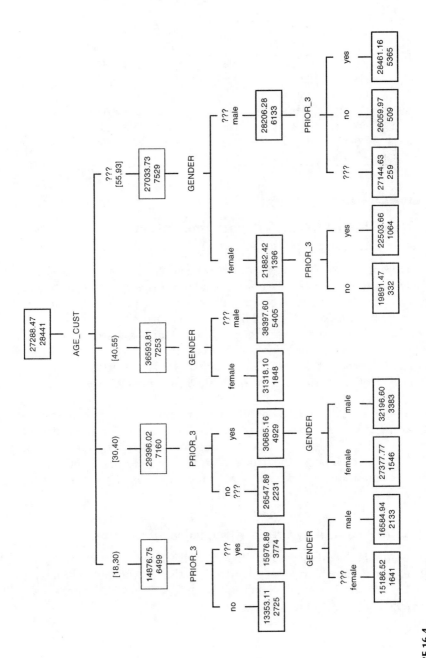

FIGURE 16.4
AGE_CUST, GENDER, and PRIOR_3 regression tree for LIFE_DOL.

TABLE 16.5

Regression Tree Imputation Estimates for Missing Values of
LIFE_DOL

AGE_CUST Class	GENDER	PRIOR_3	Class Size	Imputation Estimates
??? or [55, 93]	Female	No	55	$19,891.47
??? or [55, 93]	Female	Yes	105	$22,503.66
??? or [55, 93]	Male	No	57	$26,059.97
??? or [55, 93]	Male	Yes	254	$28,461.16
??? or [55, 93]	???	No	58	$26,059.97
??? or [55, 93]	???	Yes	162	$28,461.16
Total			691	

16.6 CHAID Most Likely Category Imputation for a Categorical Variable

CHAID for imputation of a categorical variable is very similar to the CHAID with a continuous variable, except for slight changes in assessment and interpretation. CHAID with a continuous variable assigns a mean value to an imputation class. In contrast, CHAID with a categorical variable assigns the predominant or most likely category to an imputation class. CHAID with a continuous variable provides PVE. In contrast, with a categorical variable, CHAID provides the measure proportion of total correct classifications (PTCC)* to identify the matching variable(s) whose imputation estimates are preferred.

As noted in CHAID with a continuous variable, there is a similar note that a large PTCC value does not necessarily guarantee reliable imputation estimates, and the largest PTCC value does not necessarily guarantee the best imputation estimates. The data analyst may have to perform the analysis with imputations based on several of the large PTCC value-matching variables.

16.6.1 CHAID Most Likely Category Imputation for GENDER

I wish to impute the missing values for GENDER. I perform a CHAID most likely category imputation using GENDER as the dependent variable and AGE_CUST as the matching variable. The AGE_CUST CHAID tree is in Figure 16.5. The PTCC value is 68.7%.

---

* PTCC is calculated with the percentage of observations in each of the end nodes of the tree that fall in the modal category. The weighted sum of these percentages over all end nodes of the tree is PTCC. A given node is weighted by the number of observations in the node relative to the total size of the tree.

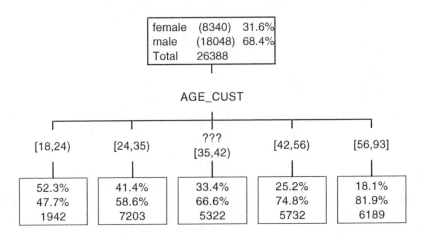

FIGURE 16.5
AGE_CUST CHAID tree for GENDER.

The AGE_CUST CHAID tree is read as follows:

1. The top box indicates the incidences of females and males are 31.6% and 68.4%, respectively, based on the available sample of 26,388 (nonmissing) observations for GENDER.

2. The CHAID creates five nodes with respect to AGE_CUST. Node 1 consists of 1,942 individuals whose ages are in interval [18, 24). The incidences of female and male are 52.3% and 47.7%, respectively. Node 2 consists of 7,203 individuals whose ages are in interval [24, 35). The incidences of female and male are 41.4% and 58.6%, respectively.

3. The remaining nodes are interpreted similarly.

4. Note: CHAID places the missing values in the middle-age node. Compare the LIFE_DOL CHAID: The missing ages are placed in the oldest-age/missing node.

I perform two additional CHAID most likely category imputations (Figures 16.6 and 16.7) for GENDER using the individual matching variables PRIOR_3 and LIFE_DOL, respectively. The PTCC values are identical, 68.4%. Thus, I select the imputation estimates for GENDER based on AGE_CUST as its PTCC value is the largest (68.7%), albeit not noticeably different from the other PTCC values.

The AGE_CUST CHAID most likely category imputation estimates for GENDER are the most likely, that is, the largest percentage categories of the nodes in Figure 16.5: female (52.3%), male (58.6%), male (66.6%), male (74.8%), and male (82.9%). I replace an individual's missing GENDER value with the predominant category of the imputation class to which the individual matches. There are 2,744 individuals with GENDER missing; their AGE_CUST distribution is in Table 16.6. The individuals whose ages are in the interval [18, 24) are

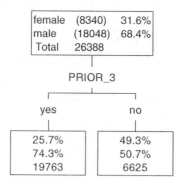

FIGURE 16.6
PRIOR_3 CHAID tree for GENDER.

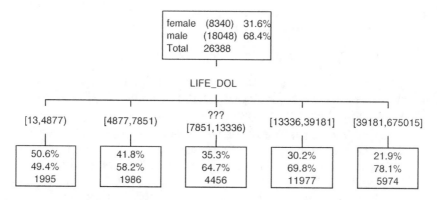

FIGURE 16.7
LIFE_DOL CHAID tree for GENDER.

classified as female because the females have the largest percentage; all other individuals are classified as male. It is not surprising that most of the classifications are males, given the large (68.4%) incidence of males in the sample.

Comparative note: Traditional mean imputation can be performed with only matching variable PRIOR_3. (Why?) CHAID most likely category imputation conveniently offers three choices of imputation estimates and a guideline to select the best.

16.6.2 Classification Tree Imputation for GENDER

I can selectively add matching variables* to the preferred single-variable CHAID tree—generating a classification tree—to increase the reliability of

---

* Here again is a great place to discuss the relationship between increasing the number of variables in a (tree) model and its effects on bias and stability of the estimates determined by the model.

TABLE 16.6

CHAID Imputation Estimates for Missing Values of GENDER

AGE_CUST Class	Class Size	Imputation Estimate
[18, 24)	182	Female
[24, 35)	709	Male
??? or [35, 42)	628	Male
[42, 56)	627	Male
[56, 93]	598	Male
Total	2,744	

the imputation estimates (i.e., increase the PTCC value). Extending the AGE_CUST tree, I obtain the classification tree (Figure 16.8), for GENDER based on AGE_CUST, PRIOR_3, and LIFE_DOL. The PTCC value for this tree is 79.3%, which represents an increase of 15.4% (= 10.6%/68.7%) over the AGE_CUST-PTCC value.

The AGE_CUST-PRIOR_3-LIFE_DOL classification tree is read as follows:

1. Extending the GENDER tree with the addition of PRIOR_3 and LIFE_DOL, I obtain a classification tree with 12 end nodes.

2. Node 1 consists of 1,013 individuals whose ages are in the interval [18, 24) *and* who have *not* made a prior purchase in the past 3 months (PRIOR_3 = no). The female and male incidences are 59.0% and 41.0%, respectively.

3. Node 2 consists of 929 individuals whose ages are in the interval [18, 24) *and* who have made a prior purchase in the past 3 months (PRIOR_3 = yes). The female and male incidences are 44.9% and 55.1%, respectively.

4. The remaining nodes are interpreted similarly.

5. Node 4 has no predominant category as the GENDER incidences are equal to 50%.

6. Nodes 1, 7, and 9 have female as the predominant category; all remaining nodes have male as the predominant category.

The AGE_CUST-PRIOR_3-LIFE_DOL classification tree imputation estimates for GENDER are the most likely categories of the nodes. I replace an individual's missing GENDER value with the predominant category of the imputation class to which the individual belongs. The distribution of missing GENDER values (Table 16.7) falls within all 12 nodes. Individuals in nodes 1, 7, and 9 are classified as females. For individuals in node 4, I flip a coin. All other individuals are classified as males.

Comparative note: Traditional logistic regression-based imputation for GENDER based on three matching variables (AGE_CUST, LIFE_DOL, and one dummy variable for PRIOR_3) results in a complete-case sample size of

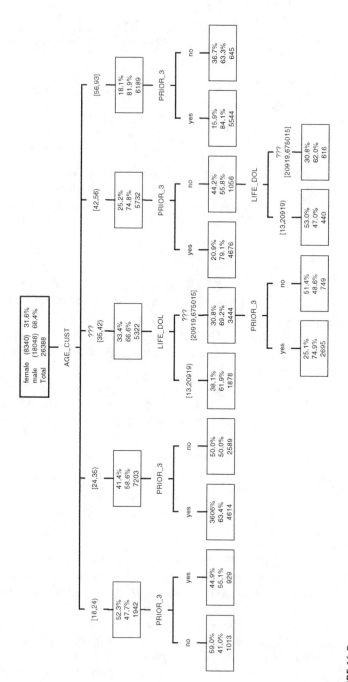

FIGURE 16.8
AGE_CUST-PRIOR_3-LIF_DOL Classification Tree for GENDER

TABLE 16.7

Classification Tree Imputation Estimates for Missing Values of GENDER

Node	AGE_CUST Class	PRIOR_3	LIFE_DOL Class	Class Size	Imputation Estimates
1	[18, 24)	No	—	103	Females
2	[18, 24)	Yes	—	79	Males
3	[24, 35)	Yes	—	403	Males
4	[24, 35)	No	–	306	Females/Males
5	??? or [35, 42)	–	[13, 20919)	169	Males
6	??? or [35, 42)	Yes	??? or [20919, 675015]	163	Males
7	??? or [35, 42)	No	??? or [20919, 675015]	296	Females
8	[42, 56)	Yes	—	415	Males
9	[42, 56)	No	[13, 20919)	70	Females
10	[42, 56)	No	[20919, 675015]	142	Males
11	[56, 93]	Yes	—	449	Males
12	[56, 93]	No	—	149	Males
Total				2,744	

26,219, which represents a barely noticeable 0.6% (= 169/26,388) loss of information from the CHAID analysis sample.

16.7 Summary

It is rare to find a dataset that has no missing data values. A given is that the data analyst first tries to recover or minimize the loss of information from the incomplete data. I illustrated briefly the popular missing data methods, which include complete-case and available-case analyses, and mean-value and regression-based imputation methods. All these methods have at least one version of the missing data assumptions: MAR and MCAR, which are difficult and impossible to test for validity, respectively.

I remarked that the conventional wisdom of missing data solutions is that their performance is, at best, satisfactory, especially for big data applications. Experts in missing data admonish us to remember that imputation is seductive and dangerous; therefore, the best solution of the missing data problem is not to have missing data. I strongly urged that imputation of big data applications must be used with restraint, and their findings must be used judiciously.

Then, I recovered from the naysayers to advance the proposed CHAID imputation method. I presented CHAID as an alternative method for mean-value/regression-based imputation. The justification of CHAID for missing data is that CHAID creates optimal homogeneous groups that can be used as trusty imputation classes, which ensure the reliability of the imputation estimates. This renders CHAID as a reliable method of mean-value/regression-based imputation. Moreover, the CHAID imputation method has salient features commensurate with the best of what EDA offers.

I illustrated the CHAID imputation method with a database catalogue case study. I showed how CHAID—for both a continuous and a categorical variable to be imputed—offers imputations based on several individual matching variables and rules for selecting the preferred CHAID mean-value imputation estimates and the preferred CHAID regression tree imputation estimates.

References

1. Cohen, J., and Cohen, P., *Applied Multiple Regression and Correlation Analysis for the Behavioral Sciences*, Erlbaum, Hillsdale, NJ, 1987.
2. Rubin, D.B., Inference and missing data, *Biometrika*, 63, 581–592, 1976.
3. Allison, P.D., *Missing Data*, Sage, Thousand Oaks, CA, 2002, 2.
4. Dempster, A.P., and Rubin, D.B., Overview, In Madow, W.G., Okin, I., and Rubin, D.B., Eds., *Incomplete Data in Sample Surveys, Vol. 2: Theory and Annotated Bibliography*, Academic Press, New York, 1983, 3.

17

Identifying Your Best Customers: Descriptive, Predictive, and Look-Alike Profiling*

17.1 Introduction

Marketers typically attempt to improve the effectiveness of their campaigns by targeting their best customers. Unfortunately, many marketers are unaware that typical target methods develop a descriptive profile of their target customer—an approach that often results in less-than-successful campaigns. The purpose of this chapter is to illustrate the inadequacy of the *descriptive* approach and to demonstrate the benefits of the correct *predictive profiling* approach. I explain the predictive profiling approach and then expand the approach to look-alike profiling.

17.2 Some Definitions

It is helpful to have a general definition of each of the three concepts discussed in this chapter. Descriptive profiles report the characteristics of a group of individuals. These profiles *do not allow* for drawing inferences about the group. The value of a descriptive profile lies in its definition of the target group's salient characteristics, which are used to develop an effective marketing strategy.

Predictive profiles report the characteristics of a group of individuals. These profiles do allow for drawing inferences about a specific behavior, such as response. The value of a predictive profile lies in its predictions of the behavior of individuals in a target group; the predictions are used in producing a list of likely responders to a marketing campaign.

A look-alike profile is a predictive profile based on a group of individuals who look like the individuals in a target group. When resources do not allow

---

* This chapter is based on an article with the same title in *Journal of Targeting, Measurement and Analysis for Marketing*, 10, 1, 2001. Used with permission.

for the gathering of information on a target group, a predictive profile built on a surrogate or "look-alike" group provides a viable approach for predicting the behavior of the individuals in the target group.

17.3 Illustration of a Flawed Targeting Effort

Consider a hypothetical test mailing to a sample of 1,000 individuals conducted by Cell-Talk, a cellular phone carrier promoting a new bundle of phone features. Three hundred individuals responded, yielding a 30% response rate. (The offer also included the purchase of a cellular phone for individuals who do not have one, but now want one because of the attractive offer.) Cell-Talk analyzes the responders and profiles them in Tables 17.1 and 17.2 by using variables GENDER and OWN_CELL (current cellular phone ownership), respectively. Ninety percent of the 300 responders are males, and 55% already own a cellular phone. Cell-Talk concludes the typical responder is a male and owns a cellular phone.

Cell-Talk plans to target the next "features" campaign to males and to owners of cellular phones. The effort is sure to fail. The reason for the poor prediction is the profile of their best customers (responders) is descriptive, not predictive. That is, the descriptive responder profile describes responders without regard to responsiveness; therefore, the profile does not imply the best customers are responsive.*

Using a descriptive profile for predictive targeting draws a false implication of the descriptive profile. In our example, the descriptive profile of "90% of the responders are males" does not imply 90% of males are responders, or even that males are more likely to respond.† In addition, "55% of the responders who own cellular phones" does not imply 55% of cellular phone owners are responders, or even that cellular phone owners are more likely to respond.

The value of a descriptive profile lies in its definition of the best customers' salient characteristics, which are used to develop an effective marketing strategy. In the illustration, knowing the target customer is a male and owns a cellular phone, I instruct Cell-Talk to position the campaign offer with a man wearing a cellular phone on his belt instead of a woman reaching for a cellular phone in her purse. Accordingly, a descriptive profile tells how to talk to the target audience. As discussed in the next section, a predictive profile helps find the target audience.

---

* A descriptive responder profile may also describe a typical nonresponder. In fact, this is the situation in Tables 17.1 and 17.2.
† More likely to respond than a random selection of individuals.

TABLE 17.1

Responder and Nonresponder Profile Response Rates by GENDER

	Responders		Nonresponders		
GENDER	Count	%	Count	%	Response Rate %
Female	30	10	70	10	30
Male	270	90	630	90	30
Total	300	100	700	100	

17.4 Well-Defined Targeting Effort

A predictive profile describes responders *with regard* to responsiveness, that is, in terms of variables that discriminate between responders and nonresponders. Effectively, the discriminating or predictive variables produce *varied* response rates and imply an expectation of responsiveness. To clarify this, consider the response rates for GENDER in Table 17.1. The response rates for both males and females are 30%. Accordingly, GENDER does not discriminate between responders and nonresponders (in terms of responsiveness). Similar results for OWN_CELL are in Table 17.2.

Hence, GENDER and OWN_CELL have no value as predictive profiles. Targeting of males and current cellular phone owners by Cell-Talk is expected to generate the average or sample response rate of 30%. In other words, this profile in a targeting effort will not produce more responders than will a random sample.

I now introduce a new variable, CHILDREN, which it is hoped has predictive value. CHILDREN is defined as "yes" if an individual belongs to a household with children and "no" if an individual does not belong to a household with children. Instead of discussing CHILDREN using a tabular display (such as in Tables 17.1 and 17.2), I prefer the user-friendly visual display of CHAID trees.

Response rates are best illustrated by use of CHAID tree displays. I review the GENDER and OWN_CELL variables in the tree displays in Figures 17.1 and 17.2, respectively. From this point, I refer only to the tree in this discussion, underscoring the utility of a tree as a profiler and reducing the details of tree building to nontechnical summaries.

The GENDER tree in Figure 17.1 is read as follows:

1. The top box indicates that for the sample of 1,000 individuals, there are 300 responders and 700 nonresponders. The response rate is 30% and nonresponse rate is 70%.

TABLE 17.2

Responder and Nonresponder Profile Response Rates by OWN_CELL

OWN_CELL	Responders		Nonresponders		Response Rate %
	Count	%	Count	%	
Yes	165	55	385	55	30
No	135	45	315	45	30
Total	300	100	700	100	

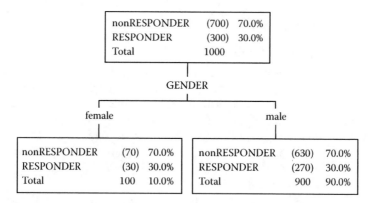

FIGURE 17.1
GENDER tree.

2. The left box represents 100 *females,* consisting of 30 responders and 70 nonresponders. The response rate among the 100 females is 30%.

3. The right box represents 900 *males,* consisting of 270 responders and 630 nonresponders. The response rate among the 900 males is 30%.

The OWN_CELL tree in Figure 17.2 is read as follows:

1. The top box indicates that for the sample of 1,000 individuals, there are 300 responders and 700 nonresponders. The response rate is 30% and nonresponse rate is 70%.

2. The left box represents 550 individuals who *own* a cell phone. The response rate among these individuals is 30%.

3. The right box represents 450 individuals who *do not own* a cell phone. The response rate among these individuals is 30%.

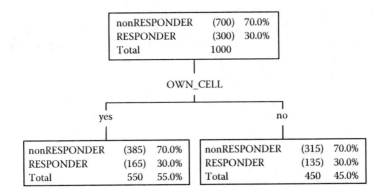

FIGURE 17.2
OWN_CELL tree.

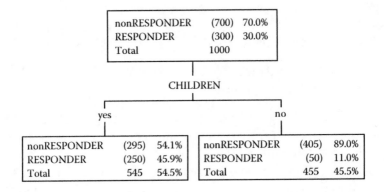

FIGURE 17.3
CHILDREN tree.

The new variable CHILDREN is defined as presence of children in the household (yes/no). The CHILDREN tree in Figure 17.3 is read as follows:

1. The top box indicates that for the sample of 1,000 individuals, there are 300 responders and 700 nonresponders. The response rate is 30% and nonresponse rate is 70%.
2. The left box represents 545 individuals belonging to households *with children*. The response rate among these individuals is 45.9%.
3. The right box represents 455 individuals belonging to households *with no children*. The response rate among these individuals is 11.0%.

CHILDREN has value as a predictive profile as it produces varied response rates of 45.9% and 11.0% for CHILDREN equal to "yes" and "no," respectively.

If Cell-Talk targets individuals belonging to households with children, the expected response rate is 45.9%. This represents a *profile lift* of 153. (*Profile lift* is defined as profile response rate 45.9% divided by sample response rate 30% multiplied by 100.) Thus, a targeting effort to the predictive profile is expected to produce 1.53 times more responses than expected from a random solicitation.

17.5 Predictive Profiles

Using additional variables, I can grow a single-variable tree into a full tree with many interesting and complex predictive profiles. Although actual building of a full tree is beyond the scope of this chapter, suffice it to say that *a tree is grown to create end-node profiles (segments) with the greatest variation in response rates across all segments.* A tree has value as a *set of predictive profiles* to the extent (1) the number of segments with response rates greater than the sample response rate is "large," and (2) the corresponding profile (segment) lifts are "large."*

Consider the full tree defined by GENDER, OWN_CELL, and CHILDREN in Figure 17.4. The tree is read as follows:

1. The top box indicates that for the sample of 1,000 individuals, there are 300 responders and 700 nonresponders. The response rate is 30% and nonresponse rate is 70%.

2. I reference the end-node segments from left to right, 1 through 7.

3. Segment 1 represents 30 *females* who *own* a cellular phone and belong to households *with children*. The response rate among these individuals is 50.0%.

4. Segment 2 represents 15 *females* who *do not own* a cellular phone and belong to households *with children*. The response rate among these individuals is 100.0%.

5. Segment 3 represents 300 *males* who *own* a cellular phone and belong to households *with children*. The response rate among these individuals is 40.0%.

6. Segment 4 represents 200 *males* who *do not own* a cellular phone and belong to households *with children*. The response rate among these individuals is 50.0%.

---

* The term *large* is subjective. Accordingly, the tree-building process is subjective, which is an inherent weakness in CHAID trees.

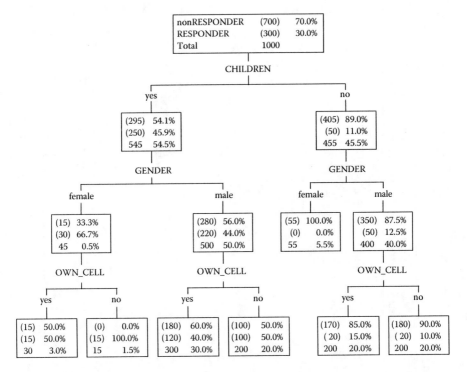

FIGURE 17.4
Full tree defined by GENDER, OWN_CELL, and CHILDREN.

7. Segment 5 represents 55 *females* who belong to households *with no children*. The response rate among these individuals is 0.0%.

8. Segment 6 represents 200 *males* who *own* a cellular phone and belong to households *with no children*. The response rate among these individuals is 15.0%.

9. Segment 7 represents 200 *males* who *do not own* a cellular phone and belong to households *with no children*. The response rate among these individuals is 10.0%.

I provide a summary of the segment response rates in the *gains chart* in Table 17.3. The construction and interpretation of the gains chart are as follows:

1. Segments are descendingly ranked by segment response rate.

2. In addition to the descriptive statistics (size of segment, number of responses, segment response rate), various calculated statistics are posted in the chart. They include the self-explanatory cumulative responses, segment response rate, and cumulative response rate.

TABLE 17.3

Gains Chart for Tree Defined by GENDER, OWN CELL, and CHILDREN

Segment[a]	Size of Segment	Number of Responses	Cumulative Responses	Segment Response Rate	Cumulative Rate	Cumulative Lift
2 OWN_CELL, no GENDER, female	15	15	15	100.0%	100.0%	333
1 CHILDREN, yes OWN_CELL, yes GENDER, female	30	15	30	50.0%	66.7%	222
3 CHILDREN, yes OWN_CELL, no GENDER, male	200	100	130	50.0%	53.1%	177
4 OWN_CELL, yes GENDER, male	300	120	250	40.0%	45.9%	153
6 CHILDREN, yes OWN_CELL, yes GENDER, male	200	30	280	15.0%	37.6%	125
7 CHILDREN, no OWN_CELL, no GENDER, male	200	20	300	10.0%	31.7%	106
5 CHILDREN, no GENDER, female CHILDREN, no	55	0	300	0.0%	30.0%	100
	1,000	300		30.0%		

[a] Segments are ranked by response rates.

3. The statistic posted in the rightmost column is the cumulative lift. *Cumulative lift* is defined as cumulative response rate divided by the sample response rate multiplied by 100. It measures how many more responses are obtained by targeting various levels of aggregated segments over a random solicitation. Cumulative lift is discussed in detail further in this section.

4. CHAID trees unvaryingly identify *sweet spots*—segments with above-average response rates* that account for small percentages of the sample. Under the working assumption that the sample is random and accurately reflects the population under study, sweet spots account for small percentages of the population. There are two sweet spots: Segment 2 has a response rate of 100% and accounts for only 1.5% (= 15/1,000) of the sample/population; segment 1 has a response rate of 50% and accounts for only 3.0% (= 30/1,000) of the sample/population.

5. A targeting strategy to a single sweet spot is limited as it is effective for only solicitations of mass products to large populations. Consider sweet spot 2 with a population of 1.5 million. Targeting this segment produces a campaign of size 22,500 with an expected yield of 22,500 responses. Mass product large campaigns have low breakeven points, which make sweet-spot targeting profitable.

6. Reconsider sweet spot 2 in a moderate-size population of 100,000. Targeting this segment produces a campaign of size 1,500 with an expected yield of 1,500 responses. However, small campaigns for mass products have high breakeven points, which render such campaigns neither practical nor profitable. In contrast, upscale product small campaigns have low breakeven points, which make sweet-spot targeting (e.g., to potential Rolls-Royce owners) both practical and profitable.

7. For moderate-size populations, the targeting strategy is to solicit an aggregate of several top consecutive responding segments to yield a campaign of a cost-efficient size to ensure a profit. Here, I recommend a solicitation consisting of the top three segments, which would account for 24.5% (= (15 + 30 + 200)/1,000) of the population with an expected yield of 53.1%. (See the cumulative response rate for the segment 3 row in Table 17.3.) Consider a population of 100,000. Targeting the aggregate of segments 2, 1, and 3 yields a campaign of size 24,500 with an expected 11,246 (= 53.1%*24,500) responses. Assuming a product offering with a good profit margin, the aggregate targeting approach should be successful.

8. With respect to the cumulative lift, Cell-Talk can expect the following:

---

* Extreme small segment response rates (close to both 0% and 100%) reflect another inherent weakness of CHAID trees.

a. Cumulative lift of 333 by targeting the top segment, which accounts for only 1.5% of the population. This is sweet-spot targeting as previously discussed.

b. Cumulative lift of 222 by targeting the top two segments, which account for only 4.5% (= (15 + 30)/1,000) of the population. This is effectively an enhanced sweet-spot targeting because the percentage of the aggregated segments is small.

c. Cumulative lift of 177 by targeting the top three segments, which account for 24.5% of the population. This is the recommended targeting strategy as previously discussed.

d. Cumulative lift of 153 by targeting the top four segments, which account for 54.5% ((15 + 30 + 200 + 300)/1,000) of the population. Unless the population is not too large, a campaign targeted to the top four segments may be cost prohibitive.

17.6 Continuous Trees

So far, the profiling uses only categorical variables, that is, variables that assume two or more discrete values. Fortunately, trees can accommodate continuous variables, or variables that assume many numerical values, which allows for developing expansive profiles.

Consider INCOME, a new variable. The INCOME tree in Figure 17.5 is read as follows:

1. Tree notation: Trees for a continuous variable denote the continuous values in ranges: a closed interval or a left-closed/right-open interval. The former is denoted by [x, y], indicating all values between and including x and y. The latter is denoted by [x, y), indicating all values greater than or equal to x and less than y.

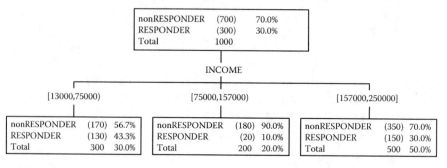

FIGURE 17.5
INCOME tree.

2. The top box indicates that for the sample of 1,000 individuals, there are 300 responders and 700 nonresponders. The response rate is 30% and nonresponse rate is 70%.

3. Segment 1 represents 300 individuals with income in the interval [$13,000, $75,000). The response rate among these individuals is 43.3%.

4. Segment 2 represents 200 individuals with income in the interval [$75,000, $157,000). The response rate among these individuals is 10.0%.

5. Segment 3 represents 500 individuals with income in the interval [$157,000, $250,000]. The response rate among these individuals is 30.0%.

The determination of the number of nodes and the range of an interval are based on a computer-intensive iterative process, which tests all possible numbers and ranges. That is, a tree is grown to create nodes/segments with the greatest variation in response rates across all segments.

The full tree with the variables GENDER, OWN_CELL, CHILDREN, and INCOME is displayed in Figure 17.6. The gains chart of this tree is in Table 17.4. Cell-Talk can expect a cumulative lift of 177, which accounts for 24.5% of the population, by targeting the top three segments.

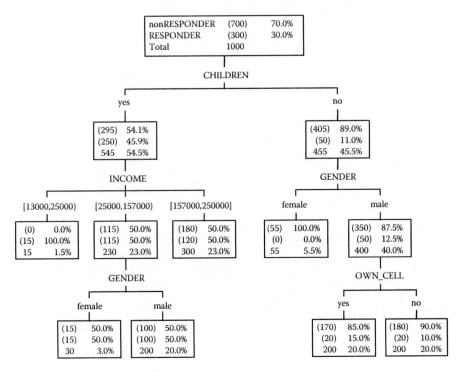

FIGURE 17.6
Full tree defined by GENDER, OWN_CELL, CHILDREN, and INCOME.

TABLE 17.4

Gains Chart for Tree Defined by GENDER, OWN CELL, CHILDREN, and INCOME

Segment[a]	Size of Segment	Number of Responses	Cumulative Responses	Segment Response Rate	Cumulative Response Rate	Cumulative Lift
1 INCOME, [13000,25000) CHILDREN, yes GENDER, male	15	15	15	100.0%	100.0%	333
3 INCOME, [25000,157000) CHILDREN, yes GENDER, female	200	100	115	50.0%	53.5%	178
2 INCOME, [25000,157000) CHILDREN, yes	30	15	130	50.0%	53.1%	177
4 INCOME [157000,250000] CHILDREN, yes OWN_CELL, yes GENDER, male	300	120	250	40.0%	45.9%	153
6 CHILDREN, no OWN_CELL, no GENDER, male	200	30	280	15.0%	37.6%	125
7 CHILDREN, no GENDER, female	200	20	300	10.0%	31.7%	106
5 CHILDREN, no	55	0	300	0.0%	30.0%	100
	1,000	300		30.0%		

[a] Segments are ranked by response rates.

It is interesting to compare this tree, which includes INCOME, to the tree without INCOME in Figure 17.4. Based on Tables 17.3 and 17.4, these two trees have the same performance statistics, at least for the top three segments, a cumulative lift of 177, which accounts for 24.5% of the population.

This raises some interesting questions. Does INCOME add any noticeable predictive power? In other words, how important is INCOME? Which tree is better? Which set of variables is best? There is a simple answer to these questions (and many more tree-related questions): *An analyst can grow many equivalent trees to explain the same response behavior. The tree that suits the analyst is the best (at least for that analyst).* Again, detailed answers to these questions (and more) are beyond the scope of this chapter.

17.7 Look-Alike Profiling

In the Cell-Talk illustration, Cell-Talk requires predictive profiles to increase the response to its campaign for a new bundle of features. They conduct a test mailing to obtain a group of their best customers—responders of the new bundle offer—on which to develop the profiles.

Now, consider that Cell-Talk wants predictive profiles to target a solicitation based on a rental list of names, for which only demographic information is available, and the ownership of a cellular phone is not known. The offer is a discounted rate plan, which should be attractive to cellular phone owners with high monthly usage (around 500 minutes of use per month).

Even though Cell-Talk does not have the time or money to conduct another test mailing to obtain their target group of responders with high monthly usage, they can still develop profiles to help in their targeting efforts as long as they have a notion of what their target group looks like. Cell-Talk can use a look-alike group—individuals who look like individuals in the target group—as a substitute for the target group. This substitution allows Cell-Talk to develop look-alike profiles: profiles that identify individuals (in this case, persons on the rental list) who are most likely to look like individuals in the target group.

The construction of the look-alike group is important. The greater the similarity between the look-alike group and target group, the greater the reliability of the resultant profiles will be.

Accordingly, the definition of the look-alike group should be as precise as possible to ensure that the look-alikes are good substitutes for the target individuals. The definition of the look-alike group can include as many variables as needed to describe pertinent characteristics of the target group. Note, the definition always involves at least one variable that is not available on the solicitation file or, in this case, the rental list. If all the variables are available, then there is no need for look-alike profiles.

Cell-Talk believes the target group looks like their current upscale cellular phone subscribers. Because cellular conversation is not inexpensive, Cell-Talk assumes that heavy users must have a high income to afford the cost of cellular use. Accordingly, Cell-Talk defines the look-alike group as individuals with a cellular phone (OWN_CELL = yes) and INCOME greater than $175,000.

Look-alike profiles are based on the following assumption: Individuals who look like individuals in a target group have levels of responsiveness similar to the group. Thus, the look-alike individuals serve as surrogates or would-be responders.

Keep in mind that individuals identified by look-alike profiles are expected probabilistically to look like the target group but not expected necessarily to respond. In practice, it has been shown that the look-alike assumption is tenable, as solicitations based on look-alike profiles produce noticeable response rates.

Look-alike profiling via tree analysis identifies variables that discriminate between look-alike individuals and non-look-alike individuals (the balance of the population without the look-alike individuals). Effectively, the discriminating variables produce varied look-alike rates. I use the original sample data in Table 17.1 along with INCOME to create the LOOK-ALIKE variable required for the tree analysis. LOOK-ALIKE equals 1 if an individual has OWN_CELL = yes and INCOME greater than $175,000; otherwise, LOOK-ALIKE equals 0. There are 300 look-alikes and 700 non-look-alikes, resulting in a sample look-alike rate of 30.0%.* These figures are reflected in the top box of the look-alike tree in Figure 17.7. (Note: The sample look-alike rate and the original sample response rate are equal; this is purely coincidental.)

The gains chart for the look-alike tree is in Table 17.5. Targeting the top segment yields a cumulative lift of 333 (with a 30% depth of population). This means the predictive look-alike profile is expected to identify 3.33 times more individuals who look like the target group than expected from random selection (of 30% of rented names).

A closer look at the look-alike tree raises a question. INCOME is used in defining both the LOOK-ALIKE variable and the profiles. Does this indicate that the tree is poorly defined? No. For this particular example, INCOME is a wanted variable. Without INCOME, this tree could not guarantee that the identified *males with children* have high incomes, a requirement for being a look-alike.

---

* The sample look-alike rate sometimes needs to be adjusted to equal the incidence of the target group in the population. This incidence is rarely known and must be estimated.

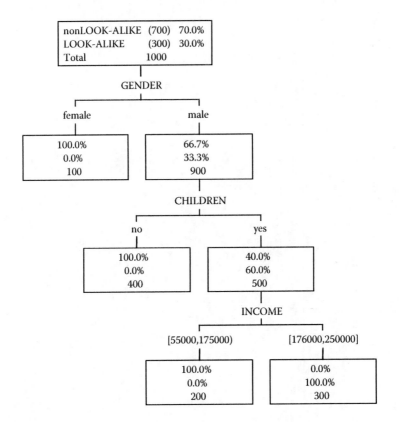

FIGURE 17.7
Look-alike tree.

17.8 Look-Alike Tree Characteristics

It is instructive to discuss a noticeable characteristic of look-alike trees. In general, upper segment rates in a look-alike tree are quite large and often reach 100%. Similarly, lower segment rates are quite small and often fall to 0%. I observe these patterns in Table 17.5. There is one segment with a 100% look-alike rate and three segments with a 0% look-alike rate.

The implication is as follows:

1. It is easier to identify an individual who looks like someone with predefined characteristics (e.g., gender, and children) than someone who behaves in a particular manner (e.g., responds to a solicitation).

2. The resultant look-alike rates are biased estimates of target response rates to the extent the defined look-alike group differs from the target

TABLE 17.5

Gains Chart for Look-Alike Tree Defined by GENDER, CHILDREN, and INCOME

Segment[a]	Size of Segment	Number of Responses	Cumulative Responses	Segment Response Rate	Cumulative Response Rate	Cumulative Lift
4 INCOME, [176000,250000) CHILDREN, yes GENDER, male	300	300	300	100.0%	100.0%	333
2 CHILDREN, no GENDER, male	400	0	300	0.0%	42.9%	143
1 GENDER, female	100	0	300	0.0%	37.5%	125
3 INCOME [55000,175000) CHILDREN, yes GENDER, male	200	0	300	0.0%	30.0%	100
	1,000	300		30.0%		

[a] Segments are ranked by look-alike rates.

group. Care should be exercised in defining the look-alike group because it is easy to include individuals inadvertently unwanted.

3. The success of a solicitation based on look-alike profiles, in terms of the actual responses obtained, depends on the disparity of the defined look-alike group and target group and the tenability of the look-alike assumption.

17.9 Summary

Marketers typically attempt to improve the effectiveness of their campaigns by targeting their best customers. However, targeting only the best customers (responders) based on their characteristics is sure to fail. The descriptive profile, which represents responders without reference to nonresponders, is a nonpredictive profile. Therefore, targeting with a descriptive profile does not provide any assurance that the best customers will respond to a new campaign. The value of the descriptive profile lies in its definition of the salient characteristics of the target group, which are used to develop an effective marketing strategy.

I contrasted descriptive and predictive profiles. A predictive profile describes responders with regard to responsiveness, that is, in terms of variables that discriminate between responders and nonresponders. The predictive profile is used for finding responders to a new campaign, after which the descriptive profile is used to communicate effectively with those customers.

Then, I introduced the tree analysis method of developing a set of complex and interesting predictive profiles. With an illustration, I presented the gains chart as the standard report of the predictive power of the tree-based predictive profiles, for example, the expected response rates on implementing the profiles in a solicitation.

Last, I expanded the tree-based approach of predictive profiling to look-alike profiling, a reliable method when actual response information is not available. A look-alike profile is a predictive profile based on a group of individuals who look like the individuals in a target group, thus serving as surrogate responders. I addressed a warning that the look-alike rates are biased estimates of target response rates because the look-alike profiles are about surrogate responders, not actual responders.

18

Assessment of Marketing Models*

18.1 Introduction

Marketers use decile analysis to assess their models in terms of classification or prediction accuracy. The uninformed marketers do not know that additional information from the decile analysis can be extracted to supplement the model assessment. The purpose of this chapter is to present two additional concepts of model assessment—precision and separability—and to illustrate these concepts by further use of the decile analysis.

I begin the discussion with the traditional concepts of accuracy for response and profit models and illustrate the basic measures of accuracy. Then, I introduce the accuracy measure used in marketing, known as Cum Lift. The discussion of Cum Lift is set in the context of a decile analysis, which is the usual approach marketers use to evaluate the performance of response and profit models. I provide a systematic procedure for conducting a decile analysis with an illustration.

I continue with the illustration to present the new concepts, precision, and separability. Last, I provide guidelines for using all three measures in assessing marketing models.

18.2 Accuracy for Response Model

How well does a response model correctly classify individuals as responder and nonresponder? The traditional measure of accuracy is the proportion of total correct classifications (PTCC), which is calculated from a simple cross tabulation.

Consider the classification results in Table 18.1 of a response model based on the validation sample consisting of 100 individuals with a 15% response

---

* This chapter is based on an article with the same title in *Journal of Targeting, Measurement and Analysis for Marketing*, 7, 3, 1998. Used with permission.

TABLE 18.1

Classification Results of Response Model

		Predicted		
		Nonresponder	Responder	Total
Actual	Nonresponder	74	11	85
	Responder	2	13	15
Total		76	24	100

rate. The "Total" column indicates there are 85 actual nonresponders and 15 actual responders in the sample. The "Total" row indicates the model predicts 76 nonresponders and 24 responders. The model correctly classifies 74 nonresponders and 13 responders. Accordingly, the PTCC is (74 + 13)/100 = 87%.

Although PTCC is frequently used, it may not be appropriate for the given situation. For example, if the assessment criterion imposes a penalty for misclassifications,* then PTCC must be either modified or discarded for a more relevant measure.

Marketers have defined their own measure of accuracy for response models: Cum Lift. They use response models to identify those individuals most likely to respond to a solicitation. They create a solicitation list of the most likely individuals to obtain an advantage over a random selection of individuals. The Cum Lift indicates how many more responses can be expected with a selection based on a model over the expected responses from a random selection (no model). Specifically, the Cum Lift is an index of the expected response rate with a selection based on a model compared with the expected response rate based on a random selection. Before I illustrate the calculation of the Cum Lift, I provide a companion exposition of accuracy for the profit model.

18.3 Accuracy for Profit Model

How well does a profit model correctly predict an individual's profit value? There are several measures of prediction accuracy, all of which are based on the concept of *error*, namely, actual profit minus predicted profit. The mean squared error (MSE) is by far the most popular measure, but it is flawed, thus necessitating three alternative measures. I briefly review the four error measures.

---

* For example, there is a $2 loss if a responder is classified as a nonresponder and a $4 loss if a nonresponder is classified as a responder.

1. MSE is the mean of individual squared errors. It gives greater importance to larger errors and tends to underestimate the predictive
   accuracy of the model. This point is illustrated in the discussion that
   follows.

2. MPE is the mean of individual percentage errors. It measures the
   bias in the estimates. Percentage error for an individual is the error
   divided by the actual profit multiplied by 100.

3. MAPE is the mean of individual absolute percentage errors. It disre-
   gards the sign of the error.

4. MAD is the mean of individual absolute deviations (errors). It disre-
   gards the sign of the error.

Consider the prediction results in Table 18.2 of the profit model (not
shown); the validation sample consists of 30 individuals with a mean profit
of $6.76. The first two columns after the ID number column are the actual
profit and predicted profit produced by the profit model, respectively. The
remaining columns in order from left to right are "Error," "Squared Error,"
"Percentage Error," "Absolute Error," and "Absolute Percentage Error." The
bottom row "MEAN" consists of the means, based on the 30 individuals, for
the last four columns on the right. The mean values are 23.22, 52.32%, 2.99,
and 87.25%, for MSE, MPE, MAD, and MAPE, respectively. These measures
are only indicators of a good model; "smaller" values tend to correspond to
"better" models.

To highlight the sensitivity of MSE to far-out values, I calculate MSE, MPE,
MAD, and MAPE based on the sample without the large error of 318.58 corresponding to individual 26. The adjusted mean values for MSE, MPE, MAD,
and MAPE are 13.03, 49.20%, 2.47, and 85.30%, respectively. The sensitivity
of MSE is clear as it is dramatically reduced by almost 50%, whereas MPE,
MAD, and MAPE remain relatively stable.

Except for the occasional need for an individual-level profit accuracy
assessment (requiring one of the four error measures), marketers use their
own measure of accuracy—Cum Lift—for profit models. When Cum Lift is
used for a profit model—Cum Lift (profit)—it is very similar to the Cum
Lift for a response model, except for slight changes in assessment and interpretation. Marketers use profit models to identify individuals contributing
maximum profit from a solicitation and create a solicitation list of those
individuals to obtain an advantage over a random selection. The Cum Lift
(profit) indicates how much more profit can be expected with a selection
based on a model over the expected profit from a random selection with no
profit model. Specifically, the Cum Lift (profit) is an index of the expected
profit with a selection based on a model compared with the expected profit
based on a random selection.

TABLE 18.2

Profit Model: Four Measures of Errors

ID #	PROFIT	Predicted PROFIT	Error	Squared Error	Percentage Error	Absolute Error	Absolute Percentage Error
1	0.60	0.26	0.34	0.12	132.15%	0.34	132.15%
2	1.60	0.26	1.34	1.80	519.08%	1.34	519.08%
3	0.50	0.26	0.24	0.06	93.46%	0.24	93.46%
4	1.60	0.26	1.34	1.80	519.08%	1.34	519.08%
5	0.50	0.26	0.24	0.06	93.46%	0.24	93.46%
6	1.20	0.26	0.94	0.89	364.31%	0.94	364.31%
7	2.00	1.80	0.20	0.04	11.42%	0.20	11.42%
8	1.30	1.80	-0.50	0.25	27.58%	0.50	27.58%
9	2.50	1.80	0.70	0.50	39.27%	0.70	39.27%
10	2.20	3.33	-1.13	1.28	33.97%	1.13	33.97%
11	2.40	3.33	-0.93	0.87	27.96%	0.93	27.96%
12	1.20	3.33	-2.13	4.54	63.98%	2.13	63.98%
13	3.50	4.87	-1.37	1.87	28.10%	1.37	28.10%
14	4.10	4.87	-0.77	0.59	15.78%	0.77	15.78%
15	5.10	4.87	0.23	0.05	4.76%	0.23	4.76%
16	5.70	6.40	-0.70	0.50	11.00%	0.70	11.00%
17	3.40	7.94	-4.54	20.62	57.19%	4.54	57.19%
18	9.70	7.94	1.76	3.09	22.14%	1.76	22.14%
19	8.60	7.94	0.66	0.43	8.29%	0.66	8.29%
20	4.00	9.48	-5.48	30.01	57.80%	5.48	57.80%
21	5.50	9.48	-3.98	15.82	41.97%	3.98	41.97%
22	10.50	9.48	1.02	1.04	10.78%	1.02	10.78%
23	17.50	11.01	6.49	42.06	58.88%	6.49	58.88%
24	13.40	11.01	2.39	5.69	21.66%	2.39	21.66%
25	4.50	11.01	-6.51	42.44	59.14%	6.51	59.14%
26	30.40	12.55	17.85	318.58	142.21%	17.85	142.21%
27	12.40	15.62	-3.22	10.40	-20.64%	3.22	20.64%
28	13.40	17.16	-3.76	14.14	21.92%	3.76	21.92%
29	26.20	17.16	9.04	81.71	52.67%	9.04	52.67%
30	7.40	17.16	-9.76	95.28	56.88%	9.76	56.88%
			MEAN	23.22	52.32%	2.99	87.25%
			MEAN without ID 26	13.03	49.20%	2.47	85.30%

18.4 Decile Analysis and Cum Lift for Response Model

The decile analysis is a tabular display of model performance. I illustrate the construction and interpretation of the response decile analysis in Table 18.3. (The response model on which the decile analysis is based is not shown.)

1. Score the sample (i.e., calibration or validation file) using the response model under consideration. Every individual receives a model score, Prob_est, the estimated probability of response of the model.
2. Rank the scored file, in descending order by Prob_est.
3. Divide the ranked and scored file into 10 equal groups. The *Decile* variable is created, which takes on 10 ordered "values": top (1), 2, 3, 4, 5, 6, 7, 8, 9, and bottom (10). The top decile consists of the best 10% of individuals most likely to respond; decile 2 consists of the next 10% of individuals most likely to respond, and so on for the remaining deciles. Accordingly, the decile separates and orders the individuals on an ordinal scale ranging from most to least likely to respond.
4. *Number of individuals* is the number of individuals in each decile, 10% of the total size of the file.
5. *Number of responses (actual)* is the actual—not predicted—number of responses in each decile. The model identifies 911 actual responders in the top decile. In decile 2, the model identifies 544 actual responders, and so on for the remaining deciles.

TABLE 18.3

Response Decile Analysis

Decile	Number of Individuals	Number of Responders	Decile Response Rate	Cumulative Response Rate	Cumulative Lift
Top	7,410	911	12.3%	12.3%	294
2	7,410	544	7.3%	9.8%	235
3	7,410	437	5.9%	8.5%	203
4	7,410	322	4.3%	7.5%	178
5	7,410	258	3.5%	6.7%	159
6	7,410	188	2.5%	6.0%	143
7	7,410	130	1.8%	5.4%	129
8	7,410	163	2.2%	5.0%	119
9	7,410	124	1.7%	4.6%	110
Bottom	7,410	24	0.3%	4.2%	100
Total	74,100	3,101	4.2%		

6. *Decile response rate* is the actual response rate for each decile group. It is *Number of responses* divided by *Number of individuals* for each decile group. For the top decile, the response rate is 12.3% (= 911/7,410). For the second decile, the response rate is 7.3% (= 544/7,410), and so on for the remaining deciles.

7. *Cumulative response rate* for a given depth of file (the aggregated or cumulative deciles) is the response rate among the individuals in the cumulative deciles. For example, the cumulative response rate for the top decile (10% depth of file) is 12.3% (= 911/7,410). For the top two deciles (20% depth of file), the cumulative response rate is 9.8% (= [911 + 544]/[7410 + 7410]), and so on for the remaining deciles.

8. *Cum Lift* for a given depth of file is the *Cumulative response rate* divided by the overall response rate of the file multiplied by 100. It measures how much better one can expect to do with the model than without a model. For example, a Cum Lift of 294 for the top decile means that when soliciting to the top 10% of the file based on the model, one can expect 2.94 times the total number of responders found by randomly soliciting 10% of the file. The Cum Lift of 235 for the top two deciles means that when soliciting to 20% of the file based on the model, one can expect 2.35 times the total number of responders found by randomly soliciting 20% of the file without a model, and so on for the remaining deciles.

Rule: The larger the Cum Lift value is, the better the accuracy for a given depth of file will be.

18.5 Decile Analysis and Cum Lift for Profit Model

Calculation of the profit decile analysis is very similar to that of the response decile analysis with "response" and "response rates" replaced by "profit" and "mean profit," respectively. I illustrate the construction and interpretation of the profit decile analysis in Table 18.4.

1. Score the sample (i.e., calibration or validation file) using the profit model under consideration. Every individual receives a model score, Pred_est, the predicted profit of the model.

2. Rank the scored file in descending order by Pred_est.

3. Divide the ranked and scored file into 10 equal groups, producing the *decile* variable. The "top" decile consists of the best 10% of individuals contributing maximum profit; decile 2 consists of the next 10% of individuals contributing maximum profit, and so on for the

TABLE 18.4

Profit Decile Analysis

Decile	Number of Individuals	Total Profit	Decile Mean Profit	Cumulative Mean Profit	Cumulative Lift
Top	3	$47.00	$15.67	$15.67	232
2	3	$60.30	$20.10	$17.88	264
3	3	$21.90	$7.30	$14.36	212
4	3	$19.40	$6.47	$12.38	183
5	3	$24.00	$8.00	$11.51	170
6	3	$12.70	$4.23	$10.29	152
7	3	$5.80	$1.93	$9.10	135
8	3	$5.80	$1.93	$8.20	121
9	3	$2.70	$0.90	$7.39	109
Bottom	3	$3.30	$1.10	$6.76	100
Total	30	$202.90	$6.76		

remaining deciles. Accordingly, the decile separates and orders the individuals on an ordinal scale ranging from maximum to minimum contribution of profit.

4. *Number of individuals* is the number of individuals in each decile, 10% of the total size of the file.

5. *Total profit (actual)* is the actual—not predicted—total profit in each decile. The model identifies individuals contributing $47 profit in the top decile. In decile 2, the model identifies individuals contributing $60.30 profit, and so on for the remaining deciles.

6. *Decile mean profit* is the actual mean profit for each decile group. It is *Total profit* divided by *Number of individuals* for each decile group. For the top decile, the actual mean profit is $15.67 (= $47/3). For the second decile, the value is $20.10 (= $60.30/3), and so on for the remaining deciles.

7. *Cumulative mean profit* for a given depth of file (the aggregated or cumulative deciles) is the mean profit among the individuals in the cumulative deciles. For example, the cumulative mean profit for the top decile (10% depth of file) is $15.67 (= $47/3). For the top two deciles (20% depth of file), the cumulative response rate is $17.88 (= [$47 + $60.30]/[3 + 3]), and so on for the remaining deciles.

8. *Cum Lift* for a given depth of file is the *Cumulative mean profit* divided by the overall profit of the file multiplied by 100. It measures how much better one can expect to do with the model than without a model. For example, a Cum Lift of 232 for the top decile means that when soliciting to the top 10% of the file based on the model, one

can expect 2.32 times the total profit found by randomly soliciting 10% of the file. The Cum Lift of 264 for the top two deciles means that when soliciting to 20% of the file based on the model, one can expect 2.64 times the total profit found by randomly soliciting 20% of the file, and so on for the remaining deciles. Note that the nondecreasing profit values throughout the decile suggest that something is "wrong" with this model, for example, an important predictor variable is not included or a predictor variable in the model needs to be reexpressed.

Rule: The larger the Cum Lift value is, the better the accuracy for a given depth of file will be.

18.6 Precision for Response Model

How close are the predicted probabilities of response to the true probabilities of response? Closeness or precision of the response cannot directly be determined because an individual's true probability is not known—if it were, then a model would not be needed. I recommend and illustrate the method of *smoothing* to provide estimates of the true probabilities. Then, I present the HL index as the measure of precision for response models.

Smoothing is the averaging of values within "neighborhoods." In this application of smoothing, I average actual responses within "decile" neighborhoods formed by the model. Continuing with the response model illustration, the actual response rate for a decile, column 4 in Table 18.3, is the estimate of true probability of response for the group of individuals in that decile.

Next, I calculate the mean predicted probabilities of response based on the response model scores (Prob_est) among the individuals in each decile. I insert these predicted means in column 4 in Table 18.5. Also, I insert the actual response rate (fourth column from the left in Table 18.3) in column 3 in Table 18.5. I can now determine response model precision.*

Comparing columns 3 and 4 (in Table 18.5) is informative. I see that for the top decile, the model underestimates the probability of response: 12.3% actual versus 9.6% predicted. Similarly, the model underestimates for deciles 2 through 4. Decile 5 is perfect. Going down deciles 6 through bottom, one can see clearly that the model is overestimating. This type of evaluation for precision is perhaps too subjective. An objective summary measure of precision is needed.

---

* This assessment is considered at a 10% level of smooth. A ventile-level analysis with the scored, ranked file divided into 20 groups provides an assessment of model precision at a 5% level of smooth. There is no agreement on a reliable level of smooth among statisticians.

TABLE 18.5

Response Model: HL and CV Indices

Decile	Column 1 Number of Individuals	Column 2 Number of Responders	Column 3 Decile Response Rate (Actual)	Column 4 Prob_Est (Predicted)	Column 5 Square of (Column 3 - Column 4) Times Column 1	Column 6 Column 4 Times (1 - Column 4)	Column 7 Column 5 Divided by Column 6
Top	7,410	911	12.3%	9.6%	5.40	0.086	62.25
2	7,410	544	7.3%	4.7%	5.01	0.044	111.83
3	7,410	437	5.9%	4.0%	2.68	0.038	69.66
4	7,410	322	4.3%	3.7%	0.27	0.035	7.49
5	7,410	258	3.5%	3.5%	0.00	0.033	0.00
6	7,410	188	2.5%	3.4%	0.60	0.032	18.27
7	7,410	130	1.8%	3.3%	1.67	0.031	52.25
8	7,410	163	2.2%	3.2%	0.74	0.031	23.92
9	7,410	124	1.7%	3.1%	1.45	0.030	48.35
Bottom	7,410	24	0.3%	3.1%	5.81	0.030	193.40
Total	74,100	3,101	4.2%				
		Separability CV	80.23			Precision HL	587.40

I present the HL index (the Hosmer-Lemeshow goodness-of-fit measure) as the measure of precision. The calculations for the HL index are illustrated in Table 18.5 for the response model illustration:

1. Columns 1, 2, and 3 are available from the response decile analysis.
2. Calculate the mean predicted probability of response for each decile from the model scores, Prob_est (column 4).
3. Calculate column 5: Take the difference between column 3 and column 4. Square the results. Then, multiply by column 1.
4. Column 6: Column 4 times the quantity 1 minus column 4.
5. Column 7: Column 5 divided by column 6.
6. HL index: Sum of the 10 elements of column 7.

Rule: The smaller the HL index value is, the better the precision will be.

18.7 Precision for Profit Model

How close are the predicted profits to the true profits? Just as in the discussion of precision for the response model, closeness cannot directly be determined because an individual's true profit is not known, and I recommend and illustrate the method of smoothing to provide estimates of the true profit values. Then, I present the SWMAD (smooth weighted mean of the absolute deviation) index as the measure of precision for a profit model.

To obtain the estimates of the true profit values, I average actual profit within "decile" neighborhoods formed by the profit model. Continuing with the profit model illustration, the mean actual profit for a decile, the fourth column from the left in Table 18.4, is the estimate of true mean profit for the group of individuals in that decile. Next, I calculate the mean predicted profit based on the model scores (Pred_est) among the individuals in each decile. I insert these predicted means in column 2 in Table 18.6. Also, I insert the mean actual profit (fourth column from the left in Table 18.4) in column 1 in Table 18.6. I can now determine model precision.

Comparing columns 1 and 2 (in Table 18.6) is informative. The ranking of the deciles based on the mean actual profit values is not strictly descending—not a desirable indicator of a good model. The mean profit values for the top and second deciles are reversed, $15.67 and $20.10, respectively. Moreover, the third-largest decile mean profit value ($8.00) is in the fifth decile. This type of evaluation is interesting, but a quantitative measure for this "out-of-order" ranking is preferred.

TABLE 18.6

Profit Model: SWMAD and CV Indices

Decile	Column 1 Decile Mean Profit (Actual)	Column 2 Decile Mean Pred_Est (Predicted Profit)	Column 3 Absolute Error	Column 4 Rank of Decile Actual Profit	Column 5 Rank of Decile Predicted Profit	Column 6 Absolute Difference Between Ranks	Column 7 Wt	Column 8 Weighted Error
Top	$15.67	$17.16	$1.49	2	1	1.0	1.10	1.64
2	$20.10	$13.06	$7.04	1	2	1.0	1.10	7.74
3	$7.30	$10.50	$3.20	4	3	1.0	1.10	3.52
4	$6.47	$8.97	$2.50	5	4	1.0	1.10	2.75
5	$8.00	$7.43	$0.57	3	5	2.0	1.20	0.68
6	$4.23	$4.87	$0.63	6	6	0.0	1.00	0.63
7	$1.93	$3.33	$1.40	7.5	7	0.5	1.05	1.47
8	$1.93	$1.80	$0.14	7.5	8	0.5	1.05	0.15
9	$0.90	$0.26	$0.64	10	9.5	0.5	1.05	0.67
Bottom	$1.10	$0.26	$0.84	9	9.5	0.5	1.05	0.88
Separability	95.86					SUM	10.8	20.15
CV						Precision	SWMAD	1.87

18.7.1 Construction of SWMAD

I present the measure SWMAD for the precision of a profit model: a weighted mean of the absolute deviation between smooth decile actual and predicted profit values; the weights reflect discordance between the rankings of the smooth decile actual and predicted values. The steps for the calculation of SWMAD for the profit model illustration are given next and are reported in Table 18.6:

1. Column 1 is available from the profit decile analysis.
2. Calculate the mean predicted profit for each decile from the model scores, Pred_est (column 2).
3. Calculate column 3: Take the absolute difference between column 1 and column 2.
4. Column 4: Rank the deciles based on actual profit (column 1); assign the lowest rank value to the highest decile mean actual profit value. Tied ranks are assigned the mean of the corresponding ranks.
5. Column 5: Rank the deciles based on predicted profit (column 2); assign the lowest rank value to the highest decile mean predicted profit value. Tied ranks are assigned the mean of the corresponding ranks.
6. Column 6: Take the absolute difference between column 4 and column 5.
7. Column 7: The weight variable (Wt) is defined as column 6 divided by 10 plus 1.
8. Column 8 is column 3 times column 7.
9. Calculate SUMWGT, the sum of the 10 values of column 7.
10. Calculate SUMWDEV, the sum of the 10 values of column 8.
11. Calculate SWMAD: SUMWDEV/SUMWGT.

Rule: The smaller the SWMAD value is, the better the precision will be.

18.8 Separability for Response and Profit Models

How different are the individuals across the deciles in terms of likelihood to respond or contribution to profit? Is there a real variation or separation of individuals as identified by the model? I can measure the variability across the decile groups by calculating the traditional coefficient of variation (CV) among the decile estimates of true probability of response for the response model and among the decile estimates of true profit for the profit model.

I illustrate the calculation of CV with the response and profit model illustrations. CV (response) is the standard deviation of the 10 smooth values of column 3 in Table 18.5 divided by the mean of the 10 smooth values and multiplied by 100. CV (response) is 80.23 in Table 18.5. CV (profit) is the standard deviation of the 10 smooth values of column 1 in Table 18.6 divided by the mean 10 smooth values and multiplied by 100. CV (profit) is 95.86 in Table 18.6.

Rule: The larger the CV value is, the better the separability will be.

18.9 Guidelines for Using Cum Lift, HL/SWMAD, and CV

The following are guidelines for selecting the best model based on the three assessment measures Cum Lift, HL/SWMAD, and CV:

1. In general, a good model has large HL/SWMAD and CV values.

2. If maximizing response rate/mean profit is not the objective of the model, then the best model is among those with the smallest HL/SWMAD values and largest CV values. Because small HL/SWMAD values do not necessarily correspond with large CV values, the data analyst must decide on the best balance of small HL/SWMAD and large CV values for declaring the best model.

3. If maximizing response rate/mean profit is the objective, then the best model has the largest Cum Lift. If there are several models with comparable largest Cum Lift values, then the model with the "best" HL/SWMAD-CV combination is declared the best model.

4. If decile-level response/profit prediction is the objective of the model, then the best model has the smallest HL/SWMAD value. If there are several models with comparable smallest HL/SWMAD values, then the model with the largest CV value is declared the best.

5. The measure of separability CV itself has no practical value. A model that is selected solely on the largest CV value will not necessarily have good accuracy or precision. Separability should be used in conjunction with the other two measures of model assessment as discussed.

18.10 Summary

The traditional measures of model accuracy are a PTCC and MSE or a variant of mean error for response and profit models, respectively. These measures have limited value in marketing. Marketers have their own measure of model accuracy—Cum Lift—that takes into account the way they implement the model. They use a model to identify individuals most likely to respond or contribute profit and create a solicitation list of those individuals to obtain advantage over a random selection. The Cum Lift is an index of the expected response/profit with a selection based on a model compared with the expected response/profit with a random selection (no model). A maxim of Cum Lift is, the larger the Cum Lift value, the better the accuracy.

I discussed the Cum Lift by illustrating the construction of the decile analysis for both response and profit models. Then, using the decile analysis as a backdrop, I presented two additional measures of model assessment—HL/SWMAD for response/profit precision and the traditional CV for separability for response and profit models. Because the true response/profit values are unknown, I estimated the true values by smoothing at the decile level. With these estimates, I illustrated the calculations for HL and SWMAD. For the HL/SWMAD rule, the smaller the HL/SWMAD values are, the better the precision will be.

Separability addresses the question of how different the individuals are in terms of likelihood to respond or contribute to profit across the deciles. I used traditional CV as the measure of separability among the estimates of true response/profit values. Thus, a maxim of CV is that the larger the CV value is, the better the separability will be.

Last, I provided guidelines for using all three measures together in selecting the best model.

19

Bootstrapping in Marketing: A New Approach for Validating Models*

19.1 Introduction

Traditional validation of a marketing model uses a holdout sample consisting of individuals who are not part of the sample used in building the model itself. The validation results are probably biased and incomplete unless a resampling method is used. This chapter points to the weaknesses of the traditional validation and then presents a bootstrap approach for validating response and profit models, as well as measuring the efficiency of the models.

19.2 Traditional Model Validation

The data analyst's first step in building a marketing model is to split randomly the original data file into two mutually exclusive parts: a calibration sample for developing the model and a validation or holdout sample for assessing the reliability of the model. If the analyst is lucky to split the file to yield a holdout sample with *favorable* characteristics, then a better-than-true biased validation is obtained. If unlucky and the sample has *unfavorable* characteristics, then a worse-than-true biased validation is obtained. Lucky or not, or even if the validation sample is a true reflection of the population under study, a single sample cannot provide a measure of variability that would otherwise allow the analyst to assert a level of confidence about the validation.

In sum, the traditional single-sample validation provides neither assurance that the results are not biased nor any measure of confidence in the results. These points are made clear with an illustration using a response model (RM), but all results and implications apply equally to profit models.

---

* This chapter is based on an article with the sa me title in *Journal of Targeting, Measurement and Analysis for Marketing*, 6, 2, 1997. Used with permission.

19.3 Illustration

As marketers use the Cum Lift measure from a decile analysis to assess the goodness of a model, the validation of the model* consists of comparing the Cum Lifts from the calibration and holdout decile analyses based on the model. It is expected that shrinkage in the Cum Lifts occurs: Cum Lifts from the holdout sample are typically smaller (less optimistic) than those from the calibration sample from which they were originated. The Cum Lifts on a fresh holdout sample, which does not contain the calibration idiosyncrasies, provide a more realistic assessment of the quality of the model. The calibration Cum Lifts inherently capitalize on the idiosyncrasies of the calibration sample due to the modeling process, which favors large Cum Lifts. If both the Cum Lift shrinkage and the Cum Lift values themselves are acceptable, then the model is considered successfully validated and ready to use; otherwise, the model is reworked until successfully validated.

Consider a RM that produces the validation decile analysis in Table 19.1 based on a sample of 181,100 customers with an overall response rate of 0.26%. (Recall from Chapter 18 that the Cum Lift is a measure of predictive power; it indicates the expected gain from a solicitation implemented with a model over a solicitation implemented without a model.) The Cum Lift for the top decile is 186; this indicates that when soliciting to the top decile—the top 10% of the customer file identified by the RM model—there is an expected 1.86 times the number of responders found by randomly soliciting 10% of the file (without a model). Similar to that for the second decile, the Cum Lift 154 indicates that when soliciting to the top two deciles—the top 20% of the customer file based on the RM model—there is an expected 1.54 times the number of responders found by randomly soliciting 20% of the file.

As luck would have it, the data analyst finds two additional samples on which two additional decile analysis validations are performed. Not surprisingly, the Cum Lifts for a given decile across the three validations are somewhat different. The reason for this is the expected sample-to-sample variation, attributed to chance. There is a large variation in the top decile (range is 15 = 197 - 182) and a small variation in decile 2 (range is 6 = 154 - 148). These results in Table 19.2 raise obvious questions.

---

* Validation of any response or profit model built from any modeling technique (e.g., discriminant analysis, logistic regression, neural network, genetic algorithms, or CHAID [chi-squared automatic interaction detection]).

TABLE 19.1

Response Decile Analysis

Decile	Number of Individuals	Number of Responders	Decile Response Rate	Cumulative Response Rate	Cum Lift
Top	18,110	88	0.49%	0.49%	186
2	18,110	58	0.32%	0.40%	154
3	18,110	50	0.28%	0.36%	138
4	18,110	63	0.35%	0.36%	137
5	18,110	44	0.24%	0.33%	128
6	18,110	48	0.27%	0.32%	123
7	18,110	39	0.22%	0.31%	118
8	18,110	34	0.19%	0.29%	112
9	18,110	23	0.13%	0.27%	105
Bottom	18,110	27	0.15%	0.26%	100
Total	181,100	474	0.26%		

TABLE 19.2

Cum Lifts for Three Validations

Decile	First Sample	Second Sample	Third Sample
Top	186	197	182
2	154	153	148
3	138	136	129
4	137	129	129
5	128	122	122
6	123	118	119
7	118	114	115
8	112	109	110
9	105	104	105
Bottom	100	100	100

19.4 Three Questions

With many decile analysis validations, the expected sample-to-sample variation within each decile points to the uncertainty of the Cum Lift estimates. If there is an observed large variation for a given decile, there is less

confidence in the Cum Lift for that decile; if there is an observed small variation, there is more confidence in the Cum Lift. Thus, there are the following
questions:

1. With many decile analysis validations, how can an *average* Cum Lift
   (for a decile) be defined to serve as an *honest* estimate of the Cum
   Lift? In addition, how many validations are needed?

2. With many decile analysis validations, how can the *variability* of an
   honest estimate of Cum Lift be assessed? That is, how can the standard error (a measure of precision of an estimate) of an honest estimate of the Cum Lift be calculated?

3. With only a single validation dataset, can an honest Cum Lift estimate and its standard error be calculated?

The answers to these questions and more lie in the bootstrap methodology.

19.5 The Bootstrap

The bootstrap is a computer-intensive approach to statistical inference [1].
It is the most popular resampling method, using the computer to resample
extensively the sample at hand* [2]. By random selection with replacement
from the sample, some individuals occur more than once in a *bootstrap* sample, and some individuals occur not at all. Each same-size bootstrap sample
will be slightly different from the others. This variation makes it possible
to induce an empirical sampling distribution[†] of the desired statistic, from
which estimates of bias and variability are determined.

The bootstrap is a flexible technique for assessing the accuracy[‡] of any statistic. For well-known statistics, such as the mean, the standard deviation,
regression coefficients, and R-squared, the bootstrap provides an alternative
to traditional parametric methods. For statistics with unknown properties,
such as the median and Cum Lift, traditional parametric methods do not
exist; thus, the bootstrap provides a viable alternative over the inappropriate
use of traditional methods, which yield questionable results.

The bootstrap falls also into the class of nonparametric procedures. It
does not rely on unrealistic parametric assumptions. Consider testing the
significance of a variable[§] in a regression model built using ordinary least

---

* Other resampling methods include, for example, the jackknife, infinitesimal jackknife, delta
  method, influence function method, and random subsampling.
[†] A sampling distribution can be considered as the frequency curve of a sample statistic from
  an infinite number of samples.
[‡] Accuracy includes bias, variance, and error.
§ That is, is the coefficient equal to zero?

squares estimation. Say the error terms do not have a normal distribution, a clear violation of the ordinary least squares assumptions [3]. The significance testing may yield inaccurate results due to the model assumption not being met. In this situation, the bootstrap is a feasible approach in determining the significance of the coefficient without concern for any assumptions. As a nonparametric method, the bootstrap does not rely on theoretical derivations required in traditional parametric methods. I review the well-known parametric approach to the construction of confidence intervals (CIs) to demonstrate the utility of the bootstrap as an alternative technique.

19.5.1 Traditional Construction of Confidence Intervals

Consider the parametric construction of a confidence interval for the population mean. I draw a random sample A of five numbers from a population. Sample A consists of (23, 4, 46, 1, 29). The sample mean is 20.60, the sample median is 23, and the sample standard deviation is 18.58.

The parametric method is based on the central limit theorem, which states that the theoretical sampling distribution of the sample mean is normally distributed with an analytically defined standard error [4]. Thus, the 100(1 - a)% CI for the mean is:

SAMPLE MEAN VALUE $\pm$ $|Z_{a/2}|$*STANDARD ERROR

where Sample Mean value is simply the mean of the five numbers in the sample.

$|Z_{a/2}|$ is the value from the standard normal distribution for a 100(1 - a)% CI. The $|Z_{a/2}|$ values are 1.96, 1.64, and 1.28 for 95%, 90%, and 80% CIs, respectively. Standard Error (SE) of the sample mean has the analytic formula: SE = the sample standard deviation divided by the square root of the sample size.

An often used term is the *margin of error*, defined as $|Z_{a/2}|$*SE.

For sample A, SE equals 8.31, and the 95% CI for the population mean is between 4.31 (= 20.60 - 1.96*8.31) and 36.89 (= 20.60 + 1.96*8.31). This is commonly, although not quite exactly, stated as follows: There is 95% confidence that the population mean lies between 4.31 and 36.89. The statistically correct statement for the 95% CI for the unknown population mean is as follows: If one repeatedly calculates such intervals from, say, 100 independent random samples, 95% of the constructed intervals would contain the true population mean. Do not be fooled: Once the confidence interval is constructed, the true mean is either in the interval or not in the interval. Thus, 95% confidence refers to the procedure for constructing the interval, not the observed interval itself. This parametric approach for the construction of confidence intervals for statistics, such as the median and the Cum Lift, does not exist because the theoretical sampling distributions (which provide the standard errors) of the median and Cum Lift are not known. If confidence intervals for the median and Cum Lift are desired, then approaches that rely on a resampling methodology like the bootstrap can be used.

19.6 How to Bootstrap

The key assumption of the bootstrap* is that the sample is the best estimate[†] of the unknown population. Treating the sample as the population, the analyst repeatedly draws same-size random samples with replacement from the original sample. The analyst estimates the sampling distribution of the desired statistic from the many bootstrap samples and is able to calculate a bias-reduced bootstrap estimate of the statistic and a bootstrap estimate of the SE of the statistic.

The bootstrap procedure consists of 10 simple steps.

1. State desired statistic, say, Y.
2. Treat sample as population.
3. Calculate Y on the sample/population; call it SAM_EST.
4. Draw a bootstrap sample from the population, that is, a random selection with replacement of size n, the size of the original sample.
5. Calculate Y on the bootstrap sample to produce a pseudovalue; call it BS_1.
6. Repeat steps 4 and 5 "m" times.[‡]
7. From steps 1 to 6, there are BS_1, BS_2, ..., BS_m.
8. Calculate the bootstrap estimate of the statistic[§]:

$$BS_{est}(Y) = 2*SAM\_EST - mean(BS_i).$$

---

* This bootstrap method is the normal approximation. Others are percentile, B-C percentile, and percentile-t.
[†] Actually, the sample distribution function is the nonparametric maximum likelihood estimate of the population distribution function.
[‡] Studies showed the precision of the bootstrap does not significantly increase for m > 250.
[§] This calculation arguably ensures a bias-reduced estimate. Some analysts question the use of the bias correction. I feel that this calculation adds precision to the decile analysis validation when conducting small solicitations and has no noticeable effect on large solicitations.

9. Calculate the bootstrap estimate of the standard error of the statistic:

$$SE_{BS}(Y) = \text{standard deviation of } (BS_i).$$

10. The 95% bootstrap confidence interval is

$$BS_{est}(Y) \pm |Z_{0.025}| {}^{*}SE_{BS}(Y).$$

19.6.1 Simple Illustration

Consider a simple illustration.* I have a sample B from a population (no reason to assume it is normally distributed) that produces the following 11 values:

Sample B: 0.1 0.1 0.1 0.4 0.5 1.0 1.1 1.3 1.9 1.9 4.7

I want a 95% confidence interval for the population standard deviation. If I knew the population was normal, I would use the parametric chi-square test and obtain the confidence interval:

0.93 < population standard deviation < 2.35.

I apply the bootstrap procedure on sample B:

1. The desired statistic is the standard deviation, StD.
2. Treat sample B as the population.
3. I calculate StD on the original sample/population. SAM_EST = 1.3435.
4. I randomly select 11 observations with replacement from the population. This is the first bootstrap sample.
5. I calculate StD on this bootstrap (BS) sample to obtain a pseudo-value, $BS_1 = 1.3478$.
6. I repeat steps 4 and 5 an additional 99 times.
7. I have $BS_1, BS_2, ..., BS_{100}$ in Table 19.3.
8. I calculate the bootstrap estimate of StD:

$$BS_{est}(StD) = 2 {}^{*}SAM\_EST - \text{mean}(BS_i) = 2 {}^{*}1.3435 - 1.2034 = 1.483$$

9. I calculate the bootstrap estimate of the standard error of StD:

$$SE_{BS}(StD) = \text{standard deviation } (BS_i) = 0.5008.$$

---

* This illustration draws on Sample B comes from Mosteller, F., and Tukey, J.W., Data Analysis and Regression, Addison-Wesley, Reading, MA, 139–143, 1977.

TABLE 19.3

100 Bootstrapped StDs

1.3431	0.60332	1.7076	0.6603	1.4614	1.43
1.4312	1.2965	1.9242	0.71063	1.3841	1.7656
0.73151	0.70404	0.643	1.6366	1.8288	1.6313
0.61427	0.76485	1.3417	0.69474	2.2153	1.2581
1.4533	1.353	0.479	0.62902	2.1982	0.73666
0.66341	1.4098	1.8892	2.0633	0.73756	0.69121
1.2893	0.67032	1.7316	0.60083	1.4493	1.437
1.3671	0.49677	0.70309	0.51897	0.65701	0.59898
1.3784	0.7181	1.4312	1.3985	0.6603	1.2972
0.47854	2.0658	1.7825	0.63281	1.8755	0.39384
0.69194	0.6343	1.31	1.3491	0.70079	1.3754
0.29609	1.5522	0.62048	1.8657	1.3919	1.6596
1.3726	2.0877	1.6659	1.4372	0.72111	1.4356
0.83327	1.4056	1.7404	1.796	1.7957	1.3994
1.399	1.3653	1.3726	1.2665	1.2874	1.8172
1.322	0.56569	0.74863	1.4085	1.6363	1.3802
0.60194	1.9938	0.57937	0.74117		
1.3476	0.6345	1.7188			

10. The bootstrap 95% confidence interval for the population standard
    deviation is 0.50 < population standard deviation < 2.47.

As you may have suspected, the sample was drawn from a normal population (NP). Thus, it is instructive to compare the performance of the
bootstrap with the theoretically correct parametric chi-square test. The
bootstrap confidence interval is somewhat wider than the chi-square/NP
interval in Figure 19.1. The BS confidence interval covers values between
0.50 and 2.47, which also includes the values within the NP confidence
interval (0.93, 2.35).

These comparative performance results are typical. Performance studies
indicate the bootstrap methodology provides results that are consistent with
the outcomes of the parametric techniques. Thus, the bootstrap is a reliable
approach to inferential statistics in most situations.

Note that the bootstrap estimate offers a more honest* and bias-reduced†
point estimate of the standard deviation. The original sample estimate is
1.3435, and the bootstrap estimate is 1.483. There is a 10.4% (1.483/1.3435) bias
reduction in the estimate.

---

* Due to the many samples used in the calculation.
† Attributable, in part, to sample size.

NP   0.93xxxxxxxxxxxxx2.35
BS   .50xxxxxxxxxxxxxxxxxx2.47

FIGURE 19.1
Bootstrap versus normal estimates.

19.7 Bootstrap Decile Analysis Validation

Continuing with the RM model illustration, I execute the 10-step bootstrap procedure to perform a bootstrap decile analysis validation. I use 50 bootstrap samples,* each of a size equal to the original sample size of 181,100. The bootstrap Cum Lift for the top decile is 183 and has a bootstrap standard error of 10. Accordingly, for the top decile, the 95% bootstrap confidence interval is 163 to 203. The second decile has a bootstrap Cum Lift of 151, and a bootstrap 95% confidence interval is between 137 and 165. Similar readings in Table 19.4 can be made for the other deciles. Specifically, this bootstrap validation indicates that the expected Cum Lift is 135 using the RM model to select the top 30% of the most responsive individuals from a randomly drawn sample of size 181,100 from the target population or database. Moreover, the Cum Lift is expected to lie between 127 and 143 with 95% confidence. Similar statements can be made for other depths of file from the bootstrap validation in Table 19.4.

Bootstrap estimates and confidence intervals for decile response rates can be easily obtained from the bootstrap estimates and confidence intervals for decile Cum Lifts. The conversion formula involves the following: Bootstrap decile response rate equals bootstrap decile Cum Lift divided by 100, then multiplied by overall response rate. For example, for the third decile, the bootstrap response rate is 0.351% (= (135/100)*0.26%); the lower and upper confidence interval end points are 0.330% (= (127/100)*0.26%) and 0.372% (= (143/100)*0.26%), respectively.

19.8 Another Question

Quantifying the predictive certainty, that is, constructing prediction confidence intervals is likely to be more informative to data analysts and their management than obtaining a *point estimate* alone. A single calculated value of a statistic such as Cum Lift can be regarded as a point estimate that provides a best guess of the true value of the statistic. However, there is an obvious need to quantify the certainty associated with such a point estimate. The decision maker wants the margin of error of the estimate. Plus or minus,

---

* I experience high precision in bootstrap decile validation with just 50 bootstrap samples.

TABLE 19.4

Bootstrap Response Decile Validation (Bootstrap Sample
Size n = 181,000)

Decile	Bootstrap Cum Lift	Bootstrap SE	95% Bootstrap CI
Top	183	10	(163, 203)
2	151	7	(137, 165)
3	135	4	(127, 143)
4	133	3	(127, 139)
5	125	2	(121, 129)
6	121	1	(119, 123)
7	115	1	(113, 117)
8	110	1	(108, 112)
9	105	1	(103, 107)
Bottom	100	0	(100, 100)

what value should be added/subtracted to the estimated value to yield an
interval for which there is a reasonable confidence that the true (Cum Lift)
value lies? If the confidence interval (equivalently, the standard error or margin of error) is too large for the business objective at hand, what can be done?

The answer rests on the well-known, fundamental relationship between
sample size and confidence interval length: Increasing (decreasing) sample
size increases (decreases) confidence in the estimate. Equivalently, increasing (decreasing) sample size decreases (increases) the standard error [5]. The
sample size-confidence length relationship can be used to increase confidence in the bootstrap Cum Lift estimates in two ways:

1. If ample additional customers are available, add them to the original
   validation dataset until the enhanced validation dataset size produces the desired standard error and confidence interval length.
2. Simulate or bootstrap the original validation dataset by increasing
   the bootstrap sample size until the enhanced bootstrap dataset size
   produces the desired standard error and confidence interval length.

Returning to the RM illustration, I increase the bootstrap sample size to
225,000 from the original sample size of 181,100 to produce a slight decrease
in the standard error from 4 to 3 for the cumulative top three deciles. This
simulated validation in Table 19.5 indicates that a Cum Lift of 136 centered
in a slightly shorter 95% confidence interval between 130 and 142 is expected
when using the RM model to select the top 30% of the most responsive individuals from a randomly selected sample of size 225,000 from the database.
Note that the bootstrap Cum Lift estimates also change (in this instance,
from 135 to 136) because their calculations are based on new larger samples.

TABLE 19.5

Bootstrap Response Decile Validation (Bootstrap Sample Size n = 225,000)

Decile	Bootstrap Cum Lift	Bootstrap SE	95% Bootstrap CI
Top	185	5	(163, 203)
2	149	3	(137, 165)
3	136	3	(127, 143)
4	133	2	(127, 139)
5	122	1	(121, 129)
6	120	1	(119, 123)
7	116	1	(113 ,117)
8	110	0.5	(108, 112)
9	105	0.5	(103, 107)
Bottom	100	0	(100, 100)

Bootstrap Cum Lift estimates rarely show big differences when the enhanced dataset size increases. (Why?) In the next section, I continue the discussion on the sample size-confidence length relationship as it relates to a bootstrap assessment of model implementation performance.

19.9 Bootstrap Assessment of Model Implementation Performance

Statisticians are often asked, "How large a sample do I need to have confidence in my results?" The traditional answer, which is based on parametric theoretical formulations, depends on the statistic in question, such as response rate or mean profit, and on additional required input. These additions include the following:

1. The expected value of the desired statistic.
2. The preselected level of confidence, such as the probability that the decision maker is willing to take by wrongly rejecting the null hypothesis; this may involve claiming a relationship *exists* when, in fact, it does not.
3. The preselected level of power to detect a relationship. That is, 1 minus the probability the decision maker is willing to take by wrongly accepting the null hypothesis; this may involve claiming that a relationship *does not exist* when, in fact, it does.
4. The assurance that sundry theoretical assumptions hold.

Regardless of who has built the marketing/database model, once it is ready for implementation, the question is essentially the same: How large a sample do I need to implement a solicitation based on the model to obtain a desired performance quantity? The database answer, which in this case does not require most of the traditional input, depends on one of two performance objectives. Objective 1 is to maximize the performance quantity for a specific depth of file, namely, the Cum Lift. Determining how large the sample should be—actually the smallest sample necessary to obtain the Cum Lift value—involves the concepts discussed in the previous section that correspond to the relationship between confidence interval length and sample size. The following procedure answers the sample size question for objective 1.

1. For a desired Cum Lift value, identify the decile and its confidence interval containing Cum Lift values closest to the desired value, based on the decile analysis validation at hand. If the corresponding confidence interval length is acceptable, then the size of the validation dataset is the required sample size. Draw a random sample of that size from the database.

2. If the corresponding confidence interval is too large, then increase the validation sample size by adding individuals or bootstrapping a larger sample size until the confidence interval length is acceptable. Draw a random sample of that size from the database.

3. If the corresponding confidence interval is unnecessarily small, which indicates that a smaller sample can be used to save time and cost of data retrieval, then decrease the validation sample size by deleting individuals or bootstrapping a smaller sample size until the confidence interval length is acceptable. Draw a random sample of that size from the database.

Objective 2 adds a constraint to the performance quantity of objective 1. Sometimes, it is desirable not only to reach a performance quantity, but also to discriminate among the *quality* of individuals who are selected to contribute to the performance quantity. This is particularly worth doing when a solicitation is targeted to an unknown, yet believed to be, relatively homogeneous population of finely graded individuals in terms of responsiveness or profitability.

The quality constraint imposes the restriction that the performance value of individuals within a decile is different across the deciles within the preselected depth-of-file range. This is accomplished by (1) determining the sample size that produces decile confidence intervals that do not overlap *and* (2) ensuring an overall confidence level that the individual decile confidence intervals are *jointly* valid; that is, the true Cum Lifts are "in" their confidence intervals. The former condition is accomplished by increasing sample size.

The latter condition is accomplished by employing the Bonferroni method, which allows the analyst to assert a confidence statement that multiple confidence intervals are *jointly* valid.

Briefly, the Bonferroni method is as follows: Assume the analyst wishes to combine k confidence intervals that individually have confidence levels 1 - a_1, 1 - a_2, ..., 1 - a_k. The analyst wants to make a joint confidence statement with confidence level 1 - a_j. The Bonferroni method states that the joint confidence level 1 - a_j is greater than or equal to 1 - a_1 - a_2..., -a_k. The joint confidence level is a conservative lower bound on the actual confidence level for a joint confidence statement. The Bonferroni method is conservative in the sense that it provides confidence intervals that have confidence levels larger than the actual level.

I apply the Bonferrroni method for four common individual confidence intervals: 95%, 90%, 85%, and 80%.

1. For combining 95% confidence intervals, there is at least 90% confidence that the two true decile Cum Lifts lie between their respective confidence intervals. There is at least 85% confidence that the three true decile Cum Lifts lie between their respective confidence intervals. And, there is at least 80% confidence that the four true decile Cum Lifts lie between their respective confidence intervals.

2. For combining 90% confidence intervals, there is at least 80% confidence that the two true decile Cum Lifts lie between their respective confidence intervals. There is at least 70% confidence that the three true decile Cum Lifts lie between their respective confidence intervals. And, there is at least 60% confidence that the four true decile Cum Lifts lie between their respective confidence intervals.

3. For combining 85% confidence intervals, there is at least 70% confidence that the three true decile Cum Lifts lie between their respective confidence intervals. There is at least 55% confidence that the three true decile Cum Lifts lie between their respective confidence intervals. And, there is at least 40% confidence that the four true decile Cum Lifts lie between their respective the confidence intervals.

4. For combining the 80% confidence intervals, there is at least 60% confidence that the three true decile Cum Lifts lie between their respective confidence intervals. There is at least 40% confidence that the three true decile Cum Lifts lie between their respective confidence intervals. And, there is at least 20% confidence that the four true decile Cum Lifts lie between their respective confidence intervals

The following procedure is used for determining how large the sample should be. Actually, the smallest sample necessary to obtain the desired quantity and quality of performance involves the following:

1. For a desired Cum Lift value for a preselected depth-of-file range, identify the decile and its confidence interval containing Cum Lift values closest to the desired value based on the decile analysis validation at hand. If the decile confidence intervals within the preselected depth-of-file range do not overlap (or concededly have acceptable minimal overlap), then the validation sample size at hand is the required sample size. Draw a random sample of that size from the database.

2. If the decile confidence intervals within the preselected depth-of-file range are too large or overlap, then increase the validation sample size by adding individuals or bootstrapping a larger sample size until the decile confidence interval lengths are acceptable and do not overlap. Draw a random sample of that size from the database.

3. If the decile confidence intervals are unnecessarily small and do not overlap, then decrease the validation sample size by deleting individuals or bootstrapping a smaller sample size until the decile confidence interval lengths are acceptable and do not overlap. Draw a random sample of that size from the database.

19.9.1 Illustration

Consider a three-variable RM predicting response from a population with an overall response rate of 4.72%. The decile analysis validation based on a sample size of 22,600, along with the bootstrap estimates, is in Table 19.6. The 95% margin of errors ($|Z_{a/2}|*SE_{BS}$) for the top four decile 95% confidence intervals is considered too large for using the model with confidence. Moreover, the decile confidence intervals severely overlap. The top decile has a 95% confidence interval of 160 to 119, with an expected bootstrap response rate of 6.61% (= (140/100)*4.72%; 140 is the bootstrap Cum Lift for the top decile). The top two decile levels have a 95% confidence interval of 113 to 141, with an expected bootstrap response rate of 5.99% (= (127/100)*4.72%; 127 is the bootstrap Cum Lift for the top two decile levels).

I create a new bootstrap validation sample of size 50,000. The 95% margins of error in Table 19.7 are still unacceptably large, and the decile confidence intervals still overlap. I further increase the bootstrap sample size to 75,000 in Table 19.8. There is no noticeable change in 95% margins of error, and the decile confidence intervals overlap.

I recalculate the bootstrap estimates using 80% margins of error on the bootstrap sample size of 75,000. The results in Table 19.9 show that the decile confidence intervals have acceptable lengths and are almost nonoverlapping. Unfortunately, the joint confidence levels are quite low: at least 60%, at least 40%, and at least 20% for the top two, top three, and top four decile levels, respectively.

After successively increasing the bootstrap sample size, I reach a bootstrap sample size (175,000) that produces acceptable and virtually nonoverlapping 95% confidence intervals in Table 19.10. The top four decile Cum Lift confidence intervals are (133, 147), (122, 132), (115, 124), and (109, 116), respectively. The joint confidence for the top three deciles and top four deciles is respectable levels of at least 85% and at least 80%, respectively. It is interesting to note that increasing the bootstrap sample size to 200,000, for either 95% or 90% margins of error, does not produce any noticeable improvement in the quality of performance. (Validations for bootstrap sample size 200,000 are not shown.)

19.10 Bootstrap Assessment of Model Efficiency

The bootstrap approach to decile analysis of marketing models can provide an assessment of model efficiency. Consider an alternative model A to the best model B (both models predict the same dependent variable). Model A is said to be *less efficient* than model B if either model A yields the same results as model B (equal Cum Lift margins of error, where model A sample size is larger than model B sample size) or model A yields worse results than model B (greater Cum Lift margins of error, where sample sizes for models A and B are equal).

The measure of efficiency is reported as the ratio of either of the following quantities: (1) the sample sizes necessary to achieve equal results or (2) the variability measures (Cum Lift margin of error) for equal sample size. The efficiency ratio is defined as follows: model B (quantity) over model A (quantity). Efficiency ratio values less (greater) than 100% indicate that model A is less (more) efficient than model B. In this section, I illustrate how a model with unnecessary predictor variables is less efficient (has larger prediction error variance) than a model with the right number of predictor variables.

Returning to the previous section with the illustration of the three-variable RM (considered the best model B), I create an alternative model (model A) by adding five unnecessary predictor variables to model B. The extra variables include irrelevant variables (not affecting the response variable) and redundant variables (not adding anything to the response variable). Thus, the eight-variable model A can be considered an overloaded and noisy model, which should produce unstable predictions with large error variance. The bootstrap decile analysis validation of model A based on a bootstrap sample size of 175,000 is in Table 19.11. To facilitate the discussion, I added the Cum Lift margins of error for model B (from Table 19.10) to Table 19.11. The efficiency ratios for the decile Cum Lifts are reported in the rightmost column of Table 19.11. Note that the decile confidence intervals overlap.

TABLE 19.6

Three-Variable Response Model 95% Bootstrap Decile Validation (Bootstrap Sample Size n = 22,600)

Decile	Number of Individuals	Number of Responses	Response Rate (%)	Cum Response Rate (%)	Cum Lift	Bootstrap Cum Lift	95% Margin of Error	95% Lower Bound	95% Upper Bound
Top	2,260	150	6.64	6.64	141	140	20.8	119	160
2nd	2,260	120	5.31	5.97	127	127	13.8	113	141
3rd	2,260	112	4.96	5.64	119	119	11.9	107	131
4th	2,260	99	4.38	5.32	113	113	10.1	103	123
5th	2,260	113	5.00	5.26	111	111	9.5	102	121
6th	2,260	114	5.05	5.22	111	111	8.2	102	119
7th	2,260	94	4.16	5.07	107	107	7.7	100	115
8th	2,260	97	4.29	4.97	105	105	6.9	98	112
9th	2,260	93	4.12	4.88	103	103	6.4	97	110
Bottom	2,260	75	3.32	4.72	100	100	6.3	93	106
Total	22,600	1,067							

TABLE 19.7

Three-Variable Response Model 95% Bootstrap Decile Validation
(Bootstrap Sample Size n = 50,000)

Decile	Model Cum Lift	Bootstrap Cum Lift	95% Margin of Error	95% Lower Bound	95% Upper Bound
Top	141	140	15.7	124	156
2nd	127	126	11.8	114	138
3rd	119	119	8.0	111	127
4th	113	112	7.2	105	120
5th	111	111	6.6	105	118
6th	111	111	6.0	105	117
7th	107	108	5.6	102	113
8th	105	105	5.1	100	111
9th	103	103	4.8	98	108
Bottom	100	100	4.5	95	104

TABLE 19.8

Three-Variable Response Model 95% Bootstrap Decile Validation
(Bootstrap Sample Size n = 75,000)

Decile	Model Cum Lift	Bootstrap Cum Lift	95% Margin of Error	95% Lower Bound	95% Upper Bound
Top	141	140	12.1	128	152
2nd	127	127	7.7	119	135
3rd	119	120	5.5	114	125
4th	113	113	4.4	109	117
5th	111	112	4.5	107	116
6th	111	111	4.5	106	116
7th	107	108	4.2	104	112
8th	105	105	3.9	101	109
9th	103	103	3.6	100	107
Bottom	100	100	3.6	96	104

It is clear that model A is less efficient than model B. Efficiency ratios are less than 100%, ranging from a low 86.0% for the top decile to 97.3% for the fourth decile, with one exception for the eighth decile, which has an anomalous ratio of 104.5%. The implication is that model A predictions are less unstable than model B predictions, that is, model A has larger prediction error variance relative to model B.

The broader implication is of a warning: Review a model with too many variables for justification of the contribution of each variable in the model; otherwise, it can be anticipated that the model has unnecessarily large prediction error variance. In other words, the bootstrap methodology should

TABLE 19.9

Three-Variable Response Model 80% Bootstrap Decile Validation
(Bootstrap Sample Size n = 75,000)

Decile	Model Cum Lift	Bootstrap Cum Lift	80% Margin of Error	80% Lower Bound	80% Upper Bound
Top	141	140	7.90	132	148
2nd	127	127	5.10	122	132
3rd	119	120	3.60	116	123
4th	113	113	2.90	110	116
5th	111	112	3.00	109	115
6th	111	111	3.00	108	114
7th	107	108	2.70	105	110
8th	105	105	2.60	103	108
9th	103	103	2.40	101	106
Bottom	100	100	2.30	98	102

TABLE 19.10

Three-Variable Response Model 95% Bootstrap Decile Validation
(Bootstrap Sample Size n = 175,000)

Decile	Model Cum Lift	Bootstrap Cum Lift	95% Margin of Error	95% Lower Bound	95% Upper Bound
Top	141	140	7.4	133	147
2nd	127	127	5.1	122	132
3rd	119	120	4.2	115	124
4th	113	113	3.6	109	116
5th	111	111	2.8	108	114
6th	111	111	2.5	108	113
7th	107	108	2.4	105	110
8th	105	105	2.3	103	108
9th	103	103	2.0	101	105
Bottom	100	100	1.9	98	102

be applied during the model-building stages. A bootstrap decile analysis of model calibration, similar to bootstrap decile analysis of model validation, can be another technique for variable selection and other assessments of model quality.

19.11 Summary

Traditional validation of marketing models involves comparing the Cum Lifts from the calibration and holdout decile analyses based on the model

TABLE 19.11

Bootstrap Model Efficiency (Bootstrap Sample Size n = 175,000)

Decile	Eight-Variable Response Model A					Three-Variable Response Model B	
	Model Cum Lift	Bootstrap Cum Lift	95% Margin of Error	95% Lower Bound	95% Upper Bound	95% Margin of Error	Efficiency Ratio
Top	139	138	8.6	129	146	7.4	86.0%
2nd	128	128	5.3	123	133	5.1	96.2%
3rd	122	122	4.3	117	126	4.2	97.7%
4th	119	119	3.7	115	122	3.6	97.3%
5th	115	115	2.9	112	117	2.8	96.6%
6th	112	112	2.6	109	114	2.5	96.2%
7th	109	109	2.6	107	112	2.4	92.3%
8th	105	105	2.2	103	107	2.3	104.5%
9th	103	103	2.1	101	105	2.0	95.2%
Bottom	100	100	1.9	98	102	1.9	100.0%

under consideration. If the expected shrinkage (difference in decile Cum Lifts between the two analyses) and the Cum Lift values themselves are acceptable, then the model is considered successfully validated and ready to use; otherwise, the model is reworked until successfully validated. I illustrated with an RM case study that the single-sample validation provides neither assurance that the results are not biased nor any measure of confidence in the Cum Lifts.

I proposed the bootstrap—a computer-intensive approach to statistical inference—as a methodology for assessing the bias and confidence in the Cum Lift estimates. I introduced briefly the bootstrap, along with a simple 10-step procedure for bootstrapping any statistic. I illustrated the procedure for a decile validation of the RM in the case study. I compared and contrasted the single-sample and bootstrap decile validations for the case study. It was clear that the bootstrap provides necessary information for a complete validation: the biases and the margins of error of decile Cum Lift estimates.

I addressed the issue of margins of error (confidence levels) that are too large (low) for the business objective at hand. I demonstrated how to use the bootstrap to decrease the margins of error.

Then, I continued the discussion on the margin of error to a bootstrap assessment of model implementation performance. I addressed the issue of how large a sample is needed to implement a solicitation based on the model to obtain a desired performance quantity. Again, I provided a bootstrap procedure for determining the smallest sample necessary to obtain the desired quantity and quality of performance and illustrated the procedure with a three-variable RM.

Last, I showed how the bootstrap decile analysis can be used for an assessment of model efficiency. Continuing with the three-variable RM study, I illustrated the efficiency of the three-variable RM relative to the eight-variable alternative model. The latter model was shown to have more unstable predictions than the former model. The implication is that review is needed of a model with too many variables for justification of each variable's contribution in the model; otherwise, it can be anticipated that the model has unnecessarily large prediction error variance. The broader implication is that a bootstrap decile analysis of model calibration, similar to a bootstrap decile analysis of model validation, can be another technique for variable selection and other assessments of model quality.

References

1. Noreen, E.W., *Computer Intensive Methods for Testing Hypotheses*, Wiley, New York, 1989.
2. Efron, B., *The Jackknife, the Bootstrap and Other Resampling Plans*, SIAM, Philadelphia, 1982.
3. Draper, N.R., and Smith, H., *Applied Regression Analysis*, Wiley, New York, 1966.
4. Neter, J., and Wasserman, W., *Applied Linear Statistical Models*, Homewood, IL., Irwin, 1974.
5. Hayes, W.L., *Statistics for the Social Sciences*, Holt, Rinehart and Winston, New York, 1973.

20

Validating the Logistic Regression Model: Try Bootstrapping

20.1 Introduction

The purpose of this how-to chapter is to introduce the principal features of a bootstrap validation method for the ever-popular logistic regression model (LRM).

20.2 Logistc Regression Model

Lest the model builder forget, the LRM depends on the important assumption: The logit of Y, a binary dependent variable (typically assuming either character values "yes" and "no" or numeric values 1 and 0) is a *linear* function of the predictor variables defining LRM ($= b_0 + b_1X_1 + b_1X_2 + + b_nX_n$). Recall the transformation to convert the logit Y = 1 into the probability of Y = 1: 1 divided by exp(-LRM), where exp is the exponentiation function e^x, and e is the number 2.718281828 … .

The standard use of the Hosmer-Lemeshow (HL) test, confirming a fitted LRM is the correct model, is not reliable because it does not distinguish between nonlinearity and noise in checking the model fit [1]. Also, the HL test is sensitive to multiple datasets.

20.3 The Bootstrap Validation Method

The *bootstrap validation method* provides (1) a measure of how sensitive LRM predictions are to small changes (random perturbations) in the data and, consequently, (2) a procedure to select the best LRM, if it exists. Bootstrap samples are randomly different by way of sampling with replacement from the original (training) data. The model builder performs repetitions of logistic regression modeling based on various bootstrap samples.

1. If the bootstrapped LRMs are *stable* (in form and predictions), then the model builder picks one of the candidate (bootstrapped) LRMs as the final model.

2. If the bootstrapped LRMs are *not* stable, then the model builder can either (a) add more data to the original data or (b) take a new larger sample to serve as the training data. Thus, the model builder starts rebuilding a LRM on the appended training data or the new sample. These tasks prove quite effective in producing a stable bootstrapped LRM. If the tasks are not effective, then the data are not homogeneous and step 3 should be conducted.

3. One approach for rendering homogeneous data is discussed in Chapter 27, which addresses *overfitting*. Delivering homogeneous data is tantamount to eliminating the components of an overfitted model. The model builder may have to stretch his or her thinking from issues of overfitting to issues of homogeneity, but the effort will be fruitful.

Lest the model builder get the wrong impression about the proposed method, the bootstrap validation method is not an elixir. Due diligence dictates that a fresh dataset—perhaps one that includes ranges outside the original ranges of the predictor variables X_i—must be used on the final LRM. In sum, I confidently use the bootstrap validation method to yield reliable and robust results along with the final LRM tested on a fresh dataset.

20.4 Summary

This how-to chapter introduces a bootstrap validation method for the ever-popular logistic regression model. The method provides 1) a measure of how sensitive LRM predictions are to small changes (random perturbations) in the data, and consequently 2) a procedure to select the best LRM, if it exists

Reference

1. Hosmer, D.W., and Lemeshow, S., *Applied Logistic Regression*, Wiley, New York, 1989.

21

*Visualization of Marketing Models*Data Mining to Uncover Innards of a Model*

21.1 Introduction

Visual displays—commonly known as graphs—have been around for a long time but are currently at their peak of popularity. The popularity is due to the massive amounts of data flowing from the digital environment and the accompanying increase in data visualization software, which provides a better picture of big data than ever before. Visual displays are commonly used in the exploratory phase of data analysis and model building. An area of untapped potential for visual displays is "what the final model is doing" on implementation of its intended task of predicting. Such displays would increase the confidence in the model builder, while engendering confidence in the marketer, an end user of marketing models. The purpose of this chapter is to introduce two data mining graphical methods—star graphs and profile curves—for *uncovering the innards of a model:* a visualization of the characteristic and performance levels of the individuals predicted by the model.

21.2 Brief History of the Graph

The first visual display had its pictorial debut about 2000 BC when the Egyptians created a real estate map describing data such as property outlines and owner. The Greek Ptolemy created the first world map circa AD 150 using latitudinal and longitudinal lines as coordinates to represent the earth on a flat surface. In the fifteenth century, Descartes realized that Ptolemy's geographic map making could serve as a graphical method to identify the

---

* This chapter is based on the following: Ratner, B., Profile curves: a method of multivariate comparison of groups, *The DMA Research Council Journal*, 28–45 1999. Used with permission.

relationship between numbers and space, such as patterns [1]. Thus, the common graph was born: a horizontal line (X-axis) and a vertical line (Y-axis) intersecting perpendicularly to create a visual space that occupies numbers defined by an ordered pair of X-Y coordinates. The original Descartes graph, which has been embellished with more than 500 years of knowledge and technology, is the genesis of the discipline of *data visualization*, which is experiencing an unprecedented growth due to the advances in microprocessor technology and a plethora of visualization software.

The scientific community was slow to embrace the Descartes graph, first on a limited basis from the seventeenth century to the mid-eighteenth century, then with much more enthusiasm toward the end of the eighteenth century. [2,3] At the end of the eighteenth century, Playfield initiated work in the area of statistical graphics with the invention of the bar diagram (1786) and pie chart (1801).* Following Playfield's progress with graphical methods, Fourier presented the cumulative frequency polygon, and Quetelet created the frequency polygon and histogram. [4] In 1857, Florence Nightingale — who was a self-educated statistician — unknowingly reinvented the pie chart, which she used in her report to the royal commission to force the British army to maintain nursing and medical care to soldiers in the field. [5]

In 1977, Tukey started a revolution of numbers and space with his seminal book *Exploratory Data Analysis* (EDA) [6]. Tukey explained, by setting forth in careful and elaborate detail, the unappreciated value of numerical, counting, and graphical detective work performed by simple arithmetic and easy-to-draw pictures. Almost three decades after reinventing the concept of graph making as a means of encoding numbers for strategic viewing, Tukey's "graphic" offsprings are everywhere. They include the box-and-whiskers plot taught in early grade schools and easily generated computer-animated and interactive displays in three-dimensional space and with 64-bit color used as a staple in business presentations. Tukey, who has been called the "Picasso of statistics," has visibly left his imprint on the visual display methodology of today [6].

In the previous chapters, I have presented model-based graphical data mining methods, including smooth and fitted graphs and other Tukey-esque displays for identifying structure of data and fitting of models. Geometry-based graphical methods, which show how the dependent variable varies over the pattern of internal model variables (variables defining the model), are in an area that has not enjoyed growth and can benefit from a Tukey-esque innovation [7]. In this chapter, I introduce two data mining methods to show what the final model is doing, that is, to visualize the individuals identified by the model in terms of variables of interest—internal model variables or external model variables (variables not defining the model)—and the levels of performance of those individuals. I illustrate the methods using a response model, but they equally apply to a profit model.

---

* Michael Friendly (2008). "Milestones in the history of thematic cartography, statistical graphics, and data visualization". 13–14. Retrieved July 7, 2008.

21.3 Star Graph Basics

A table of numbers neatly shows "important facts contained in a jungle of figures" [6]. However, more often than not, the table leaves room for further untangling of the numbers. Graphs can help the data analyst out of the "numerical" jungle with a visual display of where the analyst has been and what he or she has seen. The data analyst can go only so far into the thicket of numbers until the eye-brain connection needs a *data mining* graph to extract patterns and gather insight from within the table.

A star graph* is a visual display of multivariate data, for example, a table of many rows of many variables [8]. It is especially effective for a small table, like the decile analysis table. The basics of star graph construction are as follows:

1. Identify the units of the star graphs: the *j observations* and the *k variables*. Consider a set of *j* observations. Each observation, which corresponds to a star graph, is defined by an array or row of *k* variables X.

2. There are *k* equidistant rays emitting from the center of the star for each observation.

3. The lengths of the rays correspond to the row of X values. The variables are to be measured assuming relatively similar scales. If not, the data must be transformed to induce comparable scales. A preferred method to accomplish comparable scales is *standardization*, which transforms all the variables to have the same mean value and the same standard deviation value. The mean value used is essentially arbitrary but must satisfy the constraint that the transformed standardized values are positive. The standard deviation value used is arbitrary but is typically given a value of 1. The standardized version of X, Z(X), is defined as follows: Z(X) = (X - mean(X))/standard deviation(X). If X assumes all negative values, there is a preliminary step of multiplying X by -1, producing -X. Then, the standardization is performed on -X.

4. The ends of the rays are connected to form a polygon or star for each observation.

5. A circle is circumscribed around each star. The circumference provides a reference *line*, which aids in interpreting the star. The centers of the star and circle are the same point. The radius of the circle is equal to the length of the largest ray.

6. A star graph typically does not contain labels indicating the X values. If transformations are required, then the transformed values are virtually *meaningless*.

---

* Star graphs are also known as star glyphs.

7. The assessment of the *relative differences in the shapes* of the star graphs untangles the numbers and brings out the true insights within the table.

SAS/Graph has a procedure to generate star graphs. I provide the SAS code at the end of the chapter for the illustrations presented.

21.3.1 Illustration

Marketers use models to identify potential customers. Specifically, a marketing model provides a way of identifying individuals into 10 equal-size groups (deciles), ranging from the top 10% most likely to perform (i.e., respond or contribute profit) to the bottom 10% least likely to perform. The traditional approach of profiling the individuals identified by the model is to calculate the means of variables of interest and assess their values across the deciles.

Consider a marketing model for predicting response to a solicitation. The marketer is interested in what the final model is doing, that is, what the individuals look like in the top decile, in the second decile, and so on. How do the individuals differ across the varying levels of performance (deciles) in terms of the usual demographic variables of interest: AGE, INCOME, EDUCATION, and GENDER? Answers to these and related questions provide the marketer with strategic marketing intelligence to put together an effective targeted campaign.

The means of the four demographic variables across the RESPONSE model deciles are in Table 21.1. From the tabular display of means by deciles, I conclude the following:

1. AGE: Older individuals are more responsive than younger individuals.
2. INCOME: High-income individuals are more responsive than low-income individuals.
3. EDUCATION: Individuals who attain greater education are more responsive than individuals who attain less education.
4. GENDER: Females are more responsive than males. Note: A mean GENDER of 0 and 1 implies all females and all males, respectively.

This interpretation is correct, albeit not thorough because it only considers one variable at a time, a topic covered further in the chapter. It does describe the individuals identified by the model in terms of the four variables of interest and responsiveness. However, it does not beneficially stimulate the marketer into strategic thinking for insights. A graph—*an imprint of many insights*—is needed.

TABLE 21.1

Response Decile Analysis: Demographic Means

Decile	AGE (years)	INCOME ($000)	EDUCATION (Years of Schooling)	GENDER (1 = male, 0 = female)
Top	63	155	18	0.05
2	51	120	16	0.10
3	49	110	14	0.20
4	46	111	13	0.25
5	42	105	13	0.40
6	41	95	12	0.55
7	39	88	12	0.70
8	37	91	12	0.80
9	25	70	12	1.00
Bottom	25	55	12	1.00

21.4 Star Graphs for Single Variables

The first step in constructing a star graph is to identify the units, which serve as the observations and the variables. For decile-based single-variable star graphs, the variables of interest are the j observations, and the 10 deciles (top, 2, 3, … , bot) are the k variables. For the RESPONSE model illustration, there are four star graphs in Figure 21.1, one for each demographic variable. Each star has 10 rays, which correspond to the 10 deciles.

I interpret the star graphs as follows:

1. For AGE, INCOME, and EDUCATION: There is a decreasing trend in the mean values of the variables as the individuals are successively assigned to the top decile down through the bottom decile. These star graphs display the following: Older individuals are more responsive than younger individuals. High-income earners are more responsive than low-income earners. And, individuals who attain greater education are more responsive than individuals who attain less education.

2. AGE and INCOME star graphs are virtually identical, except for the ninth decile, which has a slightly protruding vertex. The implication is that AGE and INCOME have a similar effect on RESPONSE. Specifically, a standardized unit increase in AGE and in INCOME produce a similar change in RESPONSE.

3. For GENDER: There is an increasing trend in the incidence of males, as the individuals are successively assigned to the top decile down through the bottom decile. Keep in mind GENDER is coded zero for females.

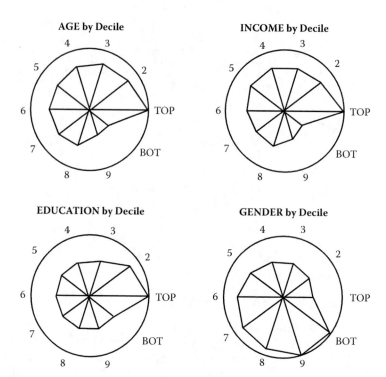

FIGURE 21.1
Star graphs for AGE, INCOME, EDUCATION, and GENDER.

In sum, the star graphs provide a unique visual display of the conclusions in Section 21.3.1. However, they are one-dimensional portraits of the effect of each variable on RESPONSE. To obtain a deeper understanding of how the model works as it assigns individuals to the decile analysis table, a full profile of the individuals using all variables considered jointly is needed. In other words, data mining to uncover the innards of the decile analysis table is the special need required. A multiple-variable star graph provides an unexampled full profile.

21.5 Star Graphs for Many Variables Considered Jointly

As with the single-variable star graph, the first step in constructing the many-variable star graphs is to identify the units. For decile-based many-variable star graphs, the 10 deciles are the j observations, and the variables of interest are the k variables. For the RESPONSE model illustration, there are 10 star graphs, one for each decile, in Figure 21.2. Each decile star has four rays, which correspond to the four demographic variables.

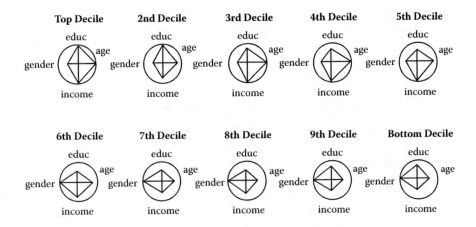

FIGURE 21.2
Star graphs for four demographic variables considered jointly.

I interpret a star graph for an array of variables in a *comparative* context. Because star graphs have no numerical labels, I assess the *shapes* of the stars by observing their *movement* within the reference circle as I go from the top to bottom deciles.

1. *Top decile star:* The rays of AGE, INCOME, and EDUCATION touch or nearly touch the circle circumference. These long rays indicate older, more educated individuals with higher income. The short ray of GENDER indicates these individuals are mostly females. The individuals in the top decile comprise the reference group for a comparative analysis of the other decile stars.

2. *Second-decile star:* Individuals in this decile are slightly younger, with less income than individuals in the top decile.

3. *Third- to fifth-decile stars:* Individuals in these deciles are less educated substantially than individuals in the top two deciles. In addition, the education level decreases as the individuals are assigned to deciles 3 through 5.

4. *Sixth-decile star:* The shape of this star makes a significant departure from the top five decile stars. This star indicates individuals who are mostly males (because the GENDER ray touches the circle), younger, less-educated individuals with less income than the individuals in the upper deciles.

5. *Seventh- to bottom decile stars:* These stars hardly move within the circle across the lower deciles (decile 6 to bottom). This indicates the individuals across the least-responsive deciles are essentially the same.

In sum, the 10 many-variable star graphs provide a lifelike animation of how the four-dimensional profiles of the individuals change as the model assigns the individuals into the 10 deciles. The first five deciles show a slight progressive decrease in EDUCATION and AGE means. Between the fifth and sixth deciles, there is a sudden change in full profile as the GENDER contribution to profile is now skewed to males. Across the bottom five deciles, there is a slight progressive decrease in INCOME and AGE means.

21.6 Profile Curves Method

I present the profile curves method as an alternative geometry-based data mining graphical method for the problem previously tackled by the star graphs, but from a slightly different concern. The star graphs provide the marketer with strategic marketing intelligence for campaign development, as obtained from an ocular inspection and complete descriptive account of their customers in terms of variables of interest. In contrast, the profile curves provide the marketer with strategic marketing intelligence for model implementation, specifically determining the number of reliable decile groups.

The profile curves are demanding to build and interpret conceptually, unlike the star graphs. Their construction is not intuitive, as they use a series of unanticipated trigonometric functions, and their display is disturbingly abstract. However, the value of profile curves in a well-matched problem solution (as presented here) can offset the initial reaction and difficulty in their use. From the discussion, the profile curves method is clearly a unique data mining method, uncovering unsuspected patterns of what the final model is doing on implementation of its intended task of predicting.

I discuss the basics of profile curves and the profile analysis, which serves as a useful preliminary step to the implementation of the profile curves method. I illustrate in the next sections profile curves and profile analysis using the RESPONSE model decile analysis table. The profile analysis involves simple pairwise scatterplots. The profile curves method requires a special computer program, which SAS/Graph has. I provide the SAS code for the profile curves for the illustration presented in the Appendix at the end of the chapter.

21.6.1 Profile Curves* Basics

Consider the curve function $f(t)$ defined in Equation (21.1) [9]:

$$f(t) = X_1/\sqrt{2} + X_2 \sin(t) + X_3 \cos(t) + X_4 \sin(2t) + X_5 \cos(2t) + \dots \quad (21.1)$$

---

* Profile curves are also known as curve plots.

where $-\pi \leq t \leq \pi$

The curve function $f(t)$ is a weighted sum of basic curves for an observation X, which is represented by many variables, that is, a multivariate data array, $X = \{X_1, X_2, X_3, \ldots, X_k\}$. The weights are the values of the X's. The basic curves are trigonometric functions sine and cosine. The plot of $f(t)$ on the Y-axis and t on the X axis for a set of multivariate data arrays (rows) of mean values for groups of individuals are called *profile curves*.

Like the star graphs, the profile curves are a visual display of multivariate data, especially effective for a small table, like the decile analysis table. Unlike the star graphs, which provide visual displays for single and many variables jointly, the profile curves only provide a visual display of the joint effects of the X variables across several groups. A profile curve for a single group is an abstract mathematical representation of the row of mean values of the variables. As such, a single group curve imparts no usable information. Information is extracted from a comparative evaluation of two or more group curves. Profile curves permit a qualitative assessment of the differences among the persons across the groups. In other words, profile curves serve as a method of multivariate comparison of groups.

21.6.2 Profile Analysis

Database marketers use models to classify customers into 10 deciles, ranging from the top 10% most likely to perform to the bottom 10% least likely to perform. To communicate effectively to the customers, database marketers combine the deciles into groups, typically three: top, middle, and bottom *decile groups*. The top group typifies high-valued/high-responsive customers to "harvest" by sending provoking solicitations for preserving their performance levels. The middle group represents medium-valued/medium-responsive customers to "retain and grow" by including them in marketing programs tailored to keep and further stimulate their performance levels. Last, the bottom group depicts a segment of minimal performers and former customers, whose performance levels can be rekindled and reactivated by creative new product and discounted offerings.

Profile analysis is used to create decile groups. Profile analysis consists of (1) calculating the means of the variables of interest and (2) plotting the means of several pairs of the variables. These profile plots suggest how the individual deciles may be combined. However, because the profiles are multidimensional (i.e., defined by many variables), the assessment of the many profile plots provides an incomplete view of the groups. If the profile analysis is fruitful, it serves as guidance for the profile curves method in determining the number of reliable decile groups.

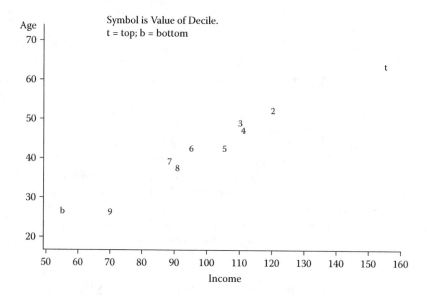

FIGURE 21.3
Plot of AGE and INCOME.

21.7 Illustration

Returning to the RESPONSE model decile analysis table (Table 21.1), I construct three profile plots in Figures 21.3 to 21.5: AGE with INCOME and with GENDER and EDUCATION with INCOME. The AGE-INCOME plot indicates that AGE and INCOME jointly decrease down through the deciles. That is, the most responsive customers are older with more income, and the least-responsive customers are younger with less income.

The AGE-GENDER plot shows that the most responsive customers are older and prominently female, and the least-responsive customers are younger and typically male. The EDUCATION-INCOME plot shows that the most responsive customers are better educated with higher incomes.

Similar plots for the other pairs of variables can be constructed, but the task of interpreting all the pairwise plots is formidable.*

The three profile plots indicate various candidate decile group compositions:

1. The AGE-INCOME plot suggests defining the groups as follows:
   a. Top group: top decile
   b. Middle group: second through eighth deciles
   c. Bottom group: ninth and bottom deciles

---

* Plotting three variables at a time can be done by plotting a pair of variables for each decile value of the third variable, clearly a challenging effort.

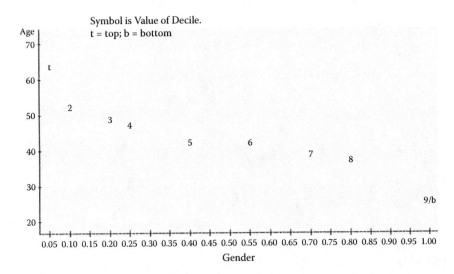

FIGURE 21.4
Plot of AGE and GENDER by decile.

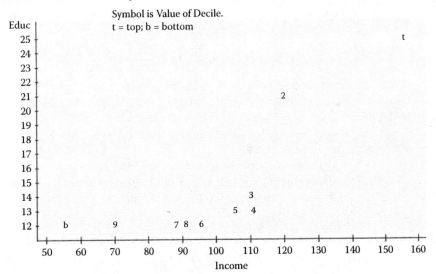

FIGURE 21.5
Plot of EDUCATION and INCOME by Decile.

2. The AGE-GENDER plot does not reveal any grouping.

3. The EDUCATION-INCOME plot indicates the following:

   a. Top group: top decile

   b. Middle group: second decile

   c. Bottom group: third through bottom deciles. Actually, this group can be divided into two subgroups: deciles 3 to 8 and deciles 9 and bottom.

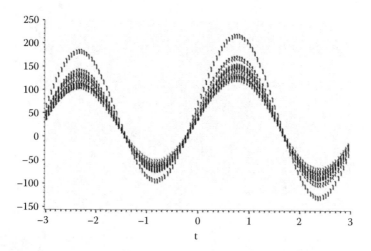

FIGURE 21.6
Profile curves: all deciles.

In this case, the profile analysis is not fruitful. It is not clear how to define the best decile grouping based on these findings. Additional plots would probably produce other inconsistent findings.

21.7.1 Profile Curves for RESPONSE Model

The profile curves method provides a graphical presentation (Figure 21.6) of the joint effects of the four demographic variables across all the deciles based on the decile analysis in Table 21.1. Interpretation of this graph is part of the following discussion, in which I illustrate the strategy for creating reliable decile groups with profile curves.

Under the working assumption that the top, middle, and bottom decile groups exist, I create profile curves for the top, fifth, and bottom deciles (Figure 21.7) as defined in Equations (21.2), (21.3), and (21.4), respectively.

$$f(t)_{top\_decile} = 63/\sqrt{2} + 155\sin(t) + 18\cos(t) + .05\sin(2t) \qquad (21.2)$$

$$f(t)_{5th\_decile} = 42/\sqrt{2} + 105\sin(t) + 13\cos(t) + .40\sin(2t) \qquad (21.3)$$

$$f(t)_{bottom\_decile} = 25/\sqrt{2} + 55\sin(t) + 12\cos(t) + 1.00\sin(2t) \qquad (21.4)$$

The upper, middle, and lower profile curves correspond to the rows of means for the top, fifth, and bottom deciles, respectively, which are reproduced for ease of discussion in Table 21.2. The three profile curves form two "hills." Based on a subjective assessment,* I declare that the profile curves

---

* I can test for statistical difference (see Andrew's article [9]); however, I am doing a visual assessment, which does not require any statistical rigor.

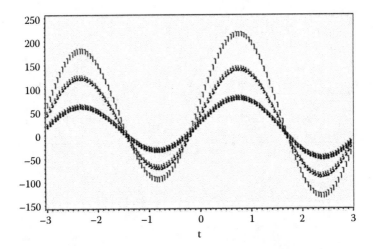

FIGURE 21.7
Profile curves for top, fifth, and bottom deciles.

are different in terms of slope of the hill. The implication is that the individuals in each decile are different—with respect to the four demographic variables considered jointly—from the individuals in each of the other two deciles; consequently, the deciles cannot be combined.

The graph of the three profile curves in Figure 21.7 exemplifies how the profile curves method works. Large variation among rows corresponds to a set of profile curves that greatly departs from a common shape. Disparate profile curves indicate that the rows should remain separate, not combined. Profile curves that slightly depart from a common shape indicate that the rows can be combined to form a more reliable row. Accordingly, I restate my preemptive implication: The graph of three profile curves indicates that individuals across the three deciles are diverse with respect to the four demographic variables considered jointly, and the deciles cannot be aggregated to form a homogeneous group.

When the rows are obviously different, as in the case in Table 21.2, the profile curves method serves as a confirmatory method. When the variation in the rows is not apparent, as is likely with a large number of variables, noticeable row variation is harder to discern, in which case the profile curves method serves as an exploratory tool.

21.7.2 Decile Group Profile Curves

I have an initial set of decile groups: The top group is the top decile; the middle group is the fifth decile; and the bottom group is the bottom decile. I have to assign the remaining deciles to one of the three groups. Can I include the second decile in the top group? The answer lies in the graph for the top and second deciles in Figure 21.8, from which I observe that the top and

TABLE 21.2

Response Decile Analysis: Demographic Means for Top, Fifth, and
Bottom Deciles

Decile	AGE (Years)	INCOME ($000)	EDUCATION (Years of Schooling)	GENDER (1 = male, 0 = female)
Top	63	155	18	0.05
5	42	105	13	0.40
Bottom	25	55	12	1.00

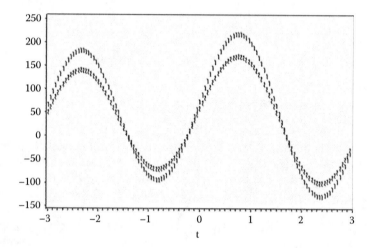

FIGURE 21.8
Profile curves for top and second deciles.

second decile profile curves are different. Thus, the top group remains with only the top decile. The second decile is assigned to the middle group; this assignment will be discussed further.

Can I add the ninth decile to the bottom group? From the graph for the ninth and bottom deciles in Figure 21.9, I observe that the two profile curves are not different. Thus, I add the ninth decile to the bottom group. Can the eighth decile also be added to the bottom group (now consisting of the ninth and bottom deciles)? I observe in the graph for the eighth to bottom deciles in Figure 21.10 that the eighth decile profile curve is somewhat different from the ninth and bottom decile profile curves. Thus, I do not include the eighth decile in the bottom group; it is placed in the middle group.

To ensure that the middle group can be defined by combining deciles 2 through 8, I generate the corresponding graph in Figure 21.11. I observe a bold common curve formed by the seven profile curves tightly stacked

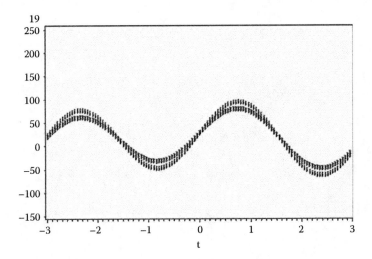

FIGURE 21.9
Profile curves for ninth to bottom deciles.

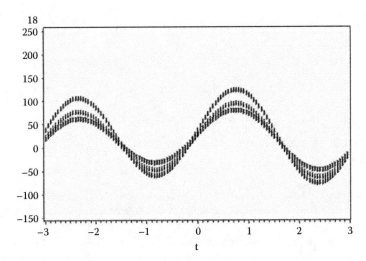

FIGURE 21.10
Profile curves for eighth to bottom deciles.

together. This suggests that the individuals across the deciles are similar. I conclude that the group comprised of deciles 2 through 8 is homogeneous.

Two points are worth noting: First, the density of the common curve of the middle group is a measure of homogeneity among the individuals across these middle group deciles. Except for a little "daylight" at the top of the

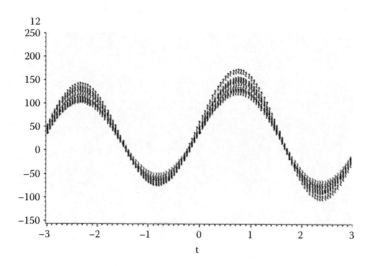

FIGURE 21.11
Profile curves second through eighth deciles.

right-side hill, the curve is solid. This indicates that nearly all the individuals within the deciles are alike. Second, if the curve had a *pattern* of daylight, then the middle group would be divided into subgroups defined by the pattern.

In sum, I define the final decile groups as follows:

1. Top group: top decile
2. Middle group: second through eighth deciles
3. Bottom group: ninth and bottom deciles

The sequential use of profile curves (as demonstrated here) serves as a general technique for multivariate profile analysis. In some situations when the profile data appear obvious, a single graph with all 10 decile curves may suffice. For the present illustration, the single graph in Figure 21.6 shows patterns of daylight that suggest the final decile groups. A closer look at the two hills confirms the final decile groups.

21.8 Summary

After taking a quick look at the history of the graph, from its debut as an Egyptian map to the visual displays of today in three dimensions with 64-bit color, I focused on two underused methods of displaying multivariate data.

I proposed the star graphs and profile curves as data mining methods of visualizing what the final model is doing on implementation of its intended task of predicting.

I presented the basics of star graph construction along with an illustration of a response model with four demographic variables. The utility of the star graph was made apparent when I compared its results with those of traditional profiling, such as examining the means of variables of interest across the deciles. Traditional profiling provides good information, but it is neither compelling nor complete in its examination of how the response model works. It is incomplete in two respects: (1) It only considers one variable at a time without any display, and (2) it does not beneficially stimulate the marketer into strategic thinking for insights. The star graph with its unique visual display does stimulate strategic thinking. However, like traditional profiling, the star graph only considers one variable at a time.

Accordingly, I extended the single-variable star graph application to the many-variable star graph application. Continuing with the response illustration, I generated many-variable star graphs, which clearly provided the "beneficial stimulation" desired. The many-variable star graphs displayed a full profile of the individuals using all four demographic variables considered jointly for a complete understanding of how the response model works as it assigns individuals to the response deciles.

Last, I presented the unique data mining profile curves method as an alternative method for the problem previously tackled by the star graphs, but from a slightly different concern. The star graph provides the marketer with strategic intelligence for campaign development. In contrast, the profile curves provide the marketer with strategic information for model implementation, specifically determining the number of reliable decile groups.

A maiden encounter with profile curves can be unremittingly severe on the data miner; the curves are demanding not only in their construction but also in their interpretation. I demonstrated with the response model illustration that the long-term utility of profile curves offsets the initial affront to the analytical senses. The response model profile curves, which accompanied a detailed analysis and interpretation, clearly showed the determination of the reliable decile groups.

References

1. Descartes, R., *The Geometry of Rene Descartes*, Dover, New York, 1954.
2. Costigan-Eaves, P., Data graphics in the 20th century: a comparative and analytical survey, PhD thesis, Rutgers University, Rutgers, NJ, 1984.
3. Funkhouser, H.G., Historical development of the graphical representation of statistical data, *Osiris*, 3, 269–404, 1937.

4. Du Toit, S.H.C., Steyn, A.G.W., and Stumpf, R.H., *Graphical Exploratory Data Analysis*, Springer-Verlag, New York, 1986, 2.
5. Salsburg, D., *The Lady Tasting Tea*, Freeman, New York, 2001.
6. Tukey, J.W., *The Exploratory Data Analysis*, Addison-Wesley, Reading, MA, 5, 1977.
7. Snee, R.D., Hare, L.B., and Trout, J. R., *Experiments in Industry. Design, Analysis and Interpretation of Results,* American Society for Quality, Milwaukee, WI, 1985.
8. Friedman, H.P., Farrell, E.S., Goldwyn, R.M., Miller, M., and Siegel, J., A graphic way of describing changing multivariate patterns, In Proceedings of the Sixth Interface Symposium on Computer Science and Statistics, University of California Press, Berkeley, 1972.
9. Andrews, D.F., Plots of high-dimensional data, *Biometrics*, 28, 1972.

Appendix 1 SAS Code for Star Graphs for Each Demographic Variable about the Deciles

```
Title1 'table';
data table;
input decile age income educ gender;
cards;
1   63 155 18 0.05
2   51 120 16 0.10
3   49 110 14 0.20
4   46 111 13 0.25
5   42 105 13 0.40
6   41 095 12 0.55
7   39 088 12 0.70
8   37 091 12 0.80
9   25 070 12 1.00
10 25 055 12 1.00
;
run;
proc print;
run;
proc standard data = table out = tablez
mean = 4 std = 1;
var
```

```
age income educ gender;
Title1 'table stdz';
proc print data = tablez;
run;
proc format; value dec_fmt
1.   = 'top' 2 = ' 2 ' 3 = ' 3 ' 4 = ' 4 ' 5 = ' 5 '
6.   = ' 6 ' 7 = ' 7 ' 8 = ' 8 ' 9 = ' 9 ' 10 = 'bot';
run;
proc greplay nofs igout = work.gseg;
delete all;
run;quit;
goptions reset = all htext = 1.05 device = win
targetdevice = winprtg ftext = swissb lfactor = 3
hsize = 2 vsize = 8;
proc greplay nofs igout = work.gseg;
delete all;
run;
goptions reset = all device = win
targetdevice = winprtg ftext = swissb lfactor = 3;
title1 'AGE by Decile';
proc gchart data = tablez;
format decile dec_fmt. ;
star decile/fill = empty discrete sumvar = age
slice = outside value = none noheading ;
run;quit;
Title1 'EDUCATON by Decile';
proc gchart data = tablez;
format decile dec_fmt. ;
star decile/fill = empty discrete sumvar = educ
slice = outside value = none noheading;
run;quit;
Title1 'INCOME by Decile';
proc gchart data = tablez;
format decile dec_fmt. ;
star decile/fill = empty discrete sumvar = income
slice = outside value = none noheading;
```

```
run;quit;
Title1 'GENDER by Decile';
proc gchart data = tablez;
format decile dec_fmt.;
star decile/fill = empty discrete sumvar = gender
slice = outside value = none noheading;
run;quit;
proc greplay nofs igout = work.gseg tc = sashelp.templt template = l2r2s;
treplay 1:1 2:2 3:3 4:4;
run;quit;
```

Appendix 2 SAS Code for Star Graphs for Each Decile about the Demographic Variables

```
data table;
input decile age income educ gender;
cards;
1  63 155 18 0.05
2  51 120 16 0.10
3  49 110 14 0.20
4  46 111 13 0.25
5  42 105 13 0.40
6  41 095 12 0.55
7  39 088 12 0.70
8  37 091 12 0.80
9  25 070 12 1.00
10 25 055 12 1.00
;
run;
proc standard data = table out = tablez
mean = 4 std = 1;
var
age income educ gender;
Title2 'table stdz';
proc print data = tablez;
```

```
run;
proc transpose data = tablez out = tablezt prefix = dec_;
var
age income educ gender;
run;
proc print data = tablezt;
run;
proc standard data = tablezt out = tableztz
mean = 4 std = 1;
Var
dec_1 - dec_10;
title2'tablezt stdz';
proc print data = tableztz;
run;
proc transpose data = tablez out = tablezt prefix = dec_;
var
age income educ gender;
run;
proc print data = tablezt;
run;
proc greplay nofs igout = work.gseg;
delete all;
run;quit;
goptions reset = all htext = 1.05 device = win
target = winprtg ftext = swissb lfactor = 3
hsize = 4 vsize = 8;
Title1 'top decile';
proc gchart data = tableztz;
star name/fill = empty sumvar = dec_1
slice = outside value = none noheading;
run;quit;
Title1 '2nd decile';
proc gchart data = tableztz;
star name/fill = empty sumvar = dec_2
slice = outside value = none noheading;
run;quit;
```

```
Title1 '3rd decile';
proc gchart data = tableztz;
star name/fill = empty sumvar = dec_3
slice = outside value = none noheading;
run;quit;
Title1 '4th decile';
proc gchart data = tableztz;
star name/fill = empty sumvar = dec_4
slice = outside value = none noheading;
run;quit;
Title1 '5th decile';
proc gchart data = tableztz;
star name/fill = empty sumvar = dec_5
slice = outside value = none noheading;
run;quit;
Title1 '6th decile';
proc gchart data = tableztz;
star name/fill = empty sumvar = dec_6
slice = outside value = none noheading;
run;quit;
Title1 '7th decile';
proc gchart data = tableztz;
star name/fill = empty sumvar = dec_7
slice = outside value = none noheading;
run;quit;
Title1 '8th decile';
proc gchart data = tableztz;
star name/fill = empty sumvar = dec_8
slice = outside value = none noheading;
run;quit;
Title1 '9th decile';
proc gchart data = tableztz;
star name/fill = empty sumvar = dec_9
slice = outside value = none noheading;
run;quit;
Title1 'bottom decile';
```

```
proc gchart data = tableztz;
star name/fill = empty sumvar = dec_10
slice = outside value = none noheading;
run;quit;
goptions hsize = 0 vsize = 0;
Proc Greplay Nofs TC = Sasuser.Templt;
Tdef L2R5 Des = 'Ten graphs: five across, two down'
1/llx = 0 lly = 51
ulx = 0 uly = 100
urx = 19 ury = 100
lrx = 19 lry = 51
2/llx = 20 lly = 51
ulx = 20 uly = 100
urx = 39 ury = 100
lrx = 39 lry = 51
3/llx = 40 lly = 51
ulx = 40 uly = 100
urx = 59 ury = 100
lrx = 59 lry = 51
4/llx = 60 lly = 51
ulx = 60 uly = 100
urx = 79 ury = 100
lrx = 79 lry = 51
5/llx = 80 lly = 51
ulx = 80 uly = 100
urx = 100 ury = 100
lrx = 100 lry = 51
6/llx = 0 lly = 0
ulx = 0 uly = 50
urx = 19 ury = 50
lrx = 19 lry = 0
7/llx = 20 lly = 0
ulx = 20 uly = 50
urx = 39 ury = 50
lrx = 39 lry = 0
8/llx = 40 lly = 0
```

```
ulx = 40 uly = 50
urx = 59 ury = 50
lrx = 59 lry = 0
9/llx = 60 lly = 0
ulx = 60 uly = 50
urx = 79 ury = 50
lrx = 79 lry = 0
10/llx = 80 lly = 0
ulx = 80 uly = 50
urx = 100 ury = 50
lrx = 100 lry = 0;
Run;
Quit;
Proc Greplay Nofs Igout = Work.Gseg
TC = Sasuser.Templt Template = L2R5;
Treplay 1:1 2:2 3:3 4:4 5:5 6:6 7:7 8:8 9:9 10:10;
run;quit;
```

Appendix 3 SAS Code for Profile Curves: All Deciles

```
Title1'table';
data table;
input decile age income educ gender;
cards;
1 63 155 18 0.05
2 51 120 16 0.10
3 49 110 14 0.20
4 46 111 13 0.25
5 42 105 13 0.40
6 41 095 12 0.55
7 39 088 12 0.70
8 37 091 12 0.80
9 25 070 12 1.00
10 25 055 12 1.00
;
```

```
run;
data table;
Set table;
x1 = age;x2 = income;x3 = educ;x4 = gender;
proc print;
run;
data table10;
sqrt2 = sqrt(2);
array f {10};
do t = -3.14 to 3.14 by.05;
do i = 1 to 10;
set table point = i;
f(i) = x1/sqrt2 + x4*sin(t) + x3*cos(t) + x2*sin(2*t);
end;
output;
label f1 = '00'x;
end;
stop;
run;
goptions reset = all device = win target = winprtg ftext = swissb lfactor
    = 3;
Title1 'Figure 21.6 Profile Curves: All Deciles';
proc gplot data = table10; plot
f1*t = 'T'
f2*t = '2'
f3*t = '3'
f4*t = '4'
f5*t = '5'
f6*t = '6'
f7*t = '7'
f8*t = '8'
f9*t = '9'
f10*t = 'B'
/overlay haxis = -3 -2 -1 0 1 2 3
nolegend vaxis = -150 to 250 by 50;
run;quit;
```

22

The Predictive Contribution Coefficient: A Measure of Predictive Importance

22.1 Introduction

Determining the most important predictor variables in a regression model is a vital element in the interpretation of the model. A general rule is to view the predictor variable with the largest standardized coefficient (SRC) as the most important variable, the predictor variable with the next-largest SRC as the next important variable, and so on. This rule is intuitive, is easy to apply, and provides practical information for understanding how the model works. Unknown to many, however, is that the rule is theoretically problematic. The purpose of this chapter is twofold: first, to discuss why the decision rule is theoretically amiss yet works well in practice; second, to present an alternative measure—the *predictive contribution coefficient* (PCC)—that offers greater utile information than the standardized coefficient as it is an assumption-free measure founded in the data mining paradigm.

22.2 Background

Let Y be a continuous dependent variable, and X_1, X_2, ... , X_i, ... , X_n be the predictor variables. The linear regression model is defined in Equation (22.1):

$$Y = b_0 + b_1{}^*X_1 + b_2{}^*X_2 + ... + b_i{}^*X_i + ... + b_n{}^*X_n \qquad (22.1)$$

The b's are the raw regression coefficients, which are estimated by the method of ordinary least squares. Once the coefficients are estimated, an individual's predicted Y value is calculated by "plugging in" the values of the predictor variables for that individual in the equation.

The interpretation of the raw regression coefficient points to the following question: How does X_i affect the prediction of Y? The answer is the predicted Y experiences an average change of b_i with a unit increase in X_i when the other X's are held constant. A common misinterpretation of the raw regression coefficient is that the predictor variable with the largest value (the sign of the coefficient is ignored) has the greatest effect on the predicted Y. Unless the predictor variables are measured in the same units, the raw regression coefficient values can be so unequal that they cannot be compared with each other. The raw regression coefficients must be standardized to import the different units of measurement of the predictor variables to allow a fair comparison. This discussion is slightly modified for the situation of a binary dependent variable: The linear regression model is the logistic regression model, and the method of maximum likelihood is used to estimate the regression coefficients, which are linear in the logit of Y.

The standardized regression coefficients (SRCs) are just plain numbers, like "points," allowing material comparisons among the predictor variables. The SRC can be determined by multiplying the raw regression coefficient by a conversion factor, a ratio of a unit measure of X_i variation to a unit measure of Y variation. The SRC in an ordinary regression model is defined in Equation (22.2), where $StdDevX_i$ and StdDevY are the standard deviations for X_i and Y, respectively.

$$\text{SRC for } X_i = (StdDevX_i/StdDevY)*\text{Raw reg., coefficient for } X_i \quad (22.2)$$

For the logistic regression model, the problem of calculating the standard deviation of the dependent variable, which is logit Y, not Y, is complicated. The problem has received attention in the literature, with solutions that provide inconsistent results [1]. The simplest is the one used in the SAS system, although it is not without its problems. The StdDev for the logit Y is taken as 1.8138 (the value of the standard deviation of the standard logistic distribution). Thus, the SRC in the logistic regression model is defined in Equation (22.3).

$$\text{SRC for } X_i = (StdDevX_i/1.8138)*\text{Raw reg., coefficient for } X_i \quad (22.3)$$

The SRC can also be obtained directly by performing the regression analysis on standardized data. Recall that standardizing the dependent variable Y and the predictor variables X_is creates new variables zY and zX_i, such that their means and standard deviations are equal to zero and one, respectively. The coefficients obtained by regressing zY on the zX_is are, by definition, the SRCs.

The question of which variables in the regression model are its most important predictors, in ranked order, needs to be annotated before being answered. The importance or ranking is traditionally taken in terms of the statistical characteristic of reduction in prediction error. The usual answer is provided by the decision rule: The variable with the largest SRC is the most important variable; the variable with the next-largest SRC is the next important variable, and so on. This decision rule is correct with the *unnoted caveat that the predictor variables are uncorrelated*. There is a rank order correspondence between the SRC (the sign of the coefficient is ignored) and the reduction in prediction error in a regression model with only uncorrelated predictor variables. There is one other unnoted caveat for the proper use of the decision rule: The SRC for a dummy predictor variable (defined by only two values) is not reliable as the standard deviation of a dummy variable is not meaningful.

Regression models in virtually all applications have correlated predictor variables, which challenge the utility of the decision rule. Yet, the rule provides continuously useful information for understanding how the model works without raising sophistic findings. The reason for its utility is there is an unknown working assumption at play: The reliability of the ranking based on the SRC increases as the average correlation among the predictor variables decreases (see Chapter 13). Thus, for well-built models, which necessarily have a minimal correlation among the predictor variables, the decision rule remains viable in virtually all regression applications. However, there are caveats: Dummy variables cannot be reliably ranked, and composite variables (derived from other predictor variables) and the elemental variables (defining the composite variables) are highly correlated inherently and thus cannot be ranked reliably.

22.3 Illustration of Decision Rule

Consider RESPONSE (0 = no; 1 = yes), PROFIT (in dollars), AGE (in years), GENDER (1 = female, 0 = male) and INCOME (in thousand dollars) for 10 individuals in the small data in Table 22.1. I standardized the data to produce the standardized variables with the notation used previously: zRESPONSE, zPROFIT, zAGE, zGENDER, and zINCOME (data not shown).

I perform two ordinary regression analyses based on both the raw data and the standardized data. Specifically, I regress PROFIT on INCOME, AGE, and

TABLE 22.1

Small Data

ID#	Response	Profit	Age	Gender	Income
1	1	185	65	0	165
2	1	174	56	0	167
3	1	154	57	0	115
4	0	155	48	0	115
5	0	150	49	0	110
6	0	119	40	0	99
7	0	117	41	1	96
8	0	112	32	1	105
9	0	107	33	1	100
10	0	110	37	1	95

TABLE 22.2

PROFIT Regression Output Based on Raw Small Data

Variable	DF	Parameter Estimate	Standard Error	t Value	Pr > t	Standardized Estimate
INTERCEPT	1	37.4870	16.4616	2.28	0.0630	0.0000
INCOME	1	0.3743	0.1357	2.76	0.0329	0.3516
AGE	1	1.3444	0.4376	3.07	0.0219	0.5181
GENDER	1	−11.1060	6.7221	−1.65	0.1496	−0.1998

TABLE 22.3

PROFIT Regression Output Based on Standardized Small Data

Variable	DF	Parameter Estimate	Standard Error	t Value	Pr 1+1	Standardized Estimate
INTERCEPT	1	−4.76E−16	0.0704	0.0000	1.0000	0.0000
zINCOME	1	0.3516	0.1275	2.7600	0.0329	0.3516
zAGE	1	0.5181	0.1687	3.0700	0.0219	0.5181
zGENDER	1	−0.1998	0.1209	−1.6500	0.1496	−0.1998

GENDER and regress zPROFIT on zINCOME, zAGE, and zGENDER. The raw regression coefficients and SRCs based on the raw data are in the "Parameter Estimate" and "Standardized Estimate" columns in Table 22.2, respectively. The raw regression coefficients for INCOME, AGE, and GENDER are 0.3743, 1.3444, and -11.1060, respectively. The SRCs for INCOME, AGE, and GENDER are 0.3516, 0.5181, and -0.1998, respectively.

The raw regression coefficients and SRCs based on the standardized data are in the "Parameter Estimate" and "Standardized Estimate" columns in Table 22.3, respectively. As expected, the raw regression coefficients—which are now the SRCs—are equal to the values in the "Standardized Estimate" column; for zINCOME, zAGE, and zGENDER, the values are 0.3516, 0.5181, and -0.1998, respectively.

I calculate the average correlation, among the three predictor variables, which is a gross 0.71. Thus, the SRC provides a questionable ranking of the predictor variables for PROFIT. AGE is the most important predictor variable, followed by INCOME; the ranked position of GENDER is undetermined.

I perform two logistic regression analyses based on both the raw data and the standardized data. Specifically, I regress RESPONSE on INCOME, AGE, and GENDER and regress RESPONSE* on zINCOME, zAGE, and zGEN-DER. The raw and standardized logistic regression coefficients based on the raw data are in the "Parameter Estimate" and "Standardized Estimate" columns in Table 22.4, respectively. The raw logistic regression coefficients for INCOME, AGE, and GENDER are 0.0680, 1.7336, and 14.3294, respectively.

---

* I choose not to use zRESPONSE. Why?

TABLE 22.4

RESPONSE Regression Output Based on Raw Small Data

Variable	DF	Parameter Estimate	Standard Error	Wald Chi-Square	Pr > ChiSq	Standardized Estimate
INTERCEPT	1	−99.5240	308.0000	0.1044	0.7466	
INCOME	1	0.0680	1.7332	0.0015	0.9687	1.0111
AGE	1	1.7336	5.9286	0.0855	0.7700	10.5745
GENDER	1	14.3294	82.4640	0.0302	0.8620	4.0797

TABLE 22.5

RESPONSE Regression Output Based on Standardized Small Data

Variable	DF	Parameter Estimate	Standard Error	Wald Chi-Square	Pr > Chi-Sq	Standardized Estimate
INTERCEPT	1	−6.4539	31.8018	0.0412	0.8392	
zINCOME	1	1.8339	46.7291	0.0015	0.9687	1.0111
zAGE	1	19.1800	65.5911	0.0855	0.7700	10.5745
zGENDER	1	7.3997	42.5842	0.0302	0.8620	4.0797

The standardized logistic regression coefficients for zINCOME, zAGE, and zGENDER are 1.8339, 19.1800, and 7.3997, respectively (Table 22.5). Although this is a trivial example, it still serves valid PCC calculations.

The raw and standardized logistic regression coefficients based on the standardized data are in the "Parameter Estimate" and "Standardized Estimate" columns in Table 22.5, respectively. Unexpectedly, the raw logistic regression coefficients—which are now the standardized logistic regression coefficients—do not equal the values in the "Standardized Estimate" column. If the ranking of predictor variables is required of the standardized coefficient, this inconsistency presents no problem as the raw and standardized values produce the same rank order. If information about the expected increase in the predicted Y is required, I prefer the SRCs in the "Parameter Estimate" column as it follows the definition of SRC.

As determined previously, the average correlation among the three predictor variables is a gross 0.71. Thus, the SRC provides questionable ranking of the predictor variables for RESPONSE. AGE is the most important predictor variable, followed by INCOME; the ranked position of GENDER is undetermined.

22.4 Predictive Contribution Coefficient

The PCC is a development in the data mining paradigm. It has the hallmarks of the data mining paradigm: flexibility because it is an assumption-free measure that works equally well with ordinary and logistic regression models; practicality and innovation because it offers greater utile information than the SRC; and above all, simplicity because it is easy to understand and calculate, as is evident from the following discussion.

Consider the linear regression model built on standardized data and defined in Equation (22.4):

$$zY = b_0 + b_1{}^*zX_1 + b_2{}^*zX_2 + \ldots + b_i{}^*zX_i + \ldots + b_n{}^*zX_n \qquad (22.4)$$

The PCC for zX_i, $PCC(zX_i)$, is a measure of the contribution of zX_i relative to the contribution of the other variables to the predictive scores of the model. $PCC(zX_i)$ is defined as the average absolute ratio of the zX_i score-point contribution $(zX_i{}^*b_i)$ to the score-point contribution (total predictive score minus zX_is score-point) of the other variables. Briefly, the PCC is read as follows: The larger the $PCC(zX_i)$ value is, the more significant the part zX_i has in the predictions of the model and the more important zX_i is as a predictor variable. Exactly how the PCC works and what are its benefits over the SRC are discussed in the next section. Now, I provide justification for the PCC.

Justification of the trustworthiness of the PCC is required as it depends on the SRC, which itself, as discussed, is not a perfect measure to rank predictor variables. The effects of any "impurities" (biases) carried by the SRC on the PCC are presumably negligible due to the "wash-cycle" calculations of the PCC. The six-step cycle, which is described in the next section, crunches the actual values of the SRC such that any original bias effects are washed out.

I would like now to revisit the question of which variables in a regression model are the most important predictors. The predictive contribution decision rule is that the variable with the largest PCC is the most important variable; the variable with the next-largest PCC is the next important variable, and so on. The predictor variables can be ranked from most to least important based on the descending order of the PCC. Unlike the decision rule for reduction in prediction error, there are no presumable caveats for the predictive contribution decision rule. Correlated predictor variables, including composite and dummy variables, can be thus ranked.

22.5 Calculation of Predictive Contribution Coefficient

Consider the logistic regression model based on the standardized data in Table 22.5. I illustrate in detail the calculation of PCC(zAGE) with the necessary data in Table 22.6.

1. Calculate the TOTAL PREDICTED (logit) score for each individual in the data. For individual ID 1, the values of the standardized predictor variable, in Table 22.6, which are multiplied by the corresponding

TABLE 22.6
Necessary Data

ID #	zAGE	zINCOME	zGENDER	Total Predicted Score	zAGE Score-Point Contribution	OTHERVARS Score-Point Contribution	zAGE_othvars
1	1.7354	1.7915	-0.7746	24.3854	33.2858	-8.9004	3.7398
2	0.9220	1.8657	-0.7746	8.9187	17.6831	-8.7643	2.0176
3	1.0123	-0.0631	-0.7746	7.1154	19.4167	-12.3013	1.5784
4	0.1989	-0.0631	-0.7746	-8.4874	3.8140	-12.3013	0.3100
5	0.2892	-0.2485	-0.7746	-7.0938	5.5476	-12.6414	0.4388
6	-0.5243	-0.6565	-0.7746	-23.4447	-10.0551	-13.3897	0.7510
7	-0.4339	-0.7678	1.1619	-7.5857	-8.3214	0.7357	11.3105
8	-1.2474	-0.4340	1.1619	-22.5762	-23.9241	1.3479	17.7492
9	-1.1570	-0.6194	1.1619	-21.1827	-22.1905	1.0078	22.0187
10	-0.7954	-0.8049	1.1619	-14.5883	-15.2560	0.6677	22.8483

standardized logistic regression coefficients in Table 22.5, produce the TOTAL PREDICTED score of 24.3854.

2. Calculate the zAGE SCORE-POINT contribution for each individual in the data. For individual ID 1, the zAGE SCORE-POINT contribution is 33.2858 (= 1.7354*19.1800).

3. Calculate the other variables SCORE-POINT contribution for each individual in the data. For individual ID 1, the other variables SCORE-POINT contribution is -8.9004 (= 24.3854 - 33.2858).

4. Calculate the zAGE_OTHVARS for each individual in the data. zAGE_OTHVARS is defined as the absolute ratio of zAGE SCORE-POINT contribution to the other variables SCORE-POINT contribution. For ID 1, zAGE_OTHVARS is 3.7398 (= absolute value of 33.2858/-8.9004).

5. Calculate PCC(zAGE), the average (median) of the zAGE_OTHVARS values: 2.8787. The zAGE_OTHVARS distribution is typically skewed, which suggests that the median is more appropriate than the mean for the average.

I summarize the results of the PCC calculations after applying the five-step process for zINCOME and zGENDER (not shown).

1. AGE is ranked as the most important predictor variable, next is GENDER, and last is INCOME. Their PCC values are 2.8787, 0.3810, and 0.0627, respectively.

2. AGE is the most important predictor variable with the largest and "large" PCC value of 2.8787. The implication is AGE is clearly driving the predictions of the model. When a predictor variable has a large PCC value, it is known as a *key driver* of the model.

Note: I do not use the standardized variable names (e.g., zAGE instead of AGE) in the summary of findings. This is to reflect the issue of determining the importance of predictor variables in the content of variable, which is clearly conveyed by the original name, not the technical name, which reflects a mathematical necessity of the calculation process.

22.6 Extra Illustration of Predictive Contribution Coefficient

This section assumes an understanding of the decile analysis table, which is discussed in full detail in Chapter 18. Readers who are not familiar with the decile analysis table may still be able to glean the key points of the following discussion without reading Chapter 18.

The PCC offers greater utile information than the SRC, notwithstanding the working assumption and caveats of the SRC. Both coefficients provide an overall ranking of the predictor variables in a model. However, the PCC can extend beyond an overall ranking by providing a ranking at various levels of model performance. Moreover, the PCC allows for the identification of key drivers—salient features—which the SRC metric cannot legitimately yield. By way of continuing with the illustration, I discuss these two benefits of the PCC.

Consider the decile performance of the logistic regression model based on the standardized data in Table 22.5. The decile analysis in Table 22.7 indicates the model works well as it identifies the lowly three responders in the top three deciles.

The PCC as presented so far provides an overall model ranking of the predictor variables. In contrast, the admissible calculations of the PCC at the decile level provide a decile ranking of the predictor variables with a response modulation, ranging from most to least likely to respond. To effect a decile-based calculation of the PCC, I rewrite step 5 in Section 22.5 on the calculation of the PCC:

Step 5: Calculate PCC(zAGE)—the median of the zAGE_OTHVARS values for each decile.

The PCC decile-based small data calculations, which are obvious and trivial, are presented to make the PCC concept and procedure clear and to generate interest in its application. Each of the 10 individuals is itself a decile in which the median value is the value of the individual. However, this point is also instructional. The reliability of the PCC value is sample size dependent. In real applications, in which the decile sizes are large to ensure the reliability

TABLE 22.7

Decile Analysis for RESPONSE Logistic Regression

Decile	Number of Individuals	Number of Responses	Response Rate (%)	Cum Response Rate (%)	Cum Lift
Top	1	1	100	100.0	333
2	1	1	100	100.0	333
3	1	1	100	100.0	333
4	1	0	0	75.0	250
5	1	0	0	60.0	200
6	1	0	0	50.0	167
7	1	0	0	42.9	143
8	1	0	0	37.5	125
9	1	0	0	33.3	111
Bottom	1	0	0	30.0	100
Total	10	3			

of the median, the PCC decile analysis is quite informational regarding how the predictor variable ranking interacts across the response modulation produced by the model. The decile PCCs for the RESPONSE model in Table 22.8 are clearly daunting. I present two approaches to analyze and draw implications from the seemingly scattered arrays of PCCs left by the decile-based calculations. The first approach ranks the predictor variables by the decile PCCs for each decile. The rank values, which descend from 1 to 3, from most to least important predictor variable, respectively, are in Table 22.9. Next, the decile PCC rankings are compared with the "overall" PCC ranking.

The analysis and implications of Table 22.9 are as follows:

1. The overall PCC importance ranking is AGE, GENDER, and INCOME, descendingly. The implication is that an inclusive marketing strategy is

TABLE 22.8

Decile PCC: Actual Values

Decile	PCC(zAGE)	PCC(zGENDER)	PCC(zINCOME)
Top	3.7398	0.1903	0.1557
2	2.0176	0.3912	0.6224
3	1.5784	0.4462	0.0160
4	0.4388	4.2082	0.0687
5	11.3105	0.5313	0.2279
6	0.3100	2.0801	0.0138
7	22.8483	0.3708	0.1126
8	22.0187	0.2887	0.0567
9	17.7492	0.2758	0.0365
Bottom	0.7510	0.3236	0.0541
Overall	2.8787	0.3810	0.0627

TABLE 22.9

Decile PCC: Rank Values

Decile	PCC(zAGE)	PCC(zGENDER)	PCC(zINCOME)
Top	1	2	3
2	1	3	2
3	1	2	3
4	2	1	3
5	1	2	3
6	2	1	3
7	1	2	3
8	1	2	3
9	1	2	3
Bottom	1	2	3
Overall	1	2	3

defined by primary focus on AGE, secondary emphasis on GENDER, and incidentally calling attention to INCOME.

2. The decile PCC importance rankings are in agreement with the overall PCC importance ranking for all but deciles 2, 4, and 6.

3. For decile 2, AGE remains most important, whereas GENDER and INCOME are reversed in their importance relative to the overall PCC ranking.

4. For deciles 4 and 6, INCOME remains the least important, whereas AGE and GENDER are reversed in their importance relative to the overall PCC ranking.

5. The implication is that two decile tactics are called for, beyond the inclusive marketing strategy. For individuals in decile 2, careful planning includes particular prominence given to AGE, secondarily mentions INCOME, and incidentally addresses GENDER. For individuals in deciles 4 and 6, careful planning includes particular prominence given to GENDER, secondarily mentions AGE, and incidentally addresses INCOME.

The second approach determines decile-specific key drivers by focusing on the actual values of the obsequious decile PCCs in Table 22.8. To put a methodology in place, a measured value of "large proportion of the combined predictive contribution" is needed. Remember that AGE is informally declared a key driver of the model because of its large PCC value, which indicates that the predictive contribution of zAGE represents a large proportion of the combined predictive contribution of zINCOME and zGENDER. Accordingly, I define predictor variable X_i as a key driver as follows: X_i *is a key driver if* $PCC(X_i)$ *is greater than* $1/(k-1)$, *where k is the number of other variables in the model; otherwise,* X_i *is not a key driver.* The value $1/(k-1)$ is, of course, user defined, but I presuppose that if the score-point contribution of a single predictor variable is greater than the rough average score-point contribution of the other variables, it can safely be declared a key driver of model predictions.

The key driver definition is used to recode the actual PCC values into 0–1 values, which represent non-key driver/key driver. The result is a key driver table, which serves as a means of formally declaring the decile-specific key drivers and overall model key drivers. In particular, the table reveals key driver patterns across the decile relative to the overall model key drivers. I use the key driver definition to recode the actual PCC values in Table 22.9 into key drivers in Table 22.10.

The analysis and implications of Table 22.10 are as follows:

1. There is a single overall key driver of the model, AGE.

2. AGE is also the sole key driver for deciles top, 3, 7, 8, 9, and bottom.

TABLE 22.10

Decile PCC: Key Drivers

Decile	AGE	GENDER	INCOME
Top	1	0	0
2	1	0	1
3	1	0	0
4	0	1	0
5	1	1	0
6	0	1	0
7	1	0	0
8	1	0	0
9	1	0	0
Bottom	1	0	0
Overall	1	0	0

3. Decile 2 is driven by AGE and INCOME.

4. Deciles 4 and 6 are driven only by GENDER.

5. Decile 5 is driven by AGE and GENDER.

6. The implication is that the salient feature of an inclusive market-ing strategy is AGE. To ensure the effectiveness of the marketing strategy, the "AGE" message must be tactically adjusted to fit the individuals in deciles 1 through 6 (the typical range for model imple-mentation). Specifically, for decile 2, the marketing strategist must add INCOME to the AGE message; for deciles 4 and 6, the marketing strategist must center attention on GENDER with an undertone of AGE; and for decile 5, the marketing strategist must add a GENDER component to the AGE message.

22.7 Summary

I briefly reviewed the traditional approach of using the magnitude of the raw or SRCs in determining which variables in a model are its most impor-tant predictors. I explained why neither coefficient yields a perfect "impor-tance" ranking of predictor variables. The raw regression coefficient does not account for inequality in the units of the variables. Furthermore, the SRC is theoretically problematic when the variables are correlated, as is virtually the situation in database applications. With a small data set, I illustrated for ordinary and logistic regression models the often-misused raw regression coefficient and the dubious use of the SRC in determining the rank importance of the predictor variables.

I pointed out that the ranking based on the SRC provides useful information without raising sophistic findings. An unknown working assumption is that the reliability of the ranking based on the SRC increases as the average correlation among the predictor variables decreases. Thus, for well-built models, which necessarily have a minimal correlation among the predictor variables, the resultant ranking is accepted practice.

Then, I presented the PCC, which offers greater utile information than the standardized coefficient. I illustrated how it works, as well as its benefits over the SRC using the small data set.

Last, I provided an extra illustration of the benefits of the new coefficient over the SRC. First, it can rank predictor variables at a decile level of model performance. Second, it can identify a newly defined predictor variable type, namely, the key driver. The new coefficient allows the data analyst to determine at the overall model level and the individual decile levels which predictor variables are the predominant or key driver variables of the predictions of the model.

Reference

1. Menard, S., *Applied Logistic Regression Analysis*, Quantitative Applications in the Social Sciences Series, Sage, Thousand Oaks, CA, 1995.

23

Regression Modeling Involves Art, Science, and Poetry, Too

23.1 Introduction

The statistician's utterance "regression modeling involves art and science" implies a mixture of a *skill acquired by experience* (art) and a *technique that reflects a precise application of fact or principle* (science). The purpose of this chapter is to put forth my assertion that regression modeling involves the trilogy of art, science, and concrete poetry. With concrete poetry, the poet's intent is conveyed by graphic patterns of symbols (e.g., a regression equation) rather than by the conventional arrangement of words. As an example of the regression trilogy, I make use of a metrical "modelogue" to introduce the machine-learning technique GenIQ, an alternative to the statistical regression models. Interpreting the modelogue requires an understanding of the GenIQ Model. I provide the interpretation after presenting the Shakespearean modelogue "To Fit or Not to Fit Data to a Model."

23.2 Shakespearean Modelogue

To Fit or Not to Fit Data to a Model

To fit or not to fit data to a model—that is the question:
Whether 'tis nobler in the mind to suffer
The slings and arrows of outrageously using
The statistical regression paradigm of
Fitting data to a prespecified model, conceived and tested
Within the *small-data setting* of the day, 207 years ago,
Or to take arms against a sea of troubles
And, by opposing, move aside fitting data to a model.
Today's big data necessitates—*Let the data define the model.*

Fitting big data to a prespecified *small-framed* model
Produces a *skewed* model with
Doubtful interpretability and questionable results.
When we have shuffled off the expected coil,
There's the respect of the GenIQ Model,
A machine-learning alternative regression model
To the statistical regression model.
GenIQ is an assumption-free, free-form model that
Maximizes the Cum Lift statistic, equally, the decile table.

—Bruce "Shakespeare" Ratner

23.3 Interpretation of the Shakespearean Modelogue

At the beginning of every day for the regression modeler, whose tasks are to predict a continuous outcome (e.g., profit) and a binary outcome (e.g., yes/no response), the ordinary least squares (OLS) regression model and the logistic regression model (LRM), respectively, are likely to be put to use, giving promise of another workday of successful models. The essence of any prediction model is the fitness function, which quantifies the optimality (goodness or accuracy) of a solution (predictions). The fitness function of the OLS regression model is mean squared error (MSE), which is minimized by calculus.

Historians date calculus to the time of the ancient Greeks, circa 400 BC. Calculus started making great strides in Europe toward the end of the eighteenth century. Leibniz and Newton pulled these ideas together, and they are credited with the independent "invention" of calculus. The OLS regression model is celebrating 207 years of popularity, as the invention of the method of least squares was on March 6, 1805. The backstory of the OLS regression model is discussed next [1].

The fundamentals of the basis for least squares analysis is accredited to Carl Friedrich Gauss in 1795 at the age of 18. Gauss did not publish the method until 1809.* In 1822, Gauss was able to state that the least squares approach to regression analysis is optimal, in the sense that the best linear unbiased estimator of the coefficients is the least squares estimator. This result is known as the Gauss-Markov theorem. However, the term *method of least squares* was coined by Adrien Marie Legendre (1752–1833), appearing in *Sur la Méthode des moindres quarrés* (*On the Method of Least Squares*), the title of

---

* Gauss did not publish the method until 1809, when it appeared in Volume 2 of his work on celestial mechanics, *Theoria Motus Corporum Coelestium in sectionibus conicis solem ambientium*. Gauss, C.F., *Theoria Motus Corporum Coelestium*, 1809.

an appendix to *Nouvelles méthodes pour la détermination des orbites des comètes*. The appendix is dated March 6, 1805 [2].

The fitness function of the LRM is the likelihood function, which is maximized by calculus (i.e., the method of maximum likelihood) [3]. The logistic function has its roots spread back to the nineteenth century, when the Belgian mathematician Verhulst invented the function, which he named logistic, to describe population growth. The rediscovery of the function in 1920 is due to Pearl and Reed; the survival of the term *logistic* is due to Yule, and the introduction of the function in statistics is due to Berkson. Berkson used the logistic function in his regression model as an alternative to the normal probability probit model, usually credited to Bliss in 1934 and sometimes to Gaddum in 1933. (The probit can be first traced to Fechner in 1860.) As of 1944, Berkson's LRM was not accepted as a viable alternative to Bliss's probit. After the ideological debate about the logistic and probit had abated in the 1960s, Berkson's logistic gained wide acceptance. Berkson was much derided for coining the term *logit* by analogy to the probit of Bliss, who coined the term *probit* for "probability unit [4, 5]."

The not-yet-popular model is the GenIQ Model, a machine-learning alternative model to the statistical OLS and LRM; I conceived and developed this model in 1994. The modeling paradigm of GenIQ is to "let the data define the model," which is the complete antithesis of the statistical modeling paradigm, "fit the data to a model." GenIQ automatically (1) data mines for new variables, (2) performs variable selection, and (3) specifies the model—to "optimize the decile table"* (i.e., to fill the upper deciles with as much profit or as many responses as possible. The fitness function of GenIQ is the decile table, which is maximized by the Darwinian inspired machine-learning paradigm of genetic programming (GP). Operationally, optimizing the decile table creates the best possible descending ranking of the dependent variable (outcome) values. Thus, the prediction of GenIQ is that of identifying individuals who are most to least likely to respond (for a binary outcome) or who contribute large to small profits (for a continuous outcome). Historians traced the first use of the decile table, originally called a "gains chart," to the direct mail business of the early 1950s [6]. The gains chart is hallmarked by solicitations found inside the covers of matchbooks. More recently, the decile table has transcended the origin of the gains chart toward a generalized measure of model performance. The decile table is the model performance display used in direct, database, or telemarketing solicitations; marketing mix optimization programs; business intelligent offerings; customer relationship management (CRM) campaigns; Web- or e-mail-based broadcasts; and the like. (The term *decile* was first used by Galton in 1882 [7].) Historians cited the first experiments using GP by Stephen F. Smith (1980) and Nichael L. Cramer (1985), as described in the seminal book *Genetic Programming: On the*

---

* Optimizing the decile table is equivalent to maximizing the Cum Lift statistic.

Programming of Computers by Means of Natural Selection by John Koza (1992), who is considered the inventor of GP [8–10].

Despite the easy implementation of GenIQ (simply drop the GenIQ equation into the scoring database), it is not yet the everyday regression model because of the following two checks:

1. Unsuspected equation: GenIQ output is the visual display called a parse tree, depicting the GenIQ Model, and the GenIQ Model "equation," which is actually a computer program/code. The regression modeler, anticipating an equation of the form $Y = b_0 + b_1X_1 + b_2X_2 + \ldots + b_nX_n$, is stupefied when he or she sees unexpectedly, say, for the Pythagorean theorem $d = \sqrt{l^2 + h^2}$, where d (the diagonal), l (the length), and h (the height) are the sides of a right triangle; and SQRT is the square root function, the GenIQ computer code (Figure 23.1).

2. Ungainly interpretation: The GenIQ parse tree and computer code can be a "bit much" to grasp (for unknown models/solutions). The visual display provides the modeler with an ocular sense, a speck, of comfort and confidence for understanding and using the GenIQ Model. The GenIQ tree for the Pythagorean theorem (Figure 23.2), is not so ungraspable because the solution is well known (it is the sixth-most-famous equation [11]).

```
x1 = height;
x2 = x1*x1;
x3 = length;
x4 = x3*x3;
x5 = x2 + x3;
x6 = SQRT(x5);
diagonal = x6;
```

FIGURE 23.1
GenIQ computer code.

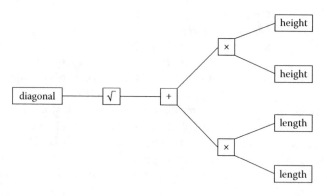

FIGURE 23.2
GenIQ tree for Pythagorean theorem.

The GenIQ tree, although not a "black box" like most other machine-learning methods, gives the modeler a graphic, albeit Picasso-like, to interpret. GenIQ, for the everyday regression model, produces a GenIQ tree defined by branches formed by predictor variables attached to various functions (like that of the Pythagorean GenIQ tree).

The GenIQ tree and computer code represent *feature 3* of GenIQ: The GenIQ Model serves as a nonstatistical, machine learning regression method that automatically *specifies the model* for the problem at hand. As well, the GenIQ tree represents *feature 1* of GenIQ: The GenIQ Model automatically *data mines for new variables*. Continuing with the Pythagorean illustration, there are four new variables (branches): new_var1 = (height × height); new_var2 = (length × length); new_var3 = (new_var1 + new_var2); and last, new_var4 = SQRT(new_var3), which is the model itself. Thus, the GenIQ Model serves as a unique data mining method that creates new variables—which cannot be intuited by the model builder—via the GP process, which evolves structure (new variables) "without explicit programming" (Adams, 1959) [12]. Moreover, appending the new variables to the dataset with the original variables for building a statistical regression model produces a *hybrid statistics-machine-learning model*, along with the regression coefficients that provide the regression modeler the necessary comfort level for model acceptance.

For the regression modeler, interpretability is all about the regression coefficients. The regression coefficients provide the key to how the model works: Which predictor variables are most important, in rank order? What effect does each predictor variable have on the dependent variable? It is not well known that the standard method of interpreting regression coefficients often leads to an incorrect interpretation of the regression model and, specifically, incorrect answers to the two questions stated in this paragraph. GenIQ is a nonstatistical machine-learning method; thus, it has no coefficients like a statistical method. But, GenIQ provides the answer to the first question by way of *feature 2* of GenIQ: The GenIQ Model provides a unique *variable selection* of important predictor variables by ranking* the relationship between each predictor variable with the dependent variable—accounting for the presence of the other predictor variables considered jointly. As for the second question, GenIQ provides the answer by analyzing the decile table (see Chapters 30 and 31).

With two checks against it: Why use GenIQ? How will GenIQ ever become popular? GenIQ is the appropriate model when the decile table is the unquestionable measure of model performance. For other instances, a trade-off has to be made between the performance and no coefficients of GenIQ versus the interpretability and use of fitness functions of statistical regression, which serve often as surrogates for optimizing the decile table. Separation anxiety for something that has been used for between 68 and 207 years is a condition

---

* The ranking is based on the mean frequency of a variable across the top 18 models in the latest generation.

that takes time to treat. Until the checks become ocularly palatable, which comes about with retraining statisticians to think out of the box, OLS and LRM will continue to be used. With the eventual recognition that "valuable information" comes in unsuspected forms (e.g., a parse tree), GenIQ will be popular.

GenIQ is the model for today's data. It can accommodate big (and small) data as it is a flexible, assumption-free, nonparametric model whose engine lets the data define the model. In stark contrast, OLS and LRM conceived, tested, and experimented on the small data setting of their day. These models are suboptimal and problematic with today's big data [13]. The paradigm of OLS and LRM is "fit the data to" an unknown, prespecified, assumption-full, parametric model, which is best for small data settings. GenIQ will become popular as today's data grow in size, necessitating that the data define the model rather than fitting "square data into a round" model.

23.4 Summary

The statistician's utterance that regression modeling involves art and science implies a mixture of a skill acquired by experience and a technique that reflects a precise application of fact or principle. I asserted the regression trilogy of art, science, and concrete poetry by bringing forth a modelogue to introduce the machine-learning technique of GenIQ, an alternative to the statistical OLS and LRM. I provided the interpretation of the poetry as only I best can, being the poet of "To Fit or Not to Fit Data to a Model" and the inventor of the GenIQ method.

References

1. Newton vs. Leibniz; The Calculus Controversy, http://www.angelfire.com/md/byme/mathsample.html, 2010.
2. Earliest Known Uses of Some of the Words of Mathematics, http://jeff560.tripod.com/m.html,2009.
3. The likelihood function represents the joint probability of observing the data that have been collected. The term joint probability means a probability that combines the contributions of all the individuals in the study.
4. Finney, D., Probit Analysis, Cambridge University, Cambridge, UK, 1947.
5. Cramer, J.S., Logit Models from Economics and Other Fields, Chapter 9, Cambridge University Press, Cambridge, UK, 2003.
6. Ratner, B., Decile analysis primer, http://www.geniq.net/DecileAnalysis Primer_2.html, 2008.

7. Ratner, B., http://www.geniq.net/res/Statistical-Terms-Who-Coined-Them-and-When.pdf, 107, 2007.
8. Smith, S.F., A Learning System Based on Genetic Adaptive Algorithms, PhD Thesis, Computer Science Department, University of Pittsburgh, December, 1980.
9. Cramer, N. L., A Representation for the Adapative Generation of Simple Sequential Programs, International Conference on Genetic Algorithms and their Applications [ICGA85], CMU, Pittsburgh, 1985.
10. Koza, J.R, Genetic Programming: On the Programming of Computers by Means of Natural Selection, MIT Press, Cambridge, MA, 1992.
11. Peter Alfeld, University of Utah, Alfred's homepage, 2010.
12. Samuel, A.L., Some studies in machine learning using the game of checkers. IBM Journal of Research and Development, 3(3), 210–229, 1959.
13. Harlow, L.L., Mulaik, S.A., and Steiger, J.H., Eds., What If There Were No Significance Tests? Erlbaum, Mahwah, NJ, 1997.

24

Genetic and Statistic Regression Models: A Comparison

24.1 Introduction

Statistical ordinary least squares (OLS) regression and logistic regression (LR) models are workhorse techniques for prediction and classification, respectively. The ordinary regression method was first published by Legendre in 1805. The LR model was developed by Berkson in 1944. Something old is not necessarily useless today, and something new is not necessarily better than something old is. Lest one forgets, consider the wheel and the written word. With respect for statistical lineage, I maintain the statistical regression paradigm, which dictates "fitting the data to model," is old, as it was developed and tested within the small data setting of yesterday, and has been shown untenable with big data of today. I further maintain the new machine-learning genetic paradigm, which "lets the data define the model," is especially effective with big data; consequently genetic models outperform the originative statistical regression models. The purpose of this chapter is to present a comparison of genetic and statistic LR in support of my assertions. (A presentation of the companion genetic OLS regression model replaces the binary dependent variable with the continuous dependent variable.)

24.2 Background

Statistical OLS regression and LR models are workhorse techniques for prediction (of a continuous dependent variable) and classification (of a binary dependent variable), respectively. The OLS regression method was first published by Legendre on March 6, 1805. The LR model was developed by Berkson in 1944. Something old (say, between 68 and 207 years ago) is not necessarily useless today, and something new is not necessarily better than something old is. Lest one forget, consider the wheel and the written word.

With respect for statistical lineage, I maintain the statistical regression paradigm, which dictates "fitting the data to model," is old as it was developed and tested within the small data* setting of yesterday and has been shown untenable with big data[†] of today. The linear nature of the regression model renders the model insupportable in that it cannot capture the underlying structure of big data. I further maintain the new machine-learning genetic paradigm, which "lets the data define the model," is especially effective with big data; consequently, genetic models outperform the originative statistical regression models.

The "engine" behind any predictive model, statistical or machine learning, linear or nonlinear, is the fitness function, also known as the objective function. The OLS regression fitness function of mean squared error is easily understood; it is defined as the average of the absolute squares of the error, actual observation minus predicted observation. The LR model fitness function is the likelihood function,[‡] an abstract mathematical expression for individuals not well versed in mathematical statistics, that is best understood with respect to the classification table, indicating the primary measure of total number of correctly classified individuals.[§] The equations to the statistical regression models are obtained by the universally routinely used calculus.

24.3 Objective

The objective of this chapter is to present the GenIQ Model as the genetic LR alternative to the statistical LR model, in support of my assertions in Section 24.2. (A presentation of the companion genetic OLS regression model replaces the binary dependent with the continuous dependent variable.)

Adding to my assertions, I bring attention to the continuing decades-long debate, *What If There Were No Significant Testing?* [1]. The debate raises doubt over the validity of the statistical regression paradigm: The regression modeler is dictated to "fit the data to" a parametric prespecified model, linear "in its coefficients," $Y = b_0 + b_1 X_1 + b_2 X_2 + \ldots + b_n X_n$. Specifically, the matter in dispute is the modeler must prespecify the *true unknowable* model, under a

---

* Hand, D.J., Daly, F., Lunn, A.D., McConway, K.J., and Ostrowski, E. (Eds.), *A Handbook of Small Data Sets*, Chapman & Hall, London, 1994. This book includes many of the actual data from the early-day efforts of developing the OLS regression method of yesterday.
[†] See Chapters 8 and 9.
[‡] Also known as the joint probability function: $P(X)*P(X)*\frac{1}{4}P(X)*\{(1 - P(X)\}*\{(1 - P(X)\} \ldots *\{(1 - P(X)\}$.
[§] Actually, there are many other measures of classification accuracy. Perhaps even more important than total classified correctly is the number of "actual" individuals classified correctly.

null hypothesis. This is undoubtedly an inappropriate way to build a model. If the hypothesis testing yields rejection of the model, then the modeler tries repeatedly until an acceptable model is finally dug up. The "try again" task is a back-to-the-future data mining procedure that statisticians deemed, over three decades ago, a work effort of finding spurious relations among variables. In light of the debated issue, perhaps something new is better: The GenIQ Model is presented next.

24.4 The GenIQ Model, the Genetic Logistic Regression

The GenIQ Model is a flexible, any-size data method guided by the paradigm of *let the data define the model*. The GenIQ Model

1. *Data mines* for new variables among the original variables.
2. *Performs* variable selection, which yields the best subset among the new and original variables.
3. *Specifies* the model to "optimize the decile table." *The decile table is the fitness function.* The fitness function is optimized via the method of *genetic programming* (GP), which executes a computer—without explicit programming—to fill up the upper deciles with as many responses as possible, equivalently, to maximize the Cum Lift. The decile table fitness function is directly appropriate for model building in industry sectors such as direct, database, or telemarketing solicitations; marketing-mix optimization programs; business-intelligent offerings; customer relationship management (CRM) campaigns; Web- or e-mail-based broadcasts; and the like. However, the decile table is becoming the universal statistic for virtually all model assessment of predictive performance.

Note: I use interchangeably the terms *optimize* and *maximize* as they are equivalent in the content of the decile table.

24.4.1 Illustration of "Filling up the Upper Deciles"

I illustrate the concept of "filling up the upper deciles." First, the decile table, the display that indicates model performance, is constructed. Construction consists of five steps: (1) apply, (2) rank, (3) divide, (4) calculate, and (5) assess (see Figure 24.1).

I present the *RESPONSE model criterion*, for a *perfect* RESPONSE model, along with the instructive decile table to show the visible representation of the concepts to which the GenIQ Model directs its efforts. Consider the RESPONSE

Model Performance Criterion

... is the DECILE ANALYSIS.

1. **Apply** model to the file (score the file).

2. **Rank** the scored file, in descending order.

3. **Divide** the ranked file into 10 equal groups.

4. **Calculate** Cum Lift.

5. **Assess** model performance. The **best** model identifies the **most** response in the **upper deciles.**

FIGURE 24.1
Construction of the decile analysis.

RESPONSE Model Criterion

How well the model **correctly classifies** response in the **upper** deciles.	Decile	Number of Individuals	Total Response
Perfect Response Model among 100 individuals	top	10	10
▶ 40 responders	2	10	10
▶ 60 nonresponsers	3	10	10
	4	10	10
	5	10	0
	6	10	0
	7	10	0
	8	10	0
	9	10	0
	bottom	10	0
	Total	100	40

FIGURE 24.2
RESPONSE model criterion.

model working with 40 responses among 100 individuals. In Figure 24.2, all 40 responses are in the upper deciles, indicating a perfect model.

In Figure 24.3, the decile table indicates that perfect models are hard to come by. Accordingly, I replace criterion with goal. Thus, one seeks a *RESPONSE model goal* for which the desired decile table is one that *maximizes* the responses in the upper deciles.

The next two decile tables are exemplary in that they show how to read the predictive performance of a RESPONSE model, stemming from any predictive modeling approach. The decile tables reflect the result from a model validated with a sample of 46,170 individuals; among these, there are 2,104 responders. In Figure 24.4, the decile table shows the decile table along with the annotation of how to read the Cum Lift of the top decile. The Cum Lift for the top decile is the top decile Cum Response Rate (18.7%) divided by

RESPONSE Model Goal

- **One seeks a model** that identifies the **maximum** responses in the **upper** deciles.

Decile	Total Response
top	max
2	max
3	max
4	
5	
6	
7	
8	
9	
bottom	

FIGURE 24.3
RESPONSE model goal.

One can expect 4.11 times the responders obtained using no model, targeting the "top" decile.

Response Decile Analysis

Decile	Number of Customers	Number of Responses	Decile Response Rate	Cum Response Rate	Cum Response Lift
top	4,617	865	18.7%	18.7%	411
2	4,617	382	8.3%	13.5%	296
3	4,617	290	6.3%	11.1%	244
4	4,617	128	2.8%	9.0%	198
5	4,617	97	2.1%	7.6%	167
6	4,617	81	1.8%	6.7%	146
7	4,617	79	1.7%	5.9%	130
8	4,617	72	1.6%	5.4%	118
9	4,617	67	1.5%	5.0%	109
bottom	4,617	43	0.9%	4.6%	100
TOTAL	46,170	2,104	4.6%		

FIGURE 24.4
The decile table.

Total-Decile Response Rate (4.6%, the average response rate of the sample) multiplied by 100, yielding 411.

The decile table in Figure 24.5 shows the "collapsed" decile table with respect to the top and second deciles combined, along with the annotation of how to read the Cum Lift of the top 20% of the sample. The Cum Lift for the

One can expect 2.96
times the responders
obtained using no
model, targeting the
"top 2" deciles.

Response Decile Analysis

Decile	Number of Customers	Number of Responses	Decile Response Rate	Cum Response Rate	Cum Response Lift
top	9,234	1,247	18.7%	18.7%	411
2			8.3%	13.5%	296
3	4,617	290	6.3%	11.1%	244
4	4,617	128	2.8%	9.0%	198
5	4,617	97	2.1%	7.6%	167
6	4,617	81	1.8%	6.7%	146
7	4,617	79	1.7%	5.9%	130
8	4,617	72	1.6%	5.4%	118
9	4,617	67	1.5%	5.0%	109
bottom	4,617	43	0.9%	4.6%	100
TOTAL	46,170	2,104	4.6%		

FIGURE 24.5
The decile table collapsed with respect to the top two deciles.

top two deciles is the "top and 2 deciles" Cum Response Rate (13.5%) divided by Total-Decile Response Rate (4.6%, the average response rate of the sample) multiplied by 100, yielding 296. The calculation of the remaining Cum Lifts follows the pattern of arithmetic tasks described.

24.5　A Pithy Summary of the Development of Genetic Programming

Unlike the statistical regression models that use calculus as their "number cruncher" to yield the model equation, the GenIQ Model uses GP as its number cruncher. As GP is relatively new, I provide a pithy summary of the development of GP.*

In 1954, GP began with the evolutionary algorithms first utilized by Nils Aall Barricelli and applied to evolutionary simulations. In the 1960s and early 1970s, evolutionary algorithms became widely recognized as a viable optimization approach. John Holland was a highly influential force behind GP during the 1970s.

The first statement of "tree-based" GP (that is, computer languages organized in tree-based structures and operated on by natural genetic operators,

---

* Genetic programming, http://en.wikipedia.org/wiki/Genetic_programming. Accordingly, I replace criterion with goal. Thus, one seeks a *RESPONSE model goal* for which the desired decile table is one that *maximizes* the responses in the upper deciles

such as reproduction, mating, and mutation) was given by Nichael L. Cramer (1985). This work was later greatly expanded by John R. Koza (1992), the leading exponent of GP who has pioneered the application of GP in various complex optimization and search problems [2.]

In the 1990s, GP was used mainly to solve relatively simple problems because the computational demand of GP could not be met effectively by the CPU power of the day. Recently, GP has produced many outstanding results due to improvements in GP technology and the exponential growth in CPU power. In 1994, the GenIQ Model, a Koza-GP machine-learning alternative model to the statistical OLS and LR models, was introduced.

Statisticians do not have a scintilla of doubt about the mathematical statistical solutions of the OLS and LR models. Statisticians may be rusty with calculus to the extent they may have to spend some time bringing back to mind how to derive the estimated regression models. Regardless, the estimated models are accepted without question (notwithstanding the effects of annoying modeling issues, e.g., multicollinearity). There is no burden of proof required for the acceptance of statistical model results.

Unfortunately, statisticians are not so kind with GP because, I dare say, they have no formal exposure to GP principles or procedures. So, I acknowledge the current presentation may not sit well with the statistical modeler. The GenIQ Model specification is set forth for the attention of statisticians, along with their acceptance of the GenIQ Model *without my obligation* to provide a primer on GP. GP methodology itself is not difficult to grasp but requires much space for a discourse, not available here. I hope this acknowledgment allows the statistical modeler *to accept assumedly* the trustworthiness of GP and the presentation of comparing genetic and statistic regression models. Note that I do present GP methodology proper in Chapter 29.

24.6 The GenIQ Model: A Brief Review of Its Objective and Salient Features

The operational objective of the GenIQ Model is to find a set of *functions* (e.g., arithmetic, trigonometric) and *variables* such that the equation (the GenIQ Model), represented symbolically as GenIQ = functions + variables, maximizes response in the upper deciles. GP determines the functions and variables (see Figure 24.6).

24.6.1 The GenIQ Model Requires Selection of Variables and Function: An Extra Burden?

The GenIQ Model requires selection of variables and function, whereas the statistic regression model requires presumably only selection of variables.

GenIQ Model

- ■ OBJECTIVE
  - – To find a set of **functions** and **variables** such that the equation (GenIQ Model)
    - ▸ GenIQ = **functions + variables**
    - ▸ **maximizes response in upper deciles**.
- ■ GP determines the functions, variables.

FIGURE 24.6
The GenIQ Model—objective and form.

The obvious question is whether the GenIQ Model places an extra burden on the modeler who uses GenIQ over the statistical model.

The selection of functions is neither a burden for the modeler nor a weakness in the GP paradigm. This is evident by the success of GP and the GenIQ Model. More to the point, as a true burden often unheeded, the statistical regression model *also requires* the selection of functions as the premodeling, mandatory exploratory data analysis (EDA) seeks the best transformations, such as log X, $1/X$, or $(5 - x)^2$. Due to the evolutionary process of GP, EDA is not needed. This is not a hyperbole; it is a salient feature of GP and consequently of GenIQ. The only GP data preparatory work called for is to eliminate impossible or improbable values (e.g., age is 120 years old, and a boy named Sue, respectively).

24.7 The GenIQ Model: How It Works

I acknowledge my assertion of "the machine-learning genetic paradigm ... is *especially effective with big data*," yet I illustrate the new technique with the smallest of data. I take poetic license in using a small dataset for the benefit of presenting effectively the GenIQ Model and thereby making it easy on the reader to follow the innards of GenIQ. That GenIQ works especially well with big data does not preclude the predictive prowess of GenIQ on small data.

I offer to view how the GenIQ Model works with a simple illustration that stresses the predictive power of GenIQ. Consider 10 customers (5 responders and 5 nonresponders) and predictor variables X1 and X2. The dependent variable is RESPONSE. The objective is to build a model that maximizes the *upper four deciles* (see Figure 24.7). Note that the dataset of size 10 is used to simulate a decile table in which each decile is of size 1.

I build two RESPONSE models, a statistical LR model and a genetic LR model (GenIQ). GenIQ identifies three of four responders, a 75% response rate for the upper four deciles. LRM identifies two of four responders, a 50% response rate for the upper four deciles (see Figure 24.8).

GenIQ Model: How It Works

GenIQ: How It Works

Consider 10 customers:

- their RESPONSE and two predictors X1 & X2

- Objective: to build a model that maximizes **upper four deciles**

i	RESPONSE	X1	X2
1	R	45	5
2	R	35	21
3	R	31	38
4	R	30	30
5	R	6	10
6	N	45	37
7	N	30	10
8	N	23	30
9	N	16	13
10	N	12	30

FIGURE 24.7
The GenIQ Model—how it works.

GenIQ = 20 + 40*(X1 + X2)
LRM   = .20 – .03*X1 + .03*X2

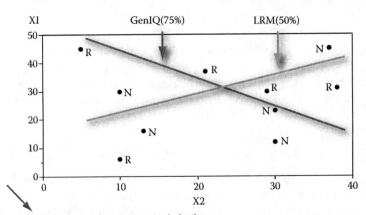

LRM yields 50% of responses in 4th decile.
GenIQ *outdoes* LRM by findings 75% of responses.

FIGURE 24.8
Genetic versus statistic logistic regression models. LRM, logistic regression model.

GenIQ = 20 + 40* (X1 + X2)

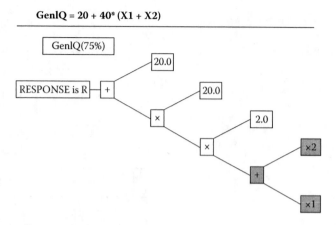

FIGURE 24.9
GenIQ parse tree.

The GenIQ Model, as a by-product of being a GP-based model, provides uniquely a visual display of its equation in the form of a *parse tree*. The parse tree for the presented GenIQ Model is in Figure 24.9. From the tree, I point out the following:

1. GenIQ has data mined a new variable: X1 + X2 as indicated in the tree by a gray-shaded branch.
2. X1 + X2 is a result of the variable selection of GenIQ, albeit it is a singleton best subset.
3. The tree itself specifies the model.

In sum, I have provided an exemplar illustration of GenIQ outperforming a LR model. Consequently, I showed the features of the GenIQ Model as described in Section 24.4: It (1) data mines for new variables, (2) performs variable selection, and (3) specifies the model to optimize the decile table (actually, the upper four deciles in this illustration).

24.7.1 The GenIQ Model Maximizes the Decile Table

I reconsider the dataset in Figure 24.7. I seek a GenIQ Model that maximizes the *full* decile table. From the modeling session of the first illustration, I obtain such a model in Figure 24.10. Before discussing the model, I explain how the model is obtained. GenIQ modeling produces, in a given session, about 5 to 10 *equivalent* models. Some models perform better, say, in the top four deciles (like the previous GenIQ Model). Some models perform better all the way from the top to the bottom deciles, the full decile table. In addition, some models have the desired discreteness of predicted scores. And, some models have undesired clumping of or gaps among the GenIQ predicted scores.

GenIQ Model Tree

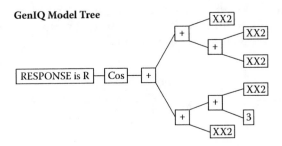

GenIQ Model - Computer Code

```
x1 = XX2;
  x2 = 3;
    x3 = XX1;
  x2 = x2 + x3;
If x1 NE 0 Then x1 = x2/x1; Else x1 = 1;
  x2 = XX2;
    x3 = XX2;
  x2 = x3 - x2;
    x3 = XX2;
  x2 = x2 + x3;
If x1 NE 0 Then x1 = x2/x1; Else x1 = 1;
  x1 = Cos(x1);
GenIQvar = x1;
```

		Table 1.		
	Scored, Ranked Data by GenIQvar			
ID	XX1	XX2	Response	GenIQvar
1	45	5	R	0.86740
2	35	21	R	0.57261
5	6	10	R	0.11528
3	31	38	R	0.05905
4	30	30	R	-0.53895
9	16	13	N	-0.86279
10	12	30	N	-0.95241
6	45	37	N	-0.96977
7	30	10	N	-0.99381
8	23	30	N	-0.99833

FIGURE 24.10
The GenIQ Model maximizes the decile table.

The GenIQer selects the model that meets the objective at hand along with desired characteristics. It is important to note that the offering by GenIQ of equivalent models is *not* spurious data mining: There is no rerunning the data through various variable selections until the model satisfies the given objective.

The GenIQ Model that maximizes the full decile table is in Figure 24.10. There are three elements of the GenIQ Model: (1) parse tree; (2) original dataset (with X1 and X2 replaced by XX1 and XX2, respectively*), to which the GenIQ Model score, the GenIQ predicted variable GenIQvar, is appended; and (3) the GenIQ Model equation, actually computer code defining the model. From the parse tree, the GenIQ Model selects the following functions: addition, subtraction, division, and the cosine. Also, GenIQ selects the two original variables XX1 and XX2 and the number 3, which in the world of GP is considered a variable.

The scored and ranked, by GenIQvar, dataset indicates a perfect ranking of the RESPONSE variable, 5 R's followed by 5 N's. This is tantamount to a max-imized decile table. Hence, the sought-after objective is met. Note this illus-tration implies the weakness of statistical regression modeling: Statistical

---

* The renaming of the X variables to XX is a necessary task as GenIQ uses X-named variables as an intermediate variable for defining the computer code of the model.

modeling offers only one model unless spurious rerunning the data through various variable selections is conducted, versus the several models of varying decile performances offered by genetic modeling.

24.8 Summary

With respect for statistical lineage, I maintained the statistical regression paradigm, which dictates fitting the data to model, is old, as it was developed and tested within the small data setting of yesterday and has been shown untenable with big data of today. I further maintained the new machine-learning genetic paradigm, which lets the data define the model, is especially effective with big data; consequently, genetic models outperform the originative statistical regression models. I presented a comparison of genetic versus statistical LR, showing great care and completeness. My comparison showed the promise of the GenIQ Model as a flexible, any-size data alternative method to the statistical regression models.

References

1. Harlow, L.L., Mulaik, S.A., & Steiger, J.H. (Eds.), *What If There Were No Significance Testing?* Erlbaum, Mahwah, NJ, 1997.
2. Koza, J.R, *Genetic Programming: On the Programming of Computers by Means of Natural Selection*, MIT Press, Cambridge, MA, 1992.

25

Data Reuse: A Powerful Data Mining Effect of the GenIQ Model

25.1 Introduction

The purpose of this chapter is to introduce the concept of *data reuse*, a powerful data mining effect of the GenIQ Model. Data reuse is appending new variables, which are found when building a GenIQ Model, to the original dataset. As the new variables are reexpressions of the original variables, the correlations among the original variables and the GenIQ data-mined variables are expectedly high. In the context of statistical modeling, the issue of such high correlations is known as *multicollinearity*. One effect of multicollinearity is unstable regression coefficients, an unacceptable result. In contrast, multicollinearity is a nonissue for the GenIQ Model because it has no coefficients. The benefit of data reuse is apparent: The original dataset is enhanced with the addition of new, predictive-full, GenIQ data-mined variables. I provide two illustrations of data reuse as a powerful data mining technique.

25.2 Data Reuse

Data reuse is appending new variables, which are found when building a GenIQ Model, to the original dataset. As the new variables are reexpressions of the original variables, the correlations among the original variables and the GenIQ data-mined variables are expectedly high. In the context of statistical modeling, the issue of such high correlations is known as multicollinearity. The effects of multicollinearity are inflated standard errors of the regression coefficients, unstable regression coefficients, lack of valid declaration of the importance of the predictor variables, and an indeterminate regression equation when the multicollinearity is severe. The simplest solution of "guess and check," although inefficient, to the multicollinearity problem is deleting suspect variables from the regression model.

Multicollinearity is a nonissue for the GenIQ Model because it has no coefficients. The benefit of data reuse is apparent: The original dataset is enhanced with the addition of new, predictive-full, GenIQ data-mined variables. The concept and benefit of data reuse are best explained by illustration. I provide two such illustrations.

25.3 Illustration of Data Reuse

To illustrate data reuse, I build an ordinary least squares (OLS) regression model with dependent variable PROFIT, predictor variables XX1 and XX2, and the data in Table 25.1.

The OLS Profit_est model in Equation (25.1) is

$$\text{Profit\_est} = 2.42925 + 0.16972 * XX1 - 0.06331 * XX2 \tag{25.1}$$

The assessment of the Profit_est model is made in terms of decile table performance. The individuals in Table 25.1 are scored and ranked by Profit_est. The resultant performance is in Table 25.2. The PROFIT ranking *is not perfect*, although three individuals are correctly placed in the proper positions. Specifically, individual ID 1 is correctly placed in the top position (top decile, decile of size 1); individuals ID 9 and ID 10 are correctly placed in the bottom two positions (ninth and bottom deciles, each decile of size 1).

25.3.1 The GenIQ Profit Model

The GenIQ Profit Model is in Figure 25.1. The assessment of the GenIQ Profit Model is made in terms of decile table performance. The individuals in

TABLE 25.1

Profit Dataset

ID	XX1	XX2	PROFIT
1	45	5	10
2	32	33	9
3	33	38	8
4	32	23	7
5	10	6	6
6	46	38	5
7	25	12	4
8	23	30	3
9	5	5	2
10	12	30	1

TABLE 25.2

Scored and Ranked Data by Profit_est

ID	XX1	XX2	PROFIT	Profit_est
1	45	5	10	9.74991
6	46	38	5	7.83047
4	32	23	7	6.40407
7	25	12	4	5.91245
2	32	33	9	5.77099
3	33	38	8	5.62417
8	23	30	3	4.43348
5	10	6	6	3.74656
9	5	5	2	2.96129
10	12	30	1	2.56660

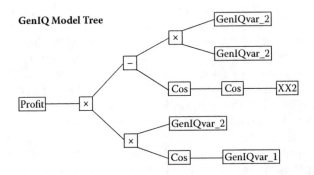

GenIQ Model – Computer Code

```
x1 = GenIQvar_1;
   x1 = Cos(x1);
      x2 = GenIQvar_2;
   x1 = x1 * x2;
      x2 = XX2;
      x2 = Cos(x2);
      x2 = Cos(x2);
         x3 = GenIQvar_2;
            x4 = GenIQvar_2;
         x3 = x3 * x4;
      x2 = x3 – x2;
   x1 = x1 * x2;
GenIQvar = x1;
```

Table 1.

Scored, Ranked Data by GenIQvar

ID	XX1	XX2	PROFIT	GenIQvar
1	45	5	10	9.27644
2	32	33	9	6.22359
3	33	38	8	0.60790
4	32	23	7	0.29207
5	10	6	6	0.09151
6	46	38	5	0.06930
7	25	12	4	–0.12350
8	23	30	3	–0.19258
9	5	5	2	–0.26727
10	12	30	1	–0.31133

FIGURE 25.1
The PROFIT GenIQ Model.

Table 25.1 are scored and ranked by GenIQvar. The resultant performance is in Table 1 of Figure 25.1. The PROFIT ranking *is perfect*. I hold off declaring GenIQ outperforms OLS.

The GenIQ Model computer code is the GenIQ Profit Model equation. The GenIQ Model tree indicates the GenIQ Profit Model has three predictor variables: the original XX2 variable and two data-reused variables, GenIQvar_1 and GenIQvar_2. Before I discuss GenIQvar_1 and GenIQvar_2, I provide a formal treatment of the fundamental characteristics of data-reused variables.

25.3.2 Data-Reused Variables

Data-reused variables, as found in GenIQ modeling, come about by the inherit mechanism of genetic programming (GP). GP starts out with an initial random set, a *genetic population,* of, say, 100 genetic models, defined by a set of predictor variables, numeric "variables" (numbers are considered variables within the GP paradigm), and a set of functions (e.g., arithmetic and trigonometric). The initial genetic population, generation 0, is subjected to GP evolutionary computation that includes mimicking the natural biological genetic operators: (1) copying/reproduction, (2) mating/sexual combination, and (3) altering/mutation. Some models are copied. Most models are mated: Two "parent" models evolve two offspring models with probabilistically greater predictive power than the parent models. Few models are altered by changing a model characteristic (e.g., replacing the addition function by the multiplication function); the altered versions of the models have probabilistically greater predictive power than the original models. The resultant genetic population of 100 models, generation 1, has probabilistically greater predictive power than the models in generation 0. The models in generation 1 are now selected for another round of copying-mating-altering (based on proportionate to total decile table fitness), resulting in a genetic population of 100 models, generation 2. Iteratively, the sequence of copying-mating-altering continues with "jumps" in the decile table fitness values until the fitness values start to flatten, that is, until no further improvement in fitness is observed; equivalently, the decile table remains stable.

To capture the data-reused variables, the iterative process is somewhat modified. When there is jump in the decile table fitness values, the GenIQer stops the process to (1) take the corresponding GenIQ Model, which is a variable itself, conventionally labeled GenIQvar_i, and (2) append the data-reused variable to the original dataset. The GenIQer captures and appends two or three GenIQvar variables. Then, the GenIQer builds the final GenIQ Model with the original variables and the appended GenIQvar variables. If an extreme jump is perchance observed in the final step of building the GenIQ Model, the GenIQer can, of course, append the corresponding

GenIQvar variable to the original dataset, then restart the final GenIQ model building.

25.3.3 Data-Reused Variables GenIQvar_1 and GenIQvar_2

GenIQvar_1 and GenIQvar_2 are displayed in Figures 25.2 and 25.3. GenIQvar_1 is defined by the original variables XX1 and XX2 and the cosine function, which is not surprising as the trigonometric functions are known to maximize the decile table fitness function. Note the bottom branch of the GenIQvar_1 tree: XX1/XX1. The GP method *edits* periodically the tree for such branches (in this case, XX1/XX1 would be replaced by 1). Also, the GP method edits to eliminate redundant branches. GenIQvar_2 is defined by GenIQvar_1, -0.345, and -0.283 and the following functions: addition, division, and multiplication. The following functions:

In sum, the seemingly simple PROFIT data in Table 25.1 is not so simple after all if one seeks to perfectly rank the PROFIT variable. Now, I declare GenIQ outperforms the logistic regression model (LRM). This exemplar illustration points to the superiority of GenIQ modeling over statistical regression modeling.

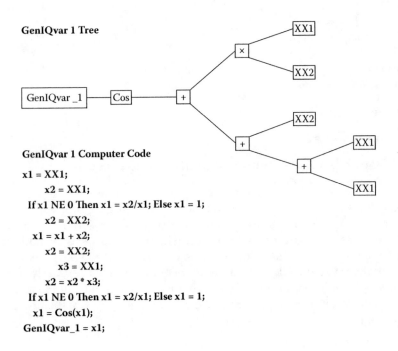

GenIQvar 1 Tree

GenIQvar 1 Computer Code

```
x1 = XX1;
    x2 = XX1;
  If x1 NE 0 Then x1 = x2/x1; Else x1 = 1;
    x2 = XX2;
  x1 = x1 + x2;
    x2 = XX2;
      x3 = XX1;
    x2 = x2 * x3;
  If x1 NE 0 Then x1 = x2/x1; Else x1 = 1;
    x1 = Cos(x1);
  GenIQvar_1 = x1;
```

FIGURE 25.2
Date-reused variable GenIQvar_1.

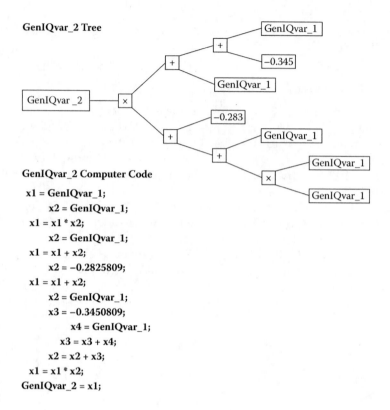

GenIQvar_2 Computer Code

```
x1 = GenIQvar_1;
    x2 = GenIQvar_1;
x1 = x1 * x2;
    x2 = GenIQvar_1;
x1 = x1 + x2;
    x2 = -0.2825809;
x1 = x1 + x2;
    x2 = GenIQvar_1;
    x3 = -0.3450809;
        x4 = GenIQvar_1;
    x3 = x3 + x4;
    x2 = x2 + x3;
x1 = x1 * x2;
GenIQvar_2 = x1;
```

FIGURE 25.3
Data-reused variable GenIQvar_2.

25.4 Modified Data Reuse: A GenIQ - Enhanced Regression Model

I modify the definition of data reuse to let the data mining prowess of GenIQ enhance the results of an already-built statistical regression model. I define data reuse as appending new variables, which are found when building a GenIQ or *any model*, to the original dataset. I build a GenIQ Model with only one predictor variable, the already-built model score. This approach of using GenIQ as a GenIQ enhancer of an existing regression model has been shown empirically to improve significantly the existent model 80% of the time.

25.4.1 Illustration of a GenIQ-Enhanced LRM

I build an LRM with dependent variable RESPONSE, predictor variables XX1 and XX2, and the data in Table 25.3.

TABLE 25.3

Response Data

ID	XXI	XX2	RESPONSE
1	31	38	Yes
2	12	30	No
3	35	21	Yes
4	23	30	No
5	45	37	No
6	16	13	No
7	45	5	Yes
8	30	30	Yes
9	6	10	Yes
10	30	10	No

TABLE 25.4

Scored and Ranked Data by Prob_of_Response

ID	XX1	XX2	RESPONSE	Prob_of_ Response
7	45	5	Yes	0.75472
10	30	10	No	0.61728
3	35	21	Yes	0.57522
5	45	37	No	0.53452
6	16	13	No	0.48164
8	30	30	Yes	0.46556
9	6	10	Yes	0.42336
1	31	38	Yes	0.41299
4	23	30	No	0.40913
2	12	30	No	0.32557

The LRM is in Equation (25.2).

$$\text{Logit of RESPONSE (= Yes)} = 0.1978 - 0.0328*XX1 + 0.0308*XX2 \quad (25.2)$$

The assessment of the RESPONSE LRM is made in terms of decile table performance. The individuals in Table 25.2 are scored and ranked by Prob_of_Response. The resultant performance is in Table 25.4. The RESPONSE ranking *is not perfect,* although three individuals are correctly placed in the proper positions. Specifically, individual ID 7 is correctly placed in the top position (top decile, decile of size 1); individuals ID 4 and ID 2 are correctly placed in the bottom two positions (ninth and bottom deciles, respectively, each decile of size 1).

Now, I build a GenIQ Model with only one predictor variable, the LRM defined in Equation (25.2). The assessment of the resultant GenIQ-enhanced

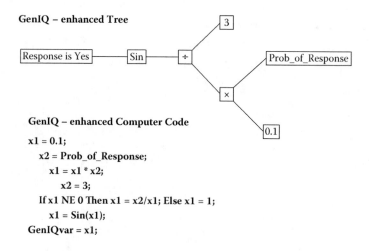

GenIQ – enhanced Tree

GenIQ – enhanced Computer Code

```
x1 = 0.1;
  x2 = Prob_of_Response;
    x1 = x1 * x2;
      x2 = 3;
  If x1 NE 0 Then x1 = x2/x1; Else x1 = 1;
      x1 = Sin(x1);
GenIQvar = x1;
```

FIGURE 25.4
The GenIQ-enhanced RESPONSE.

TABLE 25.5

Scored, Ranked Data by GenIQvar

ID	XX1	XX2	Prob_of_Response	RESPONSE	GenIQvar
8	30	30	0.46556	Yes	0.99934
9	6	10	0.42336	Yes	0.98444
3	35	21	0.57522	Yes	0.95002
7	45	5	0.75472	Yes	0.88701
1	31	38	0.41299	Yes	-0.37436
5	45	37	0.53452	No	-0.41121
6	16	13	0.48164	No	-0.51791
2	12	30	0.32557	No	-0.86183
4	23	30	0.40913	No	-0.87665
10	30	10	0.61728	No	-0.99553

RESPONSE model (Figure 25.4) is made in terms of decile table performance. The individuals in Table 25.2 are scored and ranked GenIQvar. The ranking performance is in Table 25.5. The RESPONSE ranking *is perfect*.

The GenIQ-enhanced computer code is the GenIQ-enhanced RESPONSE model equation.

The GenIQ-enhanced tree indicates the GenIQ-enhanced RESPONSE model has predictor variable Prob_of_Response and two numeric variables, 3 and 0.1, combined with the sine (Sin), division, and multiplication functions.

In sum, the GenIQ-enhanced RESPONSE Model clearly improves the already-built RESPONSE LRM. As the GenIQ Model is inherently a nonlinear model, the GenIQ-enhanced model captures nonlinear structure in the data that the statistical regression model cannot capture because it is, after all, a linear model. This exemplar illustration implies the GenIQ-enhanced model is worthy of application.

25.5 Summary

I introduced the concept of data reuse, a powerful data mining effect of the GenIQ Model. Data reuse is the appending of new variables, which are found when building a GenIQ model, to the original dataset. The benefit of data reuse is apparent: The original dataset is enhanced with the addition of new GenIQ variables, which are filled with predictive power.

I provided a formal treatment of the fundamental characteristics of data-reused variables. Then, I illustrated data reuse as a powerful data mining effect of the GenIQ Model. First, I built an OLS regression model. After discussing the results of the OLS modeling, I built a corresponding GenIQ model. Comparing and contrasting the two models, I showed the predictive power of the data reuse technique.

I modified the definition of data reuse to let the data mining prowess of GenIQ enhance the results of an already-built statistical regression model. I redefined data reuse as appending new variables, which are found when building a GenIQ or any model, to the original dataset. I built a GenIQ Model with only one predictor variable, the regression equation of an already-built LRM. This approach of using GenIQ as a GenIQ enhancer of an existing regression model showed significant improvement over the statistical model. Consequently, I illustrated the GenIQ enhancer is a power technique to extract the nonlinear structure in the data, which the statistical regression model cannot capture in its final regression equation because it is a linear model.

26

A Data Mining Method for Moderating Outliers Instead of Discarding Them

26.1 Introduction

In statistics, an outlier is an observation whose position falls outside the overall pattern of the data. Outliers are problematic: Statistical regression models are quite sensitive to outliers, which render an estimated regression model with questionable predictions. The common remedy for handling outliers is to "determine and discard" them. The purpose of this chapter is to present an alternative data mining method for moderating outliers instead of discarding them. I illustrate the data mining feature of the GenIQ Model as a method for moderating outliers with a simple, compelling presentation.

26.2 Background

In statistics, an *outlier* is an observation whose position falls outside the overall pattern of the data.* There are numerous statistical methods for identifying outliers. The most popular methods are the univariate tests.[†] There are many multivariate methods,[‡] but they are not the first choice because of the advanced expertise required to understand their underpinnings (e.g., Mahalanobis's distance). Almost all the univariate and multivariate methods are based on the assumption of normally distributed data, an untenable

---

* This is my definition of an outlier. There is no agreement for the definition of an outlier. There are many definitions, which in my opinion only reflect each author's writing style.
† Three popular classical univariate methods for normally distributed data are (a) z-score method, (b) modified z-score method, and (c) the Grubbs's test. One EDA method, which assumes no distribution of the data, is the boxplot method.
‡ The classical approach of identifying outliers is to calculate Mahalanobis's distance using robust estimators of the covariance matrix and the mean array. A popular class of robust estimators is M estimators, first introduced by Huber (Huber, P. J., Robust Estimation of a Location Parameter. Annals of Mathematical Statistics, 35:73–101, 1964).

condition to satisfy with either big or small data. If the normality assumption is not met, then the decision "there is an outlier" may be due to the nonnormality of the data rather than the presence of an outlier. There are tests for nonnormal data,* but they are difficult to use and not as powerful as the tests for normal data.

The statistical community[†] has not addressed uniting the outlier detection methodology and the "reason for the existence" of the outlier. Hence, I maintain the current approach of determining and discarding an outlier, based on application of tests *with* the untenable assumption of normality and *without* accounting for reason of existence, is wanting.

The qualified outlier is a serious matter in the context of statistical regression modeling. Statistical regression models are quite sensitive to outliers, which render an estimated regression model with questionable predictions. Without a workable robust outlier detection approach, statistical regression models go unwitting in production with indeterminable predictions. In the next section, I introduce an alternative approach to the much-needed outlier detection methodology. The alternative approach uses the bivariate graphic outlier technique, the scatterplot, and the GenIQ Model.

26.3 Moderating Outliers Instead of Discarding Them

Comparing relationships between pairs of variables, in scatterplots, is a way of drawing closer to outliers. The scatterplot is an effective nonparametric (implication: flexible), assumption-free (no assumption of normality of data) technique. Using the scatterplot in tandem with the GenIQ Model provides a perfect pair of techniques for moderating outliers instead of discarding them. I illustrate this perfect duo with a seemingly simple dataset.

26.3.1 Illustration of Moderating Outliers Instead of Discarding Them

I put forth for the illustration the dataset[‡] as described in Table 26.1.

The scatterplot is a valuable visual display to check for outliers: identify data points off the observed linear relationship between two variables under consideration. In the illustration, the scatterplot of (XX, Y) (Figure 26.1) suggests the single circled point (1, 20), labeled A, is an outlier, among the linear

---

* Popular tests for nonnormal data are given in Barnett, V., and Lewis, T., *Outliers in Statistical Data*, Wiley, New York, 1984.
[†] To the best of my knowledge.
[‡] This dataset is in Huck, S.W., Perfect correlation ¼ if not for a single outlier, *STAT*, 49, 9, 2008.

TABLE 26.1

Dataset of Paired Variables (XX, Y)

■ Consider the dataset of 101 points (XX, Y).

  – There are four "mass" points; each has 25 observations:

    • (17, 1) has 25 observations

    • (18, 2) has 25 observations

    • (19, 4) has 25 observations

    • (20, 4) has 25 observations

  – There is one "single" point.

    • (1, 20) has 1 observation

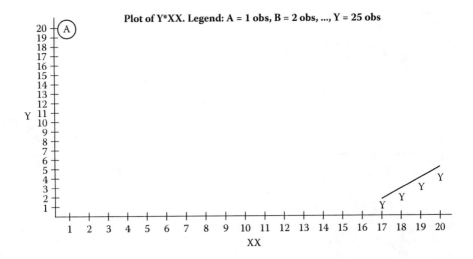

FIGURE 26.1
Scatterplot of (XX, Y).

relationship of the 100 points, consisting of the four mass points labeled Y. I assume, for a moment, the reason for existence of the outlier is sound.

The correlation coefficient of (XX, Y) is -0.41618 (see Table 26.2). The correlation coefficient value implies the strength of the linear relationship between XX and Y *only if* the corresponding scatterplot indicates an underlying linear relationship between XX and Y. Under the momentary assumption, the relationship between XX and Y is not linear; thus, the correlation coefficient value of -0.41618 is meaningless.

If a reason for the existence of outlier point (1, 20) cannot be posited, then the relationship is assumed curvilinear, as depicted in the scatterplot in Figure 26.2. In this case, there is no outlier, and the model builder seeks to reexpress the paired variables (XX, Y) to straighten the curvilinear relationship and then observe the resultant scatterplot for new outliers.

TABLE 26.2

Correlation Coefficients: N = 101 Prob > **r**
under H0: Rho=0

	XX	GenIQvar
Y	−0.41618	0.84156
	<.0001	<.0001

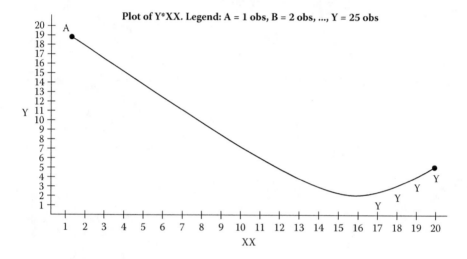

FIGURE 26.2
Scatterplot of (XX, Y) depicting a nonlinear relationship.

I have doubts about point (1, 20) as an outlier. Thus, I attempt to reexpress the paired variables (XX, Y) to straighten the curvilinear relationship. I apply the GenIQ Model to paired variables (XX, Y). The GenIQ reexpression of (XX, Y) is (GenIQvar, Y), where GenIQvar is the GenIQ data-mined transformation of XX. (The specification of the transformation is discussed in Section 26.3.2) The scatterplot of (GenIQvar, Y) (Figure 26.3) shows no outliers. More important, the scatterplot indicates a linear relationship between GenIQvar and Y and the meaningful correlation coefficient of (GenIQvar, Y), 0.84156 (see Table 26.2) quantifies the linear relationship as strong.

A posttransformation justification of point (1, 20) as a nonoutlier requires an explanation in light of the four mass points where GenIQ is −1.00, on the horizontal axis, labeled GenIQvar. The four points are in a vertical trend, which is viewed as a variation about a "true" transformed mass point. The scatterplot shows the quintessential straight line defined by two points: the true transformed point, positioned at the middle of the four Y's, (−1.00, 2.5), and the point (1.00, 20), labeled A.

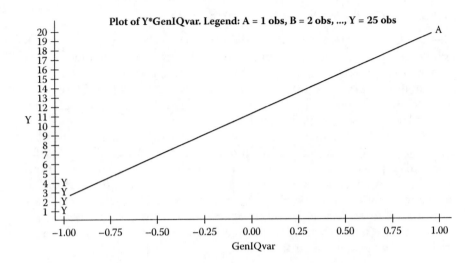

FIGURE 26.3
Scatterplot of (GenIQvar, Y).

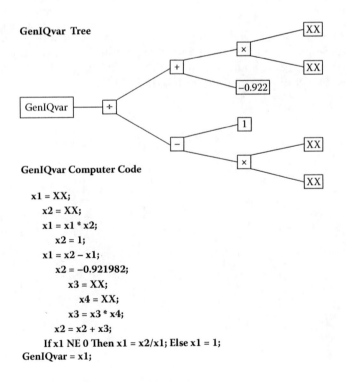

FIGURE 26.4
GenIQ Model for moderating the outlier.

26.3.2 The GenIQ Model for Moderating the Outlier

The GenIQ Model moderates the outlier point (1, 20) by reexpressing all 101 points. Comparing scatterplots in Figures 26.1 and 26.3, the outlier is repositioned from top left to top right in the scatterplot, and the remaining four mass points are repositioned from the bottom left to the bottom right in the scatterplot. The GenIQ Model transformation is defined by the GenIQvar tree and GenIQvar computer code (Figure 26.4).

The GenIQ Model, illustrated to moderate a *single* outlier, can handily serve as a multivariate method for moderating virtually *all* outliers in the data. This is possible due to the nature of the GenIQ decile table fitness function. A discussion of how optimizing the decile table fitness function brings forth a moderation of virtually all outliers is beyond the scope of this chapter. Suffice it to say that such optimization is equivalent to straightening the data by repositioning the outliers into a multilinear pattern, thus moderating the outliers instead of discarding them.

26.4 Summary

The common remedy for handling outliers is to determine and discard them. I presented an alternative data mining method for moderating outliers instead of discarding them. I illustrated the data mining feature of the GenIQ Model as the alternative method for handling outliers.

27

Overfitting: Old Problem, New Solution

27.1 Introduction

Overfitting, a problem akin to model inaccuracy, is as old as model building itself, as it is part of the modeling process. An overfitted model is one that *approaches reproducing* the training data on which the model is built. The effect of overfitting is an inaccurate model. The purpose of this chapter is to introduce a new solution, based on the data mining feature of the GenIQ Model, to the old problem of overfitting. I illustrate how the GenIQ Model identifies the complexity of the idiosyncrasies and subsequently instructs for deletion of the individuals that contribute to the complexity in the data under consideration.

27.2 Background

Overfitting, a problem akin to model inaccuracy, is as old as model building itself, as it is part of the modeling process. An overfitted model is one that *approaches reproducing* the training data on which the model is built—by capitalizing on the idiosyncrasies of the training data. The model reflects the complexity of the idiosyncrasies by including extra variables, interactions, and variable constructs. It follows that a key characteristic of an overfitted model is the model has too many variables: An overfitted model is too complex.

In another rendering, an overfitted model can be considered a *too perfect picture* of the predominant pattern in the data; the model memorizes the training data instead of capturing the desired pattern. Individuals of validation data (drawn from the population of the training data) are strangers who are unacquainted with the training data and cannot expect to fit into the model's perfect picture of the predominant pattern.

When the accuracy of a model based on the validation data is "out of the neighborhood" of the accuracy of the model based on the training data, the problem is one of overfitting. As the fit of the model *increases* by *including more information* (seemingly to be a good thing), the predictive performance of the model on the validation data *decreases*. This is the *paradox of overfitting.*

Related to overfitted models is the concept of prediction error variance. Overfitted models have large predictive error variance: The confidence interval about the prediction error is large.

27.2.1 Idiomatic Definition of Overfitting to Help Remember the Concept

A model is built to *represent* training data, not to *reproduce* training data. Otherwise, a visitor (individual's data point) from validation data will not feel at home with the model. The visitor encounters an uncomfortable fit in the model because he or she probabilistically *does not* look like a typical data point from the training data. The misfit visitor takes a poor prediction. The model is overfitted.

The underfitted model, a nonfrequenter model, has too few variables: The underfitted model is too simple. An underfitted model can be considered a *poorly rendered picture* of the predominant pattern; without recollection of the training data, the model captures poorly the desired pattern. Individuals of validation data are strangers who have no familiarity with the training data and cannot expect to fit into the model's portraiture of the predominant pattern.

As overfitted models affect (prediction) error variance, underfitted models affect error bias. *Bias* is the difference between the predicted score and the true score. Underfitted models have large error bias; predicted scores are wildly far from the true scores. Figure 27.1 is a graphical depiction of over- and underfitted models.*

Consider the two models: the simple model $g(x)$ in the left-hand graph and the *zigzag* model in right-hand graph in Figure 27.1. Clearly, I want a model that best represents the predominant pattern of the parabola as depicted by the data points indicated by circles. I fit the points with the straight-line model $g(x)$, using only one variable (too few variables). The model is visibly too simple. It does not do a good job of fitting the data and would not do well in predicting for new data points. This model is underfitted.

For the *rough* zigzag model, I fit the data to hit every data point by using too many variables. The model does a perfect job of reproducing the data points but would not do well in predicting for new data points. This model is utterly overfitted. The model does not reflect the obvious *smooth* parabolic pattern. As is plainly evident, I want a model between the $g(x)$ and zigzag models, a model that is powerful enough to represent the apparent pattern of a parabola. The conceptual building of the desired model is given to the reader.

It is "fitting" to digress here for a discussion of model accuracy. A well-fitted model is one that *faithfully represents* the sought-after predominant pattern in the data, ignoring the idiosyncrasies in the training data. A well-fitted

---

* http://www.willamette.edu/gorr/classes/cs449.html, 2010.

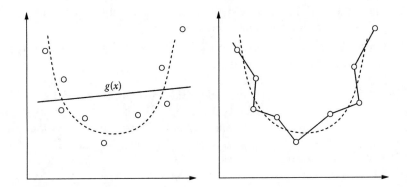

FIGURE 27.1
Over- and underfitted models.

$$\text{Model's accuracy}(\text{HOLDOUT}) \approx \text{Model's accuracy}(\text{TRAINING})$$

FIGURE 27.2
Definition of well-fitted model.

$$\text{Model Accuracy}(\text{HOLDOUT}) \leq \text{Model Accuracy}(\text{TRAINING})$$

FIGURE 27.3
Definition of overfitted model.

model is typically defined by a handful of variables. Individuals of valida-tion data, the everyman and everywoman incognizant with the training data, can expect to fit into the model's faithfully rendered picture of the pre-dominant pattern. The accuracy of the well-fitted model on validation/hold-out data is within the neighborhood of the model's accuracy based on the training data. Thus, the well-fitted model can be defined by the *approximately equal* equality, as depicted in Figure 27.2.

In contrast, the accuracy of the overfitted model on validation data will be "outside the neighborhood" of the model's accuracy based on the training data. Thus, the overfitted model can be defined by the *less than* inequality, as depicted in Figure 27.3.

27.3 The GenIQ Model Solution to Overfitting

I introduce a new solution, based on the data mining feature of the GenIQ Model, to the old problem of overfitting. The GenIQ Model is used in the following steps:

1. Identify the complexity of the idiosyncrasies, the variables, and their constructs of the idiosyncrasies.

2. Delete the individuals that contribute to the complexity of the data from the dataset under consideration.

3. A model can now be built on *clean* data to represent honestly the predominant pattern, yielding a well-fitted model.

I illustrate how the GenIQ Model identifies the complexity of the idiosyncrasies, and subsequently instructs for deletion of the individuals that contribute to the complexity in the data, from the dataset under consideration. Using the popular random-split validation, I create a variable RANDOM_ SPLIT (R-S) that randomly divides the dataset into equal halves (50%-50%).* The SAS code for the construction of RANDOM_SPLIT within a real case study dataset Overfit is

```
data Overfit;
set Overfit;
RANDOM_SPLIT = 0;
if uniform(12345) = le 0.5 then RANDOM_SPLIT = 1;
run;
```

The variables and their type (numeric or character) in the Overfit dataset are listed in Table 27.1.

There are three possible modeling events with respect to the condition of Overfit:

1. If Overfit has *no noise*, building a model with dependent variable RANDOM_SPLIT is *impossible*. The decile table has Cum Lifts equal to 100 throughout, from top to bottom deciles. Overfit is *clean* of idiosyncrasies. Accurate predictions result when building a model with Overfit data.

2. If Overfit has *negligible noise,* building a model is *most likely* possible. The decile table has Cum Lifts *within* [98, 102] in the upper deciles, say, top to third. Overfit is *almost* clean of idiosyncrasies. Highly probable accurate predictions result when building a model with Overfit data.

3. If Overfit has *unacceptable noise,* building a model is *possible*: The decile table has Cum Lifts *outside* [98, 102] in the upper deciles, say, top to third. Overfit has idiosyncrasies causing *substantial overfit-ting.* Overfit is cleaned by deleting the individuals in the deciles with Cum Lifts outside [98, 102]. As a result, a well-fitted model can be built on the clean version of Overfit. Accurate predictions result when building a model with the clean Overfit data.

---

* For unequal splits (e.g., 60%-40%), the approach is the same.

TABLE 27.1

Overfit Data: Variables, and Type

#	Variable	Type
1	RANDOM_SPLIT	Num
2	REQUESTE	Num
3	INCOME	Num
4	TERM	Num
5	APPTYPE	Char
6	ACCOMMOD	Num
7	CHILDREN	Num
S	MOVES5YR	Num
9	MARITAL	Num
10	EMPLOYME	Num
11	DIRECTDE	Char
12	CONSOLID	Num
13	NETINCOM	Num
14	EMPLOY_1	Char
15	EMAIL	Num
16	AGE	Num
17	COAPP	Num
18	GENDER	Num
19	INCOMECO	Num
20	COSTOFLI	Num
21	PHCHKHL	Char
22	PWCHKHL	Char
23	PMCHKHL	Char
24	NOCITZHL	Char
25	EMPFLGHL	Char
26	PFSFLGHL	Char
27	NUMEMPLO	Num
28	BANKAFLG	Char
29	EMPFLGML	Char
30	PFSFLGML	Char
31	CIVILSML	Char
32	TAXHL	Num
33	TAXML	Num
34	NETINCML	Num
35	LIVLOANH	Num
36	LIVCOSTH	Num
37	CARLOAN	Num
38	CARCOST	Num
39	EDLOAN	Num
40	EDCOST	Num

(Continued)

TABLE 27.1 (Continued)

Overfit Data: Variables, and Type

#	Variable	Type
41	OTLOAN	Num
42	OTCOST	Num
43	CCLOAN	Num
44	CCCOST	Num
45	EMLFLGHL	Char
46	PHONEH	Num
47	PHONEW	Num
48	PHONEC	Num
49	REQCONSR	Num
50	TIMEEMPL	Num
51	AGECOAPP	Num
52	APPLIEDY	Num
53	GBCODE	Num

27.3.1 RANDOM_SPLIT GenIQ Model

The GenIQ Model consists of two components: a tree display and computer code. Using the Overfit dataset, I build a GenIQ Model with the dependent variable RANDOM_SPLIT. The RANDOM_SPLIT GenIQ Model tree display (Figure 27.4) identifies the complexity of the idiosyncrasies (noise) in Overfit. The RANDOM_SPLIT GenIQ Model computer code is in Figure 27.5.

27.3.2 RANDOM_SPLIT GenIQ Model Decile Analysis

The decile analysis from the RANDOM_SPLIT GenIQ Model* built on the Overfit data indicates Overfit has unacceptable noise.

1. Decile table, in Table 27.2, has Cum Lifts outside [98, 102] in the top and second deciles. Individuals in these deciles cause substantial overfitting. Individuals with Cum Lifts of 103 in deciles 4, 7, and 8 are perhaps iffy, as per the *quasi N-tile analysis* in Section 27.3.3.

2. The decile table is constructed by brute (dumb) division of the data into 10 rows of *equal size regardless* of individual model scores. As model scores often spill over into lower deciles, the dumb decile table typically yields a biased assessment of model performance.

---

* FYI: The logistic regression model could be used instead of the GenIQ Model, but the results would be the identification of only linear noise, consisting of a few variables with two functions, addition and subtraction.

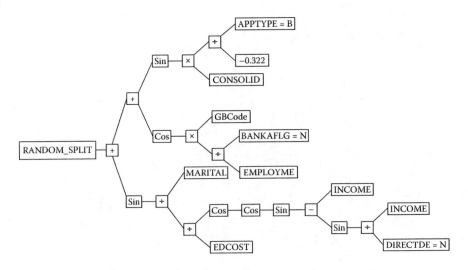

FIGURE 27.4
RANDOM_SPLIT GenIQ Model tree.

```
x1 = EDCOST;
    If DIRECTDE = "N" Then x2 = 1; Else x2 = 0;
        x3 = INCOME;
    If x2 NE 0 Then x2 = x3 / x2; Else x2 =1;
    x2 = Sin(x2);
        x3 = INCOME;
    x2 = x3 - x2; x2 = Sin(x2);
    x2 = Cos(x2); x2 = Cos(x2);
If x1 NE 0 Then x1 = x2 / x1; Else x1 =1;
    x2 = MARITAL;
If x1 NE 0 Then x1 = x2 / x1; Else x1 =1;
x1 = Sin(x1);
    x2 = EMPLOYME;
        If BANKAFLG = "N" Then x3 = 1; Else x3 = 0;
    If x2 NE 0 Then x2 = x3/x2; Else x2 = 1;
        x3 = GBCode;
x2 = x2*x3; x2 = Cos(x2);
        x3 = CON SOLID;
            x4 =-.3223163;
                If APPTYPE = "B" Then x5 = 1; Else x5=0;
            Lf x4 NE 0 Then x4 = x5/x4; Else x4 = 1;
        x3 = x3 * x4; x3 = Sin(x3);
    x2 = x2 + x3;
x1 = x1 + x2;
GenIQvar = x1;
```

FIGURE 27.5
RANDOM_SPLIT GenIQ Model computer code.

TABLE 27.2

RANDOM_SPLIT Decile Table

Decile	Predicted Random_ Split	Random_ Split Rate %	Cum Random_ Split Rate %	Cum Lift	Min score	Max score
Top	88	62.86	62.86	126	2.09	2.91
2nd	59	42.14	52.50	105	1.91	2.09
3rd	67	47.86	50.95	102	1.84	1.91
4th	73	52.14	51.25	103	1.84	1.84
5th	65	46.43	50.29	101	1.65	1.84
6th	76	54.29	50.95	102	1.14	1.65
7th	78	55.71	51.63	103	1.02	1.14
8th	69	49.29	51.34	103	0.46	1.02
9th	62	44.29	50.56	101	0.23	0.46
Bottom	63	45.00	50.00	100	-0.96	0.23

3. In Table 27.2, the top decile minimum score (2.09) spills over into the second decile. Not knowing how many model scores of 2.09 fall in the second deciles, I can only guess the percentage of data with 2.09 noise: The percentage falls in the fully open interval (0%, 20%).

4. A quasi N-tile analysis sheds light on the *spilling over of scores* (in not only the top and second deciles, but also any number of consecutive deciles).

27.3.3 Quasi N-tile Analysis

To know how many model scores fall about consecutive deciles, a quasi N-tile analysis is needed. A quasi N-tile analysis divides model scores into *score groups* or *N-tiles*, which consist of a *distinct* score for individuals within an N-tile, and *different* distinct scores across N-tiles. A quasi *smart* analysis eliminates model scores spilling over to provide an unbiased assessment of model performance.

The quasi analysis of the RANDOM_SPLIT decile table is in Table 27.3. Not knowing the distribution of the model scores, I instruct for a 20-tile analysis:

1. The smart analysis produces only six N-tiles as there are apparently only six distinct scores.

2. The top five N-tiles have Cum Lifts outside [98, 102].

3. Within these N-tiles, there are 170 (56 + 23 + 16 + 33 + 44) individuals.

4. The Cum Lift at the fifth score group is 119, after which the Cum Lifts range between [100, 102/103].

5. Deleting the 170 individuals *removes the source of noise* in Overfit data.

TABLE 27.3

Quasi 20-tile Analysis

Select Number of Tiles	20

N.TILE	NUMBER OF INDIVIDUALS	NUMBER OF Random_Split	Random_Split RATE (%)	CUM Random_Split RATE (%)	CUM% of SAMPLE	CUM LIFT (%)
Top	56	40	71.43	71.43	04.00	143
2	23	15	65.22	69.62	05.64	139
3	16	8	50.00	66.32	06.79	133
4	33	20	60.61	64.84	09.14	130
5	44	19	43.18	59.30	12.29	119
6	503	240	47.71	50.67	48.21	101

TABLE 27.4

RANDOM_SPLIT Decile Table

Decile	Predicted Random_ Split	Random_ Split Rate %	Cum Random_ Split Rate %	Cum Lift	Min score	Max score
Top	62	50.49	50.41	101	−1.26	1.33
2nd	62	50.49	50.41	101	−1.27	−1.26
3rd	61	49.67	50.27	101	−1.38	−1.27
4th	62	50.49	50.31	101	−1.54	−1.38
5th	61	49.67	50.16	100	−1.60	−1.54
6th	60	48.86	49.93	100	−3.07	−1.60
7th	62	50.49	50.00	100	−3.19	−3.07
8th	61	49.67	50.00	100	−3.28	−3.19
9th	62	50.49	50.05	100	−4.71	−3.28
Bottom	61	49.67	50.00	100	−13.44	−4.71

Overfit is now clean of noise. To test the soundness of the last assertion, I rerun GenIQ with clean Overfit data. The resultant decile table, in Table 27.4, displays Cum Lifts within [100, 101]. Hence, Overfit is clean of noise and is ready for building a well-fitted model. Note: The decile table in Table 27.4 is a *smart* decile table, which does not require my generating a corresponding quasi N-tile analysis.

27.4 Summary

Overfitting, a problem akin to model inaccuracy, is as old as model building itself because it is part of the modeling process. An overfitted model is one that approaches reproducing the training data on which the model is built—by capitalizing on the idiosyncrasies of the training data. I introduced a new solution, based on the data-mining feature of the GenIQ Model, to the old problem of overfitting. I illustrated, with a real case study, how the GenIQ Model identifies the complexity of the idiosyncrasies and instructs for deletion of the individuals that contribute to the complexity in the data to produce a dataset clean of noise. The clean dataset is ready for building a well-fitted model.

28

The Importance of Straight Data: Revisited

28.1 Introduction

The purpose of this chapter is to revisit examples discussed in Chapters 4 and 9, in which the importance of straight data is illustrated. I posited the solutions to the examples without explanation as the material needed to understand the solution was not introduced at that point.

At this point, the background required has been more than extensively covered. Thus, for completeness, I detail the posited solutions in this chapter. The solution uses the data mining feature, straightening data, of the GenIQ Model. I start with the example in Chapter 9 and conclude with the example in Chapter 4.

28.2 Restatement of Why It Is Important to Straighten Data

From Chapter 4, Section 4.2, there are five reasons why it is important to straighten data:

1. The straight-line (linear) relationship between two continuous variables, say X and Y, *is as simple as it gets*. As X increases (decreases) in its values, so does Y increase (decrease) in its values, in which case it is said that X and Y are positively correlated. Or, as X increases (decreases) in its values, so does Y decrease (increase) in its values, in which case it is said that X and Y are negatively correlated. As an example of this setting of simplicity (and everlasting importance), Einstein's E and m have a perfect positive linear relationship.

2. With linear data, the data analyst without difficulty sees *what is going on within the data*. The class of linear data is the desirable element for good model-building practice.

3. Most marketing models, belonging to the class of innumerable varieties of the statistical linear model, *require linear relationships* between a dependent variable and (a) each predictor variable in a model and

(b) *all* predictor variables considered jointly, regarding them as an array of predictor variables that have a multivariate distribution.

4. It has been shown that *nonlinear models,* which are attributed with yielding good predictions with nonstraight data, in fact *do better with straight data.*

5. I have not ignored the feature of symmetry. Not accidentally, there are theoretical reasons for *symmetry and straightness going hand in hand.* Straightening data often makes data symmetric and vice versa. Recall that symmetric data have values that are in correspondence in size and shape on opposite sides of a dividing line or middle value of the data. The iconic symmetric data profile in statistics is bell shaped.

28.3 Restatement of Section 9.3.1.1 "Reexpressing INCOME"

I envision an underlying positively sloped straight line running through the 10 points in the PROFIT-INCOME smooth plot in Figure 9.2, even though the smooth trace reveals four severe kinks. Based on the general association test with a TS (test statistic) value of 6, which is *almost* equal to the cutoff score 7, as presented in Chapter 2, I conclude there is an *almost noticeable* straight-line relationship between PROFIT and INCOME. The correlation coefficient for the relationship is a reliable $r_{PROFIT, INCOME}$ of 0.763. Notwithstanding these indicators of straightness, the relationship could use some straightening, but clearly, the bulging rule does not apply.

An alternative method for straightening data, especially characterized by nonlinearities, is the GenIQ procedure, a machine-learning, genetic-based data mining method.

28.3.1 Complete Exposition of Reexpressing INCOME

I use the GenIQ Model to reexpress INCOME. The genetic structure, which represents the reexpressed INCOME variable, labeled gINCOME, is defined in Equation (9.3):

$$gINCOME = \sin(\sin(\sin(\sin(INCOME)))*INCOME) + \log(INCOME) \quad (9.3)$$

The structure uses the nonlinear reexpressions of the trigonometric sine function (four times) and the log (to base 10) function to loosen the "kinky" PROFIT-INCOME relationship. The relationship between PROFIT and INCOME (via gINCOME) has indeed been smoothed out as the smooth trace reveals no serious kinks in Figure 9.3. Based on TS equal 6, which again is almost equal to the cutoff score of 7, I conclude there is an almost noticeable straight-line PROFIT-gINCOME relationship, a nonrandom scatter about an underlying positively sloped straight line. The correlation coefficient for the reexpressed relationship is a reliable $r_{PROFIT, gINCOME}$ of 0.894.

Visually, the effectiveness of the GenIQ procedure in straightening the data is obvious: the sharp peaks and valleys in the original PROFIT smooth plots versus the smooth wave of the reexpressed smooth plot. Quantitatively, the gINCOME-based relationship represents a noticeable improvement of 7.24% (= (0.894 - 0.763)/0.763) increase in correlation coefficient "points" over the INCOME-based relationship.

Two points of note: Recall that I previously invoked the statistical fac-toid that states a dollar-unit variable is often reexpressed with the log function. Thus, it is not surprising that the genetically evolved structure gINCOME uses the log function. With respect to logging the PROFIT vari-able, I concede that PROFIT could not benefit from a log reexpression, no doubt due to the "mini" in the dataset (i.e., the small size of the data), so I chose to work with PROFIT, not log of PROFIT, for the sake of simplicity (another EDA [exploratory data analysis] mandate, even for instructional purposes).

28.3.1.1 The GenIQ Model Detail of the gINCOME Structure

The GenIQ Model for gINCOME is in Figure 28.1.

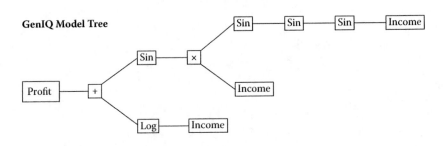

GenIQ Model Computer Code

```
x1 = INCOME;
  x1 = Log (x1);
    x2 = INCOME;
      x2 = sin (x2);
      x2 = sin (x2);
      x2 = sin (x2);
          x3 = INCOME;
      x2 = x2 + x3;
      x2 = sin (x2);
  x1 = x1 + x2;
GenIQvar = x1;
gINCOME = GenIQvar;
```

FIGURE 28.1
GenIQ Model for gINCOME.

28.4 Restatement of Section 4.6 "Data Mining the Relationship of (xx3, yy3)"

Recall, I data mine for the underlying structure of the paired variables (xx3, yy3) using a machine-learning approach under the discipline of evolutionary computation, specifically *genetic programming* (GP). The fruits of my data mining work yield the scatterplot in Figure 4.3 of Chapter 4. The data mining work is not an expenditure of preoccupied time (i.e., not waiting for time-consuming results) or mental effort as the GP-based data mining (GP-DM) is a machine-learning adaptive intelligent process, which is quite effective for straightening data. The data mining software used is the GenIQ Model, which renames the data-mined variable with the prefix GenIQvar. Data-mined (xx3, yy3) is relabeled (xx3, GenIQvar(yy3)).

28.4.1 The GenIQ Model Detail of the GenIQvar(yy3) Structure

The GenIQ Model for GenIQvar(yy3) is in Figure 28.2.

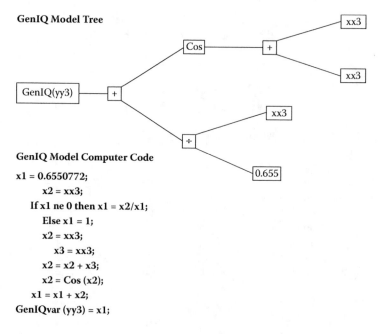

GenIQ Model Tree

GenIQ Model Computer Code

```
x1 = 0.6550772;
    x2 = xx3;
  If x1 ne 0 then x1 = x2/x1;
    Else x1 = 1;
    x2 = xx3;
      x3 = xx3;
    x2 = x2 + x3;
    x2 = Cos (x2);
  x1 = x1 + x2;
GenIQvar (yy3) = x1;
```

FIGURE 28.2
GenIQ Model for GenIQvar(yy3).

28.5 Summary

The purpose of this chapter is to revisit examples discussed in Chapters 2, 4, and 9, in which I posited the solutions to the examples without explanation, as the material needed to understand the solution was not introduced at that point. At this point, the background required has been more than extensively covered. Thus, for completeness, I detailed the posited solutions. The solutions used the data-mining feature, straightening data, of the GenIQ Model. Now, I have completed the illustrations of why it is important to straighten data.

29

The GenIQ Model: Its Definition and an Application*

29.1 Introduction

Using a variety of techniques, regression modelers build the everyday models that maximize expected response and profit based on the results of marketing programs, solicitations, and the like. Standard techniques include the statistical methods of classical discriminant analysis (DA), as well as logistic and ordinary regression. A recent addition to the regression modelers' arsenal is the machine-learning (ML) method of artificial neural networks (ANNs). Another newcomer is the GenIQ Model, a ML alternative to the statistical ordinary least squares model and the logistic regression model (LRM), is the focus of this chapter and is presented in full detail.

First, I provide background on the concept of optimization as optimization techniques provide the estimation of all models. Then, I introduce *genetic modeling*, the ML optimization approach that serves as the engine for the GenIQ Model. As the ubiquitous marketing objectives are to maximize expected response and profit for developing marketing strategies, I demonstrate how the GenIQ Model serves to meet those objectives. Actual case studies explicate further the potential of the GenIQ Model.

29.2 What Is Optimization?

Whether in business or model building, optimization is central to the decision-making process. In both theory and practice, an optimization technique involves selecting the best (or most favorable) condition within a given environment. To distinguish among available choices, an objective function (also known as a fitness function) must be predetermined. The choice, which

---

* This chapter is based on an article in *Journal of Targeting, Measurement and Analysis for Marketing,* 9, 3, 2001. Used with permission.

corresponds to the extreme value* of the objective function, is the best outcome that constitutes the details of the solution to the problem.

Modeling techniques are developed to find a specific solution to a problem. For example, in marketing, one such problem is to predict sales. The least squares regression technique is a model formulated to address sales prediction. The regression problem is formulated in terms of finding the regression equation such that the prediction errors (the difference between actual and predicted sales) are small.[†] The objective function is the prediction error, making the best equation the one that minimizes that prediction error. Calculus-based methods are used to estimate the best regression equation.

As I discuss further in the chapter, each modeling method addresses its own decision problem. The GenIQ Model addresses problems pertaining to direct, database, or telemarketing solicitations; marketing-mix optimization programs; business-intelligent offerings; customer relationship management (CRM) campaigns; Web- or e-mail-based broadcasts; and the like and uses genetic modeling as the optimization technique for its solution.

29.3 What Is Genetic Modeling?[‡]

Just as Darwin's principle of the survival of the fittest[§,¶] explains tendencies in human biology, regression modelers can use the same principle to predict the best solution to an optimization problem.[**] Each genetic model has an associated fitness function value that indicates how well the model solves, or "fits," the problem. A model with a high fitness value solves the problem

---

* If the optimization problem seeks to minimize the objective function, then the extreme value is the smallest; if it seeks to maximize, then the extreme value is the largest.
† The definition of *small* (technically called mean squared error) is the average of the squared differences between actual and predicted values.
‡ Genetic modeling as described in this chapter is formally known as genetic programming. I choose the term *modeling* instead of *programming* because the latter term, which has its roots in computer sciences, does not connote the activity of model building to data analysts with statistics or quantitative backgrounds.
§ When people hear the phrase "survival of the fittest," most think of Charles Darwin. Well, interestingly, he did not coin the term but did use it 10 years later in the fifth edition of his still-controversial *On the Origin of Species*, published in 1869.
¶ British philosopher Herbert Spence first used the phrase "survival of the fittest"—after reading Charles Darwin's *On the Origin of Species* (1859)—in his *Principles of Biology* (1864), in which he drew parallels between his own economic theories and Darwin's biological ones, writing, "This survival of the fittest is that which Mr. Darwin has called 'natural selection.'" Darwin first used Spencer's new phrase "survival of the fittest" as a synonym for "natural selection" in the fifth edition of *On the Origin of Species*, published in 1869 (this citation source was lost via "soft browse" of the Web).
** The focus of this chapter is optimization, but genetic modeling has been applied to a variety of problems: optimal control, planning, sequence induction, empirical discovery and forecasting, symbolic integration, and discovering mathematical identities.

better than a model with a lower fitness value and survives and reproduces at a high rate. Models that are less fit survive and reproduce, if at all, at a lower rate.

If two models are effective in solving a problem, then some of their parts contain probabilistically some valuable genetic material. Recombining the parts of highly fit "parent" models produces probabilistically offspring models that are better fit at solving the problem than either parent. Offspring models then become the parents of the next generation, repeating the recombination process. After many generations, an evolved model is declared the best-so-far solution of the problem.

Genetic modeling consists of the following steps [1]:

1. Define the fitness function. The fitness function allows for identifying good or bad models, after which refinements are made with the goal of producing the best model.

2. Select the set of functions (e.g., the set of arithmetic operators [addition, subtraction, multiplication, division]; log and exponential) and variables (predictors X_1, X_2, ..., X_n and numerical values) that are believed to be related to the problem at hand (the dependent variable Y).* An initial population of random models is generated using the preselected set of functions and variables.

3. Calculate the fitness of each model in the population by applying the model to a training set, a sample of individuals along with their values on the predictor variables X_1, X_2, ..., X_n and the dependent variable Y. Thus, every model has a fitness value reflecting how well it solves the problem.

4. Create a new population of models by mimicking the natural genetic operators. The genetic operators are applied to models in the current population selected with a probability based on fitness (i.e., the fitter the model, the more likely the model is to be selected).

    a. Reproduction: Copy models from the current population into the new population.

    b. Crossover: Create two offspring models for the new population by genetically recombining randomly chosen parts of two parent models from the current population.

    c. Mutation: Introduce random changes to some models from the current population into the new population.

The model with the highest fitness value produced in a generation is declared the *best-of-generation* model, which is the solution, or an approximate solution, to the problem.

---

* Effectively, I have chosen a genetic alphabet.

29.4 Genetic Modeling: An Illustration

Consider the process of building a response model, for which the dependent variable RESPONSE assumes two values: yes and no. I designate the best model as one with the highest R-squared* value. Thus, the fitness function is the formula for R-squared. (Analytical note: I am using the R-squared measure only for illustrative purposes; it is not the fitness function of the GenIQ Model. The GenIQ Model fitness function is discussed in Section 29.8.)

I have to select functions and variables that are related to the problem at hand (e.g., predicting RESPONSE). Selection is based on theoretical rationale or empirical expertise; sometimes, function selection is based on a rapid-pace trial and error.

I have two variables, X_1 and X_2, to use as predictors of RESPONSE. Thus, the variable set contains X_1 and X_2. I add the numerical value "b" to the variable set based on prior experience. I define the function set to contain the four arithmetic operations and the exponential function (exp), also based on prior experience.

Generating the initial population of random models is done with an unbiased function roulette wheel (Figure 29.1) and an unbiased function-variable roulette wheel (Figure 29.2). The slices of the function wheel are of equal size, namely, 20%. The slices of the function-variable wheel are of equal size, namely, 12.5%. Note that the division symbol % is used to denote a "protected" division. This means that division by zero, which is undefined, is set to the value 1.

To generate the first random model, I spin the function wheel. The pointer of the wheel falls on slice "+." Next, I spin the function-variable wheel, and the pointer lands on slice X_1. With two following spins of the function-variable wheel, the pointer lands on slices "X_1" and "b," successively. I decide to stop evolving the model at this point. The resultant random model 1 is depicted in Figure 29.3 as a rooted point-label tree.

I generate the second random model, in Figure 29.4, by spinning the function wheel once, then by spinning the function-variable wheel twice. The pointer lands on slices "+," X_1, and X_1, successively. Similarly, I generate three additional random models, models 3, 4, and 5 in Figures 29.5, 29.6, and 29.7, respectively. Thus, I have generated the initial population of five random models (genetic population size is five).

Each of the five models in the population is assigned a fitness value in Table 29.1 to indicate how well it solves the problem of predicting RESPONSE. Because I am using R-squared as the fitness function, I apply each model to a

---

* I know that R-squared is not an appropriate fitness function for a 0-1 dependent variable model. Perhaps I should use the likelihood function of the logistic regression model as the fitness measure for the response model or an example with a continuous (profit) variable and the R-squared fitness measure. There is more about the appropriate choice of fitness function for the problem at hand in a further section.

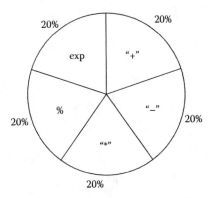

FIGURE 29.1
Unbiased function roulette wheel.

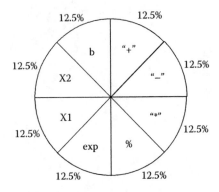

FIGURE 29.2
Unbiased function-variable roulette wheel.

Model 1: Response = b + X1

FIGURE 29.3
Random model 1.

training dataset to calculate its R-squared value. Model 1 produces the highest R-squared value, 0.52, and model 5 produces the lowest R-squared value, 0.05.

Fitness for the population itself can be calculated. The *total fitness of the population* is the sum of the fitness values among all models in the population. Here, the population total fitness is 1.53 (Table 29.1).

Model 2: Response = X1 + X1

FIGURE 29.4
Random model 2.

Model 3: Response = X1 *X1

FIGURE 29.5
Random model 3.

Model 4: Response = X1 * (b + X2)

FIGURE 29.6
Random model 4.

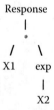

Model 5: Response = X1 * exp(X2)

FIGURE 29.7
Random model 5.

TABLE 29.1

Initial Population

	Fitness Value (R-Squared)	PTF (Fitness/Total)
Model 1	0.52	0.34
Model 2	0.41	0.27
Model 3	0.38	0.25
Model 4	0.17	0.11
Model 5	0.05	0.03
Population total fitness	1.53	

29.4.1 Reproduction

After the initial population of random models is generated, all subsequent populations of models are evolved with adaptive intelligence via the implementation of the genetic operators and the mechanism of selection *proportional to fitness* (PTF). Reproduction is the process by which models are duplicated or copied based on selection PTF. Selection PTF is defined as model fitness value divided by population total fitness (see Table 29.1). For example, model 1 has a PTF value of 0.34 (= 0.52/1.53).

Reproduction PTF means that a model with a high PTF value has a high probability of being selected for inclusion in the next generation. The reproduction operator is implemented with a biased model roulette wheel (Figure 29.8), where the slices are sized according to PTF values.

The operation of reproduction proceeds as follows: The spin of the biased model roulette wheel determines which models and how many times the models are copied. The model selected by the pointer is copied without alteration and put into the next generation. Spinning the wheel in Figure 29.8, say 100 times, produces on average the following selection: 34 copies of model 1, 27 copies of model 2, 25 copies of model 3, 11 copies of model 4, and 3 copies of model 5.

29.4.2 Crossover

The crossover (sexual recombination) operation is performed on two parent models by recombining randomly chosen parts of the two parent models; the expectation is that the offspring models are fitter than either parent model.

The crossover operation works with selection PTF. An illustration makes this operation easy to understand. Consider the parent models in Figures 29.9 and 29.10. The operation begins by randomly selecting an internal point (a function) in the tree for the crossover site.

Say, for instance, that the crossover sites are the lower "+" and "*" for parents 1 and 2, respectively.

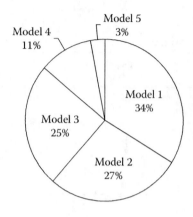

FIGURE 29.8
Biased model roulette wheel.

```
                Parent 1
                   +
                 /   \
              X2      +
                    /   \
                   b    X1
```

FIGURE 29.9
Parent 1.

```
                Parent 2
                   −
                 /   \
               *       X4
             /   \
            c    X3
```

FIGURE 29.10
Parent 2.

The crossover fragment for a parent is the subtree that has at its root the crossover site function. Crossover fragments for parents 1 and 2 are in Figures 29.11 and 29.12, respectively.

Offspring 1, in Figure 29.13, from parent 1 is produced by deleting the crossover fragment of parent 1 and then inserting the crossover fragment of parent 2 at the crossover point of parent 1. Similarly, offspring 2 (Figure 29.14) is produced.

29.4.3 Mutation

The mutation operation begins by selecting a point at random within a tree. This mutation point can be an internal point (a function) or an external

Crossover Fragment 1

FIGURE 29.11
Crossover fragment 1.

Crossover Fragment 2

FIGURE 29.12
Crossover fragment 2.

Offspring 1

FIGURE 29.13
Offspring 1.

Offspring 2

```
        –
      /   \
   +       X4
  / \
 b   X1
```

FIGURE 29.14
Offspring 2.

or terminal point (a variable or numerical value). The mutation operation either *replaces* a randomly generated function with another function (from the function set previously defined) or *inverts** the terminals of the subtree whose root is the randomly selected internal point.

For example, model I (Figure 29.15) is mutated by replacing the function "-" with "+," resulting in mutated model I.1 in Figure 29.16. Model I is also mutated by inverting the terminal points c and X3, resulting in mutated model I.2 in Figure 29.17.

---

* When a subtree has more than two terminals, the terminals are randomly permutated.

```
                Model I
                   +
                 /   \
               X1     −
                     /  \
                    c    X3
```

FIGURE 29.15
Model I for mutation.

```
                Model I.1
                   +
                 /   \
               X1     +
                     /  \
                    c    X3
```

FIGURE 29.16
Mutated model I.1.

```
                Model I.2
                   +
                 /   \
               X1     −
                     /  \
                   X3    c
```

FIGURE 29.17
Mutated model I.2.

29.5 Parameters for Controlling a Genetic Model Run

There are several control parameters that need to be set before evolving a genetic model.

1. Genetic population size, the number of models randomly generated and subsequently evolved.

2. The maximum number of generations to be run until no further improvement in the fitness function values is achieved.

3. Reproduction probability, the percentage of the population that is copied. If population size is 100 and reproduction probability is 10%, then 10 models from each generation are selected (with reselection allowed) for reproduction. Selection is based on PTF.

4. Crossover probability, the percentage of the population that is used for crossover. If population size is 100 and crossover probability is 80%, then 80 models from each generation are selected (with

reselection allowed) for crossover. Selection is based on PTF. Models are paired at random.

5. Mutation probability, the percentage of the population that is used for mutation. If population size is 100, and mutation 10%, then 10 models from each generation are selected (with reselection allowed) for mutation. Selection is based on PTF.

6. Termination criterion, the single model with the largest fitness value over all generations, the so-called best-so-far model, is declared the result of a run.

29.6 Genetic Modeling: Strengths and Limitations

Genetic Modeling has strengths and limitations, like any methodology. Perhaps the most important strength of genetic modeling is that it is a workable alternative to statistical models, which are highly parametric with sample size restrictions. Statistical models require, for their estimated coefficients, algorithms depending on smooth, unconstrained functions with the existence of derivatives (well-defined slope values). In practice, the functions (response surfaces) are noisy, multimodal, and frequently discontinuous. In contrast, genetic models are robust, assumption-free, nonparametric models and perform well on large and small samples. The only requirement is a fitness function, which can be designed to ensure that the genetic model does not perform worse than any other statistical model.

Genetic modeling has shown itself to be effective for solving large optimization problems as it can efficiently search through response surfaces of very large datasets. In addition, genetic modeling can be used to learn complex relationships, making it a viable data mining tool for rooting out valuable pieces of information.

A potential limitation of genetic modeling is in the setting of the genetic modeling parameters: genetic population size and reproduction, crossover and mutation probabilities. The parameter settings are, in part, data and problem dependent; thus, proper settings require experimentation. Fortunately, new theories and empirical studies are continually providing rules of thumb for these settings as application areas broaden. These guidelines* make genetic modeling an accessible approach to regression modelers not formally trained in genetic modeling. Even with the "correct" parameter settings, genetic models do not guarantee the optimal (best) solution. Further, genetic models are only as good as the definition of the fitness function. Precisely defining the fitness function sometimes requires expert experimentation.

---

* See http://www.geniq.net/GenIQModelFAQs.html from my Web site.

29.7 Goals of Marketing Modeling

Marketers typically attempt to improve the effectiveness of their marketing strategies by targeting their best customers or prospects. They use a model to identify individuals who are likely to respond to or generate profit* from a campaign, solicitation, and the like. The model provides, for each individual, estimates of probability of response or estimates of contribution to profit. Although the precision of these estimates is important, the performance of the model is measured at an aggregated level as reported in a decile analysis.

Marketers have defined the *Cum Lift*, which is found in the decile analysis, as the relevant measure of model performance. Based on the selection of individuals by the model, marketers create a "push-up" list of individuals likely to respond or contribute to profit to obtain an advantage over a random selection of individuals. The Cum Response Lift is an index of how many more responses are expected with a selection based on a model over the expected responses with a random selection (without a model). Similarly, the Cum Profit Lift is an index of how much more profit is expected with a selection based on a model over the profit expected with a random selection (without a model). The concept of Cum Lift and the steps of the construction in a decile analysis are described in Chapter 18.

It should be clear that a model that produces a decile analysis with more responses or profit in the upper (top, second, third, or fourth) deciles is a better model than a model with fewer responses or less profit in the upper deciles. This concept is the motivation for the GenIQ Model.

29.8 The GenIQ Response Model

The GenIQ approach to modeling is to address the ubiquitous objective concerning regression modelers across a multitude of industry sectors (e.g., direct, database, or telemarketing; business analytics; risk analytics; consumer credit; customer life cycle; financial services marketing; and the like), namely, maximizing response and profit. The GenIQ Model uses the genetic methodology to optimize *explicitly* the desired criterion: *maximize the upper deciles*. Consequently, the GenIQ Model allows regression modelers to build response and profit models in ways that are not possible with current statistical models.

The GenIQ Response Model is theoretically superior—with respect to maximizing the upper deciles—to a response model built with alternative response techniques because of the explicit nature of the fitness function. The

---

* I use the term *profit* as a stand-in for any measure of an individual's worth, such as sales per order, lifetime sales, revenue, number of visits, or number of purchases.

formulation of the fitness function is beyond the scope of this chapter. But, suffice it to say, the fitness function seeks to fill the upper deciles with as many responses as possible, equivalently, to maximize the Cum Response Lift.

Alternative response techniques, such as DA, LRM, and ANN only maximize implicitly the desired criterion. Their optimization criterion (fitness function) serves as a surrogate for the desired criterion. DA, with the assumption of bell-shaped data, is defined to maximize explicitly the ratio of between-group sum of squares to within-group sum of squares.

LRM, with the two assumptions of independence of responses and an S-shape relationship between predictor variables and response, is defined to maximize the logistic likelihood (LL) function.

The ANN, a highly parametric method, is typically defined to minimize explicitly mean squared error (MSE).

29.9 The GenIQ Profit Model

The GenIQ Profit Model is theoretically superior—with respect to maximizing the upper deciles—to the OLS regression and ANN. The GenIQ Profit Model uses the genetic methodology with a fitness function that explicitly addresses the desired modeling criterion. The fitness function is defined *to fill the upper deciles with as much profit as possible that, equivalently, maximizes the Cum Profit Lift*.

The fitness function for OLS and ANN models minimizes MSE, which serves as a surrogate for the desired criterion.

OLS regression has another weakness in a marketing application. A key assumption of the regression technique is the dependent variable data must follow a bell-shaped curve. If the assumption is violated, the resultant model may not be valid. Unfortunately, profit data are not bell shaped. For example, a 2% response rate yields 98% nonresponders with profit values of zero dollars or some nominal cost associated with nonresponse. Data with a concentration of 98% of a single value cannot be spread out to form a bell-shape distribution.

There is still another data issue when using OLS with marketing data. Lifetime value (LTV) is an important marketing performance measure. LTV is typically positively skewed. The log is the appropriate transformation to reshape positively skewed data into a bell-shaped curve. However, using the log of LTV as the dependent variable in OLS regression does not guarantee that other OLS assumptions are not violated.* Accordingly, attempts at modeling profit with ordinary regression are questionable or difficult.

---

* The error structure of the OLS equation may not necessarily be normally distributed with zero mean and constant variance, in which case the modeling results are questionable and additional transformations may be needed.

The GenIQ Response and Profit Models have no restriction on the dependent variable. The GenIQ Models produce accurate and precise predictions with a dependent variable of any shape.* This is because the GenIQ estimation is based on the genetic methodology, which is inherently nonparametric and Assumption-free.

In fact, due to its nonparametric and assumption-free estimation, the GenIQ Models place no restriction on the interrelationship among the predictor variables. The GenIQ Models are unaffected by any degree of correlation among the predictor variables. In contrast, OLS and ANN, as well as DA and LRM, can tolerate only a "moderate" degree of intercorrelation among the predictor variables to ensure a stable calculation of their models. Severe degrees of intercorrelation among the predictor variables often lead to inestimable models.

Moreover, the GenIQ Models have no restriction on sample size. The GenIQ Models can be built on small samples as well as large samples. OLS, DA, and, somewhat less, ANN and LRM[†] models require at least a "moderate" size sample.[‡]

29.10 Case Study: Response Model

Cataloguer ABC requires a response model based on a recent direct mail campaign, which produced an 0.83% response rate (dependent variable is RESPONSE). ABC's consultant built an LRM using three variables based on the techniques discussed in Chapter 8:

1. RENT_1: a composite variable measuring the ranges of rental cost[§]
2. ACCT_1: a composite variable measuring the activity of various financial accounts[¶]
3. APP_TOTL: the number of inquiries

The logistic response model is defined in Equation (29.1) as

$$\text{Logit of RESPONSE} = -1.9 + 0.19*\text{APP\_TOTL} - 0.24*\text{RENT\_1} - 0.25*\text{ACCTS\_1}$$
(29.1)

---

* The dependent variable can be bell shaped or skewed, bimodal or multimodal, and continuous or discontinuous.
† There are specialty algorithms for logistic regression with small sample size.
‡ Statisticians do not agree on how "moderate" a moderate sample size is. I drew a sample of statisticians to determine the size of a moderate sample; the average was 5,000.
§ Four categories of rental cost: less than $200 per month, $200–$300 per month, $300–$500 per month, and greater than $500 per month.
¶ Financial accounts include bank cards, department store cards, installments loans, and so on.

TABLE 29.2

LRM Response Decile Analysis

Decile	Number of Individuals	Number of Responses	Decile Response Rate	Cumulative Response Rate	Cum Lift Response
Top	1,740	38	2.20%	2.18%	264
2	1,740	12	0.70%	1.44%	174
3	1,740	18	1.00%	1.30%	157
4	1,740	12	0.70%	1.15%	139
5	1,740	16	0.90%	1.10%	133
6	1,740	20	1.10%	1.11%	134
7	1,740	8	0.50%	1.02%	123
8	1,740	10	0.60%	0.96%	116
9	1,740	6	0.30%	0.89%	108
Bottom	1,740	4	0.20%	0.83%	100
Total	17,400	144	0.83%		

The LRM response validation decile analysis in Table 29.2 shows the performance of the model over chance (i.e., no model). The decile analysis shows a model with good performance in the upper deciles: Cum Lifts for top, second, third, and fourth deciles are 264, 174, 157, and 139, respectively. Note that the model may not be as good as initially believed. There is some degree of unstable performance through the deciles; that is, the number of responses does not decrease steadily through the deciles. This unstable performance, which is characterized by "jumps" in deciles 3, 5, 6, and 8, is probably due to (1) an unknown relationship between the predictor variables and RESPONSE or (2) an important predictor variable, which is not included in this model. However, it should be pointed out that only perfect models have perfect performance throughout the deciles. Good models have some jumps, albeit minor ones.

I build a GenIQ Response Model based on the same three variables used in the LRM. The GenIQ response tree is in Figure 29.18. The validation decile analysis, in Table 29.3, shows a model with very good performance in the upper deciles: Cum Lifts for top, second, third, and fourth deciles are 306, 215, 167, and 142. In contrast with the LRM, the GenIQ Model has only two minor jumps in deciles 5 and 7. The implication is the genetic methodology has evolved a better model because it has uncovered a non-linear relationship among the predictor variables with response. This comparison between LRM and GenIQ is conservative as GenIQ used the same three predictor variables used in LRM. As I discuss in Chapter 30, the strength of GenIQ is finding its own best set of variables for the prediction task at hand.

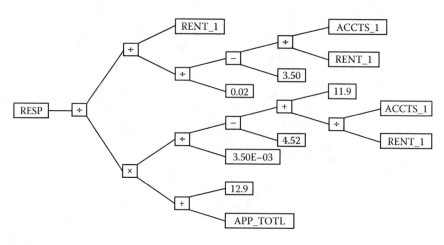

FIGURE 29.18
GenIQ response tree.

TABLE 29.3

GenIQ Response Decile Analysis

Decile	Number of Individuals	Number of Responses	Decile Response Rate	Cumulative Response Rate	Cum Lift Response
Top	1,740	44	2.50%	2.53%	306
2	1,740	18	1.00%	1.78%	215
3	1,740	10	0.60%	1.38%	167
4	1,740	10	0.60%	1.18%	142
5	1,740	14	0.80%	1.10%	133
6	1,740	10	0.60%	1.02%	123
7	1,740	12	0.70%	0.97%	117
8	1,740	10	0.60%	0.92%	111
9	1,740	8	0.50%	0.87%	105
Bottom	1,740	8	0.50%	0.83%	100
Total	17,400	144	0.83%		

The GenIQ Response Model is defined in Equation (29.2)* as:

GenIQvar_RESPONSE

$$= \frac{7.0E\text{-}5 * RENT\_1 **3}{(ACCTS\_1 - 3.50 * RENT\_1) * (12.9 + APP\_TOTL) * (ACCTS\_1 - 7.38 * RENT\_1)} \qquad (29.2)$$

---

* The GenIQ Response Model is written conveniently in a standard algebraic form as all the divisions indicated in the GenIQ response tree (e.g., ACCTS_1/RENT_1) are not undefined (division by zero). If at least one indicated division is undefined, then I would have to report the GenIQ Model computer code, as reported in previous chapters.

TABLE 29.4

Comparison: LRM and GenIQ Response

Decile	LRM	GenIQ	GenIQ Improvement over LRM
Top	264	306	16.0%
2	174	215	23.8%
3	157	167	6.1%
4	139	142	2.2%
5	133	133	−0.2%
6	134	123	−8.2%
7	123	117	−4.9%
8	116	111	−4.6%
9	108	105	−2.8%
Bottom	100	100	—

GenIQ does not outperform LRM across all the deciles in Table 29.4. However, GenIQ yields noticeable Cum Lift improvements for the important top three deciles: 16.0%, 23.8%, and 6.1%, respectively.

29.11 Case Study: Profit Model

Telecommunications Company ATMC seeks to build a zip-code-level model to predict usage, dependent variable TTLDIAL1. Based on the techniques discussed in Chapter 9, the variables used in building an ordinary regression (OLS) model are as follows:

1. AASSIS_1: composite of public assistance-related census variables
2. ANNTS_2: composite of ancestry census variables
3. FEMMAL_2: composite of gender-related variables
4. FAMINC_1: a composite variable measuring the ranges of home value*

The OLS profit (usage) model is defined in Equation (29.3) as

$$\text{TTLDIAL1} = 1.5 + \text{-}0.35{*}\text{AASSIS\_1} + 1.1{*}\text{ANNTS\_2} +$$
$$1.4{*}\text{ FEMMAL\_2} + 2.8{*}\text{FAMINC\_1} \qquad (29.3)$$

The OLS profit validation decile analysis in Table 29.5 shows the performance of the model over chance (i.e., no model). The decile analysis shows a model with good performance in the upper deciles: Cum Lifts for top, second, third, and fourth deciles are 158, 139, 131, and 123, respectively.

---

* Five categories of home value: less than $100,000, $100–$200,000, $200–$500,000, $500–$750,000, greater than $750,000.

TABLE 29.5

Decile Analysis OLS Profit (Usage) Model

Decile	Number of Customers	Total Dollar Usage	Average Usage	Cumulative Average Usage	Cum Lift Usage
Top	1,800	$38,379	$21.32	$21.32	158
2	1,800	$28,787	$15.99	$18.66	139
3	1,800	$27,852	$15.47	$17.60	131
4	1,800	$24,199	$13.44	$16.56	123
5	1,800	$26,115	$14.51	$16.15	120
6	1,800	$18,347	$10.19	$15.16	113
7	1,800	$20,145	$11.19	$14.59	108
8	1,800	$23,627	$13.13	$14.41	107
9	1,800	$19,525	$10.85	$14.01	104
Bottom	1,800	$15,428	$8.57	$13.47	100
Total	18,000	$242,404	$13.47		

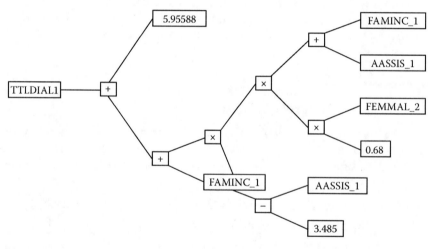

FIGURE 29.19
GenIQ Profit Tree

I build a GenIQ Profit Model based on the same four variables used in the OLS model. The GenIQ profit tree is shown in Figure 29.19. The validation decile analysis in Table 29.6 shows a model with very good performance in the upper deciles: Cum Lifts for top, second, third, and fourth deciles are 198, 167, 152, and 140, respectively. This comparison between OLS and GenIQ is conservative as GenIQ was assigned the same four predictor variables used in OLS. (Curiously, GenIQ only used three of the four variables.) As mentioned in the previous section, in which the comparison between LRM and GenIQ is made, I discuss in the next chapter why the OLS-GENIQ comparison here is conservative.

TABLE 29.6

Decile Analysis GenIQ Profit (Usage) Model

Decile	Number of Customers	Total Dollar Usage	Average Usage	Cumulative Average Usage	Cum Lift Usage
Top	1,800	$48,079	$26.71	$26.71	198
2	1,800	$32,787	$18.22	$22.46	167
3	1,800	$29,852	$16.58	$20.50	152
4	1,800	$25,399	$14.11	$18.91	140
5	1,800	$25,115	$13.95	$17.91	133
6	1,800	$18,447	$10.25	$16.64	124
7	1,800	$16,145	$8.97	$15.54	115
8	1,800	$17,227	$9.57	$14.80	110
9	1,800	$15,125	$8.40	$14.08	105
Bottom	1,800	$14,228	$7.90	$13.47	100
Total	18,000	$242,404	$13.47		

TABLE 29.7

Comparison: OLS and GenIQ Profit (Usage)

Decile	OLS	GenIQ	GenIQ Improvement Over OLS
Top	158	198	25.5%
2	139	167	20.0%
3	131	152	16.2%
4	123	140	14.1%
5	120	133	10.9%
6	113	124	9.3%
7	108	115	6.9%
8	107	110	2.7%
9	104	105	0.6%
Bottom	100	100	—

The GenIQ Profit (Usage) Model is defined in Equation (29.4) as

$$\text{GenIQvar\_TTLDIAL1} = +\ 5.95 + \text{FAMINC\_1} + (\text{FAMINC\_1}$$
$$+\ \text{AASSIS\_1})*((0.68*\text{FEMMAL\_2})*(\text{AASSIS\_1 } 3.485)) \quad (29.4)$$

GenIQ does outperform OLS across all the deciles in Table 29.7. GenIQ yields noticeable Cum Lift improvements down to the seventh decile; improvements range from 25.5% in the top decile to 6.9% in the seventh decile.

29.12 Summary

All standard statistical modeling techniques involve optimizing a fitness function to find a specific solution to a problem. The popular ordinary and logistic regression techniques, which seek accurate prediction and classification, respectively, optimize the fitness functions of MSE and the LL, respectively. Calculus-based methods are used for the optimization computations.

I presented a new modeling technique, the GenIQ Model, which seeks maximum performance (response or profit) from diversified marketing programs, solicitations, and the like. The GenIQ Model optimizes the fitness function Cum Lift. The optimization computations of GenIQ use the genetic methodology, not the usual calculus. I provided a compendious proem to genetic methodology with an illustration and inspection of its strengths and limitations.

The GenIQ Model is theoretically superior—with respect to maximizing Cum Lift—to the ordinary regression model and LRM because of its clearly and fully formulated fitness function. The GenIQ fitness function explicitly seeks to fill the upper deciles with as many responses or as much profit as possible, equivalently, to maximize the Cum Lift. Standard statistical methods only implicitly maximize the Cum Lift as their fitness functions (MSE and LL) serve as a surrogate maximizing Cum Lift.

Last, I demonstrated the potential of the new technique with response and profit model illustrations. The GenIQ Response Model illustration yields noticeable Cum Lift improvements over logistic regression for the important first three decile ranges: 16.0%, 23.8%, and 6.1%, respectively. The GenIQ Profit Model illustration yields noticeable Cum Lift improvements in ordinary regression down through the seventh decile; improvements range from 25.5% in the top decile to 6.9% in the seventh decile.

Reference

1. Koza, J., *Genetic Programming: On the Programming of Computers by Means of Natural Selection*, MIT Press, Cambridge, MA, 1992.

30

Finding the Best Variables for Marketing Models*

30.1 Introduction

Finding the best possible subset of variables to put in a model has been a frustrating exercise. Many methods of variable selection exist, but none of them is perfect. Moreover, they do not create new variables, which would enhance the predictive power of the original variables themselves. Furthermore, none uses a criterion that addresses the specific needs of marketing models. I present the GenIQ Model as a methodology that uses genetic modeling to find the best variables for marketing models. Most significant, the GenIQ Model addresses uniquely the specific requirement of marketing models, namely, to maximize the Cum Lift.

30.2 Background

The problem of finding the best subset of variables to define the best model has been extensively studied. Existing methods, based on theory, search heuristics, and rules of thumb, each use a unique criterion to build the best model. Selection criteria can be divided into two groups: one based on criteria involving classical hypothesis testing and the other involving residual error sum of squares.[†] Different criteria typically produce different subsets. The number of variables in common with the different subsets is not necessarily large, and the sizes of the subsets can vary considerably.

Essentially, the problem of variable selection is to examine certain subsets and select the subset that either maximizes or minimizes an appropriate

---

* This chapter is based on an article in *Journal of Targeting, Measurement and Analysis for Marketing*, 9, 3, 2001. Used with permission.
† Other criteria are based on information theory and Bayesian rules.

criterion. Two subsets are obvious: the best single variable and the complete set of variables. The problem lies in selecting an intermediate subset that is better than both of these extremes. Therefore, the issue is how to find the *necessary variables* among the complete set of variables by deleting both irrelevant variables (variables not affecting the dependent variable) and *redundant variables* (variables not adding anything to the dependent variable) [1].

Reviewed next are five widely used variable selection methods. The first four methods are found in major statistical software packages,* and the last is the favored rule-of-thumb approach used by many statistical modelers. The test statistic (TS) for the first three methods uses either the F statistic for a continuous dependent variable or the G statistic for a binary dependent variable (e.g., response that assumes only two values, yes/no). The TS for the fourth method is either R-squared for a continuous dependent variable or the Score statistic for a binary dependent variable. The fifth method uses the popular correlation coefficient r.

1. *Forward selection (FS):* This method adds variables to the model until no remaining variable (outside the model) can add anything significant to the dependent variable. FS begins with no variable in the model. For each variable, the TS, a measure of the contribution of the variable to the model, is calculated. The variable with the largest TS value, which is greater than a preset value C, is added to the model. Then, the TS is calculated again for the variables still remaining, and the evaluation process is repeated. Thus, variables are added to the model one by one until no remaining variable produces a TS value greater than C. Once a variable is in the model, it remains there.

2. *Backward elimination (BE):* This method deletes variables one by one from the model until all remaining variables contribute something significant to the dependent variable. BE begins with a model that includes all variables. Variables are then deleted from the model one by one until all the variables remaining in the model have TS values greater than C. At each step, the variable showing the smallest contribution to the model (i.e., with the smallest TS value that is less than C) is deleted.

3. *Stepwise (SW):* This method is a modification of the FS approach and differs in that variables already in the model do not necessarily stay. As in FS, SW adds variables to the model one at a time. Variables that have a TS value greater than C are added to the model. After a variable is added, however, SW looks at all the variables already included to delete any variable that does not have a TS value greater than C.

4. *R-squared (R-sq):* This method finds several subsets of different sizes that best predict the dependent variable. R-sq finds subsets

---

* SAS/STAT Manual. See PROC REG and PROC LOGISTIC, support.asa.com, 2011.

of variables that best predict the dependent variable based on the appropriate TS. The best subset of size k has the largest TS value. For a continuous dependent variable, TS is the popular measure R-sq., the coefficient of (multiple) determination, which measures the proportion of "explained" variance of the dependent variable by a multiple regression model. For a binary dependent variable, TS is the theoretically correct but less-known Score statistic.* R-sq finds the best one-variable model, the best two-variable model, and so forth. However, it is unlikely that one subset will stand out as clearly the best as TS values are often bunched together. For example, they are equal in value when rounded at the, say, third place after the decimal point.† R-sq generates a number of subsets of each size, which allows the modeler to select a subset, possibly using nonstatistical measures.

5. *Rule-of-thumb top k variables (Top-k):* This method selects the top ranked variables in terms of their association with the dependent variable. The association of each variable with the dependent variable is measured by the correlation coefficient r. The variables are ranked by their absolute r values,‡ from largest to smallest. The top-k ranked variables are considered the best subset. If the statistical model with the top-k variables indicates that each variable is statistically significant, then the set of k variables is declared the best subset. If any variable is not statistically significant, then the variable is removed and replaced by the next ranked variable. The new set of variables is then considered the best subset, and the evaluation process is repeated.

30.3 Weakness in the Variable Selection Methods

While the mentioned methods produce reasonably good models, each method has a drawback specific to its selection criterion. A detailed discussion of the weaknesses is beyond the scope of this chapter; however, there are two common weaknesses, which do merit attention [2, 3]. *First, the selection criteria of these methods do not explicitly address the specific needs of marketing models, namely, to maximize the Cum Lift.*

---

* R-squared theoretically is not the appropriate measure for a binary dependent variable. However, many analysts use it with varying degrees of success.
† For example, consider two TS values: 1.934056 and 1.934069. These values are equal when rounding occurs at the third place after the decimal point: 1.934.
‡ Absolute r value means that the sign is ignored. For example, if r = -0.23, then absolute r = +0.23.

Second, these methods cannot identify structure in the data. They find the best subset of variables without "digging" into the data, a feature that is necessary for finding important variables or structures. Therefore, variable selection methods without data mining capability cannot generate the enhanced best subset. The following illustration clarifies this weakness. Consider the complete set of variables, X_1, X_2, ..., X_{10}. Any of the variable selection methods in current use will only find the best combination of the original variables (say X_1, X_3, X_7, X_{10}), but can never automatically transform a variable (say transform X_1 to log X_1) if it were needed to increase the information content (predictive power) of that variable. Furthermore, none of these methods can generate a reexpression of the original variables (perhaps X_3/X_7) if the constructed variable were to offer more predictive power than the original component variables combined. In other words, current variable selection methods cannot find the enhanced best subset that needs to include transformed and reexpressed variables (possibly X_1, X_3, X_7, X_{10}, log X_1, X_3/X_7). A subset of variables without the potential of new variables offering enhanced predictive power clearly limits the modeler in building the best model.

Specifically, these methods fail to identify structure of the types discussed next.

Transformed variables with a preferred shape. A variable selection procedure should have the ability to transform an individual variable, if necessary, to induce a symmetric distribution. Symmetry is the preferred shape of an individual variable. For example, the workhorses of statistical measures—the mean and variance—are based on the symmetric distribution. A skewed distribution produces inaccurate estimates for means, variances, and related statistics, such as the correlation coefficient. Analyses based on a skewed distribution provide typically questionable findings. Symmetry facilitates the interpretation of the effect of the variable in an analysis. A skewed distribution is difficult to examine because most of the observations are bunched together at either end of the distribution.

A variable selection method also should have the ability to straighten nonlinear relationships. A linear or straight-line relationship is the preferred shape when considering two variables. A straight-line relationship between independent and dependent variables is an assumption of the popular statistical linear model. (Remember, a linear model is defined as a sum of weighted variables, such as $Y = b_0 + b_1{}^*X_1 + b_2{}^*X_2 + b_3{}^*X_3$.)* Moreover, a straight-line relationship among all the independent variables considered jointly is also a desirable property [4]. Straight-line relationships are easy to interpret: A unit of increase in one variable produces an expected constant increase in a second variable.

---

* The weights or coefficients (b_0, b_1, b_2, and b_3) are derived to satisfy some criterion, such as minimize the mean squared error used in ordinary least square regression or minimize the joint probability function used in logistic regression.

Constructed variables from the original variables using simple arithmetic functions. A variable selection method should have the ability to construct simple reexpressions of the original variables. Sum, difference, ratio, or product variables potentially offer more information than the original variables themselves. For example, when analyzing the efficiency of an automobile engine, two important variables are miles traveled and fuel used (gallons). However, it is well known that the ratio variable of miles per gallon is the best variable for assessing the performance of the engine.

Constructed variables from the original variables using a set of functions (e.g., arithmetic, trigonometric, or Boolean functions). A variable selection method should have the ability to construct complex reexpressions with mathematical functions to capture the complex relationships in the data and offer potentially more information than the original variables themselves. In an era of data warehouses and the Internet, big data consisting of hundreds of thousands to millions of individual records and hundreds to thousands of variables are commonplace. Relationships among many variables produced by so many individuals are sure to be complex, beyond the simple straight-line pattern. Discovering the mathematical expressions of these relationships, although difficult without theoretical guidance, should be the hallmark of a high-performance variable selection method. For example, consider the well-known relationship among three variables: the lengths of the three sides of a right triangle. A powerful variable selection procedure would identify the relationship among the sides, even in the presence of measurement error: The longer side (diagonal) is the square root of the sum of squares of the two shorter sides.

In sum, these two weaknesses suggest that a high-performance variable selection method for marketing models should find the best subset of variables that maximizes the Cum Lift criterion. In the sections that follow, I reintroduce the GenIQ Model of Chapter 29, this time as a high-performance variable selection technique for marketing models.

30.4 Goals of Modeling in Marketing

Marketers typically attempt to improve the effectiveness of their campaigns, solicitations, and the like by targeting their best customers or prospects. They use a model to identify individuals who are likely to respond to or generate profit* from their marketing efforts. The model provides, for each individual, estimates of probability of response and estimates of contribution to profit. Although the precision of these estimates is important, the performance of the model is measured at an aggregated level as reported in a decile analysis.

---

* I use the term *profit* as a stand-in for any measure of an individual's worth, such as sales per order, lifetime sales, revenue, number of visits, or number of purchases.

Marketers have defined the Cum Lift, which is found in the decile analysis, as the relevant measure of model performance. Based on the selection of individuals by the model, marketers create a "push-up" list to obtain an advantage over a random selection of individuals. The response Cum Lift is an index of how many more responses are expected with a selection based on a model over the expected responses with a random selection (without a model). Similarly, the profit Cum Lift is an index of how much more profit is expected with a selection based on a model over the expected profit with a random selection (without a model). The concept of Cum Lift and the steps of its construction in a decile analysis are presented in Chapter 18.

It should be clear at this point that a model that produces a decile analysis with more responses or profit in the upper (top, second, third, or fourth) deciles is a better model than one with fewer responses or less profit in the upper deciles. This concept is the motivation for the GenIQ Model. The GenIQ approach to modeling addresses specifically the objectives concerning marketers, namely, maximizing response and profit from their marketing efforts. The GenIQ Model uses the genetic methodology to optimize explicitly the desired criterion: maximize the upper deciles. Consequently, the GenIQ Model allows a regression modeler to build response and profit models in ways that are not possible with current methods.

The GenIQ Response and Profit Models are theoretically superior—with respect to maximizing the upper deciles—to response and profit models built with alternative techniques because of the explicit nature of the fitness function. The actual formulation of the fitness function is beyond the scope of this chapter; suffice it to say, the fitness function seeks to fill the upper deciles with as many responses or as much profit as possible, equivalently, to maximize response/profit Cum Lift.

Due to the explicit nature of its fitness criterion and the way it evolves models, the GenIQ Model offers high-performance variable selection for marketing models. This is apparent once I illustrate the GenIQ variable selection process in the next section.

___

30.5 Variable Selection with GenIQ

The best way of explaining variable selection with the GenIQ Model is to illustrate how GenIQ identifies structure in data. In this illustration, I demonstrate finding structure for a response model. GenIQ works equally well for a profit model, with a nominally defined profit continuous dependent variable.

Cataloguer ABC requires a response model to be built on a recent mail campaign that produced a 3.54% response rate. In addition to the RESPONSE

TABLE 30.1

Correlation Analysis: Nine Original
Variables with RESPONSE

Rank	Variable	Correlation Coefficient (r)
Top	DOLLAR_2	0.11
2	RFM_CELL	−0.10
3	PROD_TYP	0.08
4	LSTORD_M	−0.07
5	AGE_Y	0.04
6	PROMOTION	0.03
7	AVG_ORDE	0.02
8	OWN_TEL	0.10
9	FSTORD_M	0.01

dependent variable, there are nine candidate predictor variables, whose measurements were taken prior to the mail campaign.

1. AGE_Y: knowledge of customer's age (1 = if known; 0 = if not known)
2. OWN_TEL: presence of a telephone in the household (1 = yes; 0 = no)
3. AVG_ORDE: average dollar order
4. DOLLAR_2: dollars spent within last 2 years
5. PROD_TYP: number of different products purchased
6. LSTORD_M: number of months since last order
7. FSTORD_M: number of months since first order
8. RFM_CELL: recency/frequency/money cells (1 = best to 5 = worst)*
9. PROMOTION: number of promotions customer has received

To get an initial read on the information content (predictive power) of the variables, I perform a correlation analysis, which provides the correlation coefficient[†] for each candidate predictor variable with RESPONSE in Table 30.1. The top four variables in descending order of the magnitude[‡] of the strength of association are DOLLAR_2, RFM_CELL, PROD_TYP, and LSTORD_M.

I perform five logistic regression analyses (with RESPONSE) corresponding to the five variable selection methods. The resulting best subsets among the nine original variables are represented in Table 30.2. Surprisingly,

---

* RFM_CELL will be treated as a scalar variable.
† I know that the correlation coefficient with or without scatterplots is a crude gauge of predictive power.
‡ The direction of the association is not relevant. That is, the sign of the coefficient is ignored.

TABLE 30.2

Best Subsets among the Nine Original Variables

	DOLLAR_2	RFM_CELL	LSTORD_M	AGE_Y	AVG_ORDE
FS	x	x	x	x	
BE	x	x	x	x	
SW	x	x	x	x	
R-sq	x		x	x	x
Top-4	x	x	x		x
Frequency	5	4	5	4	2

TABLE 30.3

LRM Model Performance Comparison by Variable Selection Methods: Cum Lifts

Decile	FS	BE	SW	R-sq	Top-4	AVG
Top	256	256	256	239	252	252
2	204	204	204	198	202	202
3	174	174	174	178	172	174
4	156	156	156	157	154	156
5	144	144	144	145	142	144
6	132	132	132	131	130	131
7	124	124	124	123	121	123
8	115	115	115	114	113	114
9	107	107	107	107	107	107
Bottom	100	100	100	100	100	100

the forward, backward, and SW methods produced the identical subset (DOLLAR_2, RFM_CELL, LSTORD_M, AGE_Y). Because these methods produced a subset size of 4, I set the subset size to 4 for the R-sq and top-k methods. This allows for a fair comparison across all methods. R-sq and top-k produced different best subsets, which include DOLLAR_2, LSTORD_M, and AVG_ORDE. It is interesting to note that the most frequently used variables are DOLLAR_2 and LSTORD_M in the "Frequency" row in Table 30.2. The validation performance of the five logistic models in terms of Cum Lift is reported in Table 30.3. Assessment of model performance at the decile level is as follows:

1. At the top decile, R-sq produced the worst-performing model: Cum Lift 239 versus Cum Lifts 252–256 for the other models.

2. At the second decile, R-sq produced the worst-performing model: Cum Lift 198 versus Cum Lifts 202–204 for the other models.

3. At the third decile, R-sq produced the best-performing model: Cum Lift 178 versus Cum Lifts 172–174 for the other models.

Similar findings can be made at the other depths of file.

To facilitate the comparison of the five statistics-based variable selection methods and the GenIQ Model, I use a single measure of model performance for the five methods. The average performance of the five models is measured by AVG, the average of the Cum Lifts across the five methods for each decile, in Table 30.3.

30.5.1 GenIQ Modeling

This section requires an understanding of the genetic methodology and the parameters for controlling a genetic model run (as discussed in Chapter 29).

I set the parameters for controlling the GenIQ Model to run as follows:

1. Population size: 3,000 (models)
2. Number of generations: 250
3. Percentage of the population copied: 10%
4. Percentage of the population used for crossover: 80%
5. Percentage of the population used for mutation: 10%

The GenIQ-variable set consists of the nine candidate predictor variables. For the GenIQ-function set, I select the arithmetic functions (addition, subtraction, multiplication, and division); some Boolean operators (and, or, xor, greater/less than); and the log function (Ln). The log function* is helpful in symmetrizing typically skewed dollar amount variables, such as DOLLAR_2. I anticipate that DOLLAR_2 would be part of a genetically evolved structure defined with the log function. Of course, RESPONSE is the dependent variable.

At the end of the run, 250 generations of copying/crossover/mutation have evolved 750,000 (250 times 3,000) models according to selection proportional to fitness (PTF). Each model is evaluated in terms of how well it solves the problem of "filling the upper deciles with responders." Good models having more responders in the upper deciles are more likely to contribute to the next generation of models; poor models having fewer responders in the upper deciles are less likely to contribute to the next generation of models. Consequently, the last generation consists of 3,000 high-performance models, each with a fitness value indicating how well the model solves the problem. The "top" fitness values, typically the 18 largest values,† define a set of 18 "best" models with equivalent performance (filling the upper deciles with a virtually identical large number of responders).

---

* The log to the base 10 also symmetrizes dollar amount variables.
† Top fitness values typically bunch together with equivalent values. The top fitness values are considered equivalent in that their values are equal when rounded at, say, the third place after the decimal point. Consider two fitness values: 1.934056 and 1.934069. These values are equal when rounding occurs at, say, the third place after the decimal point: 1.934.

TABLE 30.4

Mean Incidence of Original Variables
across the Set of 18 Best Models

Variable	Mean Incidence
DOLLAR_2	1.43
RFM_CELL	1.37
PROD_TYP	1.22
AGE_Y	1.11
LSTORD_M	0.84
PROMOTION	0.67
AVG_ORDE	0.37
OWN_TEL	0.11

The set of variables defining one of the best models has variables in common with the set of variables defining another best model. The common variables can be considered for the best subset. The mean incidence of a variable across the set of best models provides a measure for determining the best subset. The GenIQ-selected best subset of original variables consists of variables with mean incidence greater than 0.75.* The variables that meet this cutoff score reflect an honest determination of necessary variables with respect to the criterion of maximizing the deciles.

Returning to the illustration, GenIQ provides the mean incidence of the nine variables across the set of 18 best models in Table 30.4. Thus, the GenIQ-selected best subset consists of five variables: DOLLAR_2, RFM_CELL, PROD_TYP, AGE_Y, and LSTORD_M.

This genetic-based best subset has four variables in common with the statistics-based best subsets (DOLLAR_2, RFM_CELL, LSTORD_M, and AGE_Y). Unlike the statistics-based methods, GenIQ finds value in PROD_TYP and includes it in its best subset in Table 30.5. It is interesting to note that the most frequently used variables are DOLLAR_2 and LSTORD_M, in the "Frequency" row in Table 30.5.

At this point, I can assess the predictive power of the genetic-based and statistics-based best subsets by comparing logistic regression models (LRMs) with each subset. However, after identifying GenIQ-evolved structure, I choose to make a more fruitful comparison.

30.5.2 GenIQ Structure Identification

Just as in nature, where structure is the consequence of natural selection as well as sexual recombination and mutation, the GenIQ Model evolves structure via selection PTF (natural selection), crossover (sexual recombination), and mutation. *The GenIQ fitness leads to structure, which is evolved with respect*

---

* The mean incidence cutoff score of 0.75 has been empirically predetermined.

TABLE 30.5

Best Subsets among Original Variables: Statistics- and Genetic-Based Variable Selection Methods

Method	DOLLAR_2	RFM_CELL	LSTORD_M	AGE_Y	AVG_ORDE	PROD_TYP
FS	x	x	x	x		
BE	x	x	x	x		
SW	x	x	x	x		
R-sq	x		x	x	x	
Top-4	x	x	x		x	
GenIQ	x	x	x	x		x
Frequency	6	5	6	5	2	1

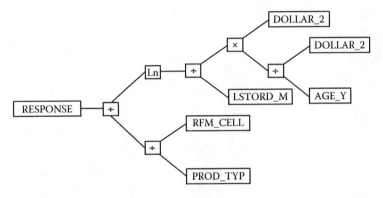

FIGURE 30.1
GenIQ Model, best 1.

to the criterion of maximizing the deciles. Important structure is found in the best models, typically the models with the four largest fitness values.

Continuing with the illustration, GenIQ has evolved several structures, or *GenIQ-constructed variables.* The GenIQ Model (Figure 30.1) has the largest fitness value and reveals five new variables, NEW_VAR1 through NEW_VAR5. Additional structures are found in the remaining three best models: NEW_VAR6 to NEW_VAR8 (in Figure 30.2), NEW_VAR9 (in Figure 30.3), and NEW_VAR10 (in Figure 30.4).

1. NEW_VAR1 = DOLLAR_2/AGE_Y; if Age_Y = 0, then NEW_VAR1 = 1
2. NEW_VAR2 = (DOLLAR_2)*NEW_VAR1
3. NEW_VAR3 = NEW_VAR2/LSTORD_M; if LSTORD_M = 0, then NEW_VAR3 = 1
4. NEW_VAR4 = Ln(NEW_VAR3); if NEW_VAR3 greater than 0, then NEW_VAR4 = 1

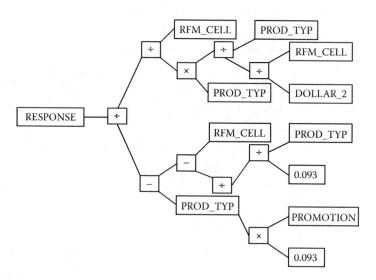

FIGURE 30.2
GenIQ Model, best 2.

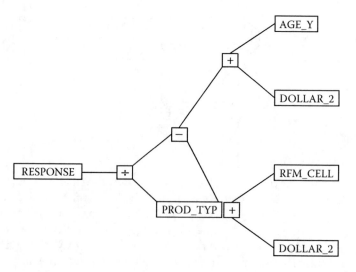

FIGURE 30.3
GenIQ Model, best 3.

5. NEW_VAR5 = RFM_CELL/PROD_TYP; if PROD_TYP = 0, then NEW_VAR5 = 1

6. NEW_VAR6 = RFM_CELL/DOLLAR_2; if DOLLAR_2 = 0, then NEW_VAR6 = 1

7. NEW_VAR7 = PROD_TYP/NEW_VAR6; if NEW_VAR6 = 0, then NEW_VAR7 = 1

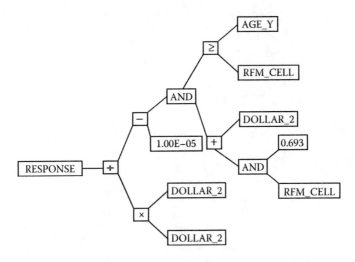

FIGURE 30.4
GenIQ Model, best 4.

8. NEW_VAR8 = NEW_VAR7*PROD_TYP
9. NEW_VAR9 = (AGE_Y/DOLLAR_2) - (RFM_CELL/DOLLAR_2); if DOLLAR_2 = 0, then NEW_VAR9 = 1
10. NEW_VAR10 = 1 if AGE_Y greater than/equal to RFM_CELL; otherwise = 0

To get a read on the predictive power of the new GenIQ-constructed variables, I perform a correlation analysis for each of the 9 original variables and the 10 new variables with RESPONSE. Some new variables have a stronger association with RESPONSE than the original variables. Specifically, the following associations (larger correlation coefficient, ignoring the sign) are observed in Table 30.6.

1. NEW_VAR7, NEW_VAR5, NEW_VAR8, and NEW_VAR1 have a stronger association with RESPONSE than the best original variable DOLLAR_2.
2. NEW_VAR10 and NEW_VAR4 fall between the second- and third-best original variables, RFM_CELL and PROD_TYP.
3. NEW_VAR2 and NEW_VAR3 are ranked 11th and 12th in importance before the last two original predictor variables, AGE_Y and PROMOTION.

30.5.3 GenIQ Variable Selection

The GenIQ-constructed variables plus the GenIQ-selected variables can be thought of as an *enhanced best subset* that reflects an honest determination

TABLE 30.6

Correlation Analysis : 9 Original and 10
GenIQ Variables with RESPONSE

Rank	Variable	Correlation Coefficient (r)
Top	NEW_VAR7	0.16
2	NEW_VAR5	0.15
3	NEW_VAR8	0.12
4	NEW_VAR1	0.12
5	DOLLAR_2	0.11
6	RFM_CELL	−0.10
7	NEW_VAR10	0.10
8	NEW_VAR4	0.10
9	PROD_TYP	0.08
10	LSTORD_M	−0.07
11	NEW_VAR2	0.07
12	NEW_VAR3	0.06
13	NEW_VAR9	0.05
14	AGE_Y	0.04
15	PROMOTION	0.03
16	NEW_VAR6	−0.02
17	AVG_ORDE	0.02
18	OWN_TEL	0.01
19	FSTORD_M	0.01

of necessary variables with respect to the criterion of maximizing the deciles. For the illustration data, the enhanced set consists of 15 variables: DOLLAR_2, RFM_CELL, PROD_TYP, AGE_Y, LSTORD_M, and NEW_VAR1 through NEW_VAR10. The predictive power of the enhanced best set is assessed by the comparison between LRMs with the genetic-based best subset and with the statistics-based best subset.

Using the enhanced best set, I perform five logistic regression analyses corresponding to the five variable selection methods. The resultant genetic-based best subsets are displayed in Table 30.7. The forward, backward, and SW methods produced different subsets (of size 4). R-sq(4) and top-4 also produced different subsets. It appears that New_VAR5 is the "most important" variable (i.e., most frequently used variable), as it is selected by all five methods (row "Frequency" in Table 30.7 equals 5). LSTORD_M is of "second importance," as it is selected by four of the five methods (row "Frequency" equals 4). RFM_CELL and AGE_Y are the "least important," as they are selected by only one method (row "Frequency" equals 1).

TABLE 30.7

Best Subsets among the Enhanced Best Subset Variables

Method	DOLLAR_ 2	RFM_ CELL	PROD_ TYP	AGE_ Y	LSTORD_ M	NEW_ VAR1	NEW_ VAR4	NEW_ VAR5
FS			x		x		x	x
BE	x			x	x			x
SW			x		x		x	x
R-sq			x		x	x		x
Top-4		x				x	x	x
Frequency	1	1	3	1	4	2	3	5

TABLE 30.8

Model Performance Comparison Based on the Genetic-Based Best Subsets: Cum Lifts

Decile	FS	BE	SW	R-sq	Top-4	AVG-g	AVG	Gain
Top	265	260	262	265	267	264	252	4.8%
2	206	204	204	206	204	205	202	1.2%
3	180	180	180	178	180	180	174	3.0%
4	166	167	167	163	166	166	156	6.4%
5	148	149	149	146	149	148	144	3.1%
6	135	137	137	134	136	136	131	3.3%
7	124	125	125	123	125	124	123	1.0%
8	116	117	117	116	117	117	114	1.9%
9	108	108	108	107	108	108	107	0.7%
Bottom	100	100	100	100	100	100	100	0.0%

To assess the gains in predictive power of the genetic-based best subset over the statistics-based best subset, I define AVG-g as the average measure of model validation performance for the five methods for each decile.

Comparison of AVG-g and AVG (average model performance based on the statistics-based set) indicates noticeable gains in predictive power obtained by the GenIQ variable selection technique in Table 30.8. The percentage gains range from an impressive 6.4% (at the fourth decile) to a slight 0.7% (at the ninth decile). The mean percentage gain for the most actionable depth of file, the top four deciles, is 3.9%.

This illustration demonstrates the power of the GenIQ variable selection technique over the current statistics-based variable selection methods. GenIQ variable selection is a high-performance method for marketing models with data mining capability. This method is significant in that it finds the best subset of variables to maximize the Cum Lift criterion.

30.6 Nonlinear Alternative to Logistic Regression Model

The GenIQ Model offers a nonlinear alternative to the inherently linear LRM. Accordingly, an LRM is a linear approximation of a potentially nonlinear response function, which is typically noisy, multimodal, and discontinuous. The LRM together with the GenIQ-enhanced best subset of variables provide an unbeatable combination of traditional statistics improved by the genetic-based machine learning of the GenIQ Model. However, this *hybrid GenIQ-LRM model* is still a linear approximation of a potentially nonlinear response function. The GenIQ Model itself—as defined by the entire tree with all its structure—is a nonlinear superstructure with a strong possibility for further improvement over the hybrid GenIQ-LRM and, of course, over LRM. Because the degree of nonlinearity in the response function is never known, the best approach is to compare the GenIQ Model with the hybrid GenIQ-LRM model. If the improvement is determined to be stable and noticeable, then the GenIQ Model should be used.

Continuing with the illustration, the GenIQ Model Cum Lifts are reported in Table 30.9. The GenIQ Model offers noticeable improvements over the performance of the hybrid GenIQ-LRM model (AVG-g). The percentage gains (fifth column from the left) range from an impressive 7.1% (at the top decile) to a respectable 1.2% (at the ninth decile). The mean percentage gain for the most actionable depth of file, the top four deciles, is 4.6%.

I determine the improvements of the GenIQ Model over the performance of the LRM (AVG). The mean percentage gain (rightmost column) for the most actionable depth of file, the top four deciles, is 8.6%, which includes a huge 12.2% in the top decile.

TABLE 30.9

Model Performance Comparison LRM and GenIQ Model: Cum Lifts

Decile	AVG-g (Hybrid)	AVG (LRM)	GenIQ	GenIQ Gain Over Hybrid	GenIQ Gain Over LRM
Top	264	252	283	7.1%	12.2%
2	205	202	214	4.4%	5.6%
3	180	174	187	3.9%	7.0%
4	166	156	171	2.9%	9.5%
5	148	144	152	2.8%	5.9%
6	136	131	139	2.5%	5.9%
7	124	123	127	2.2%	3.2%
8	117	114	118	1.3%	3.3%
9	108	107	109	1.2%	1.9%
Bottom	100	100	100	0.0%	0.0%

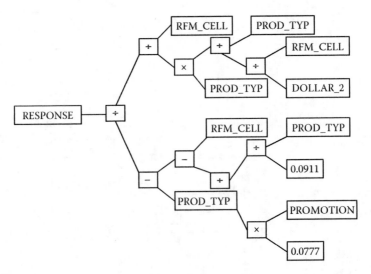

FIGURE 30.5
Best GenIQ Model, first of top four models.

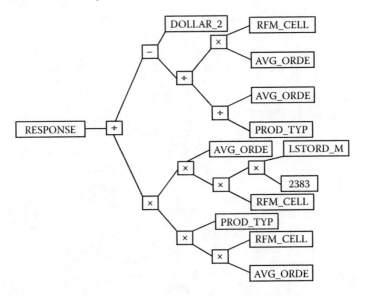

FIGURE 30.6
Best GenIQ Model, second of top four models.

Note that a set of four separate GenIQ Models is needed to obtain the reported decile performance levels because no single GenIQ Model could be evolved to provide gains for all upper deciles. The GenIQ Models that produce the top, second, third, and fourth deciles are in Figures 30.1, 30.5, 30.6, and 30.7, respectively. The GenIQ Model that produced the fifth through bottom deciles is in Figure 30.8.

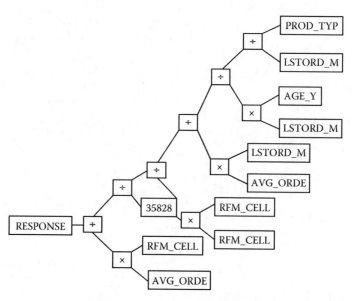

FIGURE 30.7
Best GenIQ Model, third of top four models.

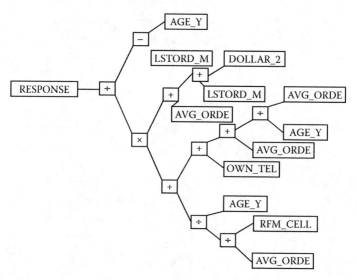

FIGURE 30.8
Best GenIQ Model, fourth of top four models.

A set of GenIQ Models is required when the response function is nonlinear with noise, multipeaks, and discontinuities. The capability of GenIQ to generate many models with desired performance gains reflects the flexibility of the GenIQ paradigm. It allows for adaptive intelligent modeling of the data to account for the variation of an apparent nonlinear response function.

This illustration shows the power of the GenIQ Model as a nonlinear alternative to the LRM. GenIQ provides a two-step procedure for response modeling. First, build the best hybrid GenIQ-LRM model. Second, select the best GenIQ Model. If the GenIQ Model offers a stable and noticeable improvement over the hybrid model, then the GenIQ Model is the preferred response model.

As previously mentioned, the GenIQ Model works equally well for finding structure in a profit model. Accordingly, the GenIQ Model is a nonlinear alternative to the ordinary least squares (OLS) regression model. The GenIQ Model offers potentially stable and noticeable improvement over OLS and the hybrid GenIQ-OLS Model.

30.7 Summary

After framing the problem of variable selection with the five popular statistics-based methods, I pointed out two common weaknesses of the methods. Each hinders its capacity to achieve the desired requirement of marketing model: neither identifying structure nor explicitly maximizing the Cum Lift criterion.

I have presented the GenIQ Model as a genetic-based approach for variable selection for marketing models. The GenIQ Response and Profit Models are theoretically superior—with respect to maximizing the upper deciles—to response and profit models built with logistic and ordinary regression models, respectively, because of the nature of their fitness function. The GenIQ fitness function seeks explicitly to fill the upper deciles with as many responses or as much profit as possible. Standard statistical methods only maximize implicitly the Cum Lift, as their fitness functions serve as a surrogate for maximizing Cum Lift.

Using a response model illustration, I demonstrated the GenIQ Model as a high-performance variable selection method with data mining capability for finding important structure to maximize the Cum Lift criterion. Starting with nine candidate predictor variables, the statistics-based variable selection methods identified five predictor variables in defining its best subsets. GenIQ also identified five predictor variables, of which four were in common with the statistics-based best subsets. In addition, GenIQ evolved 10 structures (new variables), of which 4 had a stronger association with response than the best original predictor variable. Two new variables fell between the second- and third-best original predictor variables. As a result, GenIQ created the enhanced best subset of 15 variables.

The GenIQ variable selection method outperformed the statistics-based variable selection methods. I built LRMs for the five statistics-based variable selection methods using the enhanced best subset and compared its "average" performance (AVG-g) with the average performance (AVG) of the LRM for the five statistics-based methods using the original nine variables.

Comparison of AVG-g and AVG indicated noticeable gains in predictive power: The percentage gains ranged from an impressive 6.4% to a slight 0.7%. The mean percentage gain for the most actionable depth of file, the top four deciles, was 3.9%.

Last, I advanced the GenIQ Model itself as a nonlinear alternative to the standard regression models. LRM together with the GenIQ-enhanced best subset of variables provide an unbeatable combination of traditional statistics improved by machine learning. However, this hybrid GenIQ-LRM model is still a linear approximation of a potentially nonlinear response function. The GenIQ Model itself—as defined by the entire tree with all its structure—is a nonlinear superstructure with a strong possibility for further improvement over the hybrid GenIQ-LRM. For the response illustration, the set of GenIQ Models produced noticeable improvements over the performance of the hybrid GenIQ-LRM Model. The percentage gains ranged from an impressive 7.1% to a respectable 1.2%. The mean percentage gain for the most actionable depth of file, the top four deciles, was 4.6%.

References

1. Dash, M., and Liu, H., Feature selection for classification, In *Intelligent Data Analysis*, Elsevier Science, New York, 1997.
2. Ryan, T.P., *Modern Regression Methods*, Wiley, New York, 1997.
3. Miller, A.J., *Subset Selection in Regression*, Chapman and Hall, London, 1990.
4. Fox, J., *Applied Regression Analysis, Linear Models, and Related Methods*, Sage, Thousand Oaks, CA, 1997.

31

Interpretation of Coefficient-Free Models

31.1 Introduction

The statistical ordinary least squares regression model is the reference thought of when marketers hear the words "new kind of model." Model builders use the regression concept and its prominent characteristics when judiciously evaluating an alternative modeling technique. This is because the ordinary regression paradigm is the underpinning for the solution to the ubiquitous prediction problem. Marketers with limited statistical background undoubtedly draw on their educated notions of the regression model before accepting a new technique. New modeling techniques are evaluated by the coefficients they produce. If the new coefficients impart comparable information to the prominent characteristic of the regression model—the regression coefficient—then the new technique passes the first line of acceptance. If not, the technique is summarily rejected. A quandary arises when a new modeling technique, like some machine-learning methods, produces models with no coefficients. The primary purpose of this chapter is to present a method for calculating a quasi-regression coefficient, which provides a frame of reference for evaluating and using coefficient-free models. Secondarily, the quasi-regression coefficient serves as a trusty assumption-free alternative to the regression coefficient, which is based on an implicit and hardly tested assumption necessary for reliable interpretation.

31.2 The Linear Regression Coefficient

The redoubtable regression coefficient, formally known as the ordinary least squares linear regression coefficient, linear-RC(ord), enjoys everyday use in marketing analysis and modeling. The common definition of the *linear-RC(ord) for predictor variable X* is *the predicted (expected) constant change in the dependent variable Y associated with a unit change in X*. The usual mathematical expression for the coefficient is in Equation (31.1). Although correctly stated, the definition requires commentary to have a thorough understanding for expert use. I parse the definition of linear-RC(ord) and then provide two illustrations of the statistical measure within the open work of the linear regression paradigm, which serve as a backdrop for presenting the quasi-regression coefficient.

$$\text{Linear-RC(ord)} = \frac{\text{Predicted-Change in Y}}{\text{Unit-Change in X}} \qquad (31.1)$$

Consider the simple ordinary linear regression model of Y on X based on a sample of (X_i, Y_i) points: pred_Y = a + b*X.

1. *Simple* means one predictor variable X is used.
2. *Ordinary* connotes that the dependent variable Y is continuous.
3. *Linear* has a dual meaning. Explicitly, linear denotes that the model is defined by the sum of the weighted predictor variable b*X and the constant *a*. It connotes implicitly the linearity assumption that the true relationship between Y and X is straight line.
4. *Unit change in X* means the difference of value 1 for two X values ranked in ascending order, X_r and X_{r+1}, that is, $X_{r+1} - X_r = 1$.
5. *Change in Y* means the difference in predicted Y, pred_Y, values corresponding to $(X_r, \text{pred\_}Y_r)$ and $(X_{r+1}, \text{pred\_}Y_{r+1})$: pred_$Y_{r+1}$ - pred_Y_r.
6. *Linear-RC(ord)* indicates that the expected change in Y is constant, namely, b.

31.2.1 Illustration for the Simple Ordinary Regression Model

Consider the simple ordinary (least squares) linear regression of Y on X based on the 10 observations in dataset A in Table 31.1. The satisfaction of the linearity assumption is indicated by the plot of Y versus X; there is an observed positive straight-line relationship between the two variables. The X-Y plot and the plot of residual versus predicted Y both suggest that the resultant regression model in Equation (31.2) is reliable. (These types of plots are discussed in Chapter 8. The plots for these data are not shown.) Accordingly, the model provides a level of assurance that the estimated linear-RC(ord) value of 0.7967 is a reliable point estimator of the true linear regression coefficient of X. Thus, for each unit change in X, between observed X values of 19 and 78, the expected constant change in Y is 0.7967.

$$\text{Pred\_Y} = 22.2256 + 0.7967*X \qquad (31.2)$$

TABLE 31.1

Dataset A

Y	X	pred_Y
86	78	84.3688
74	62	71.6214
66	58	68.4346
65	53	64.4511
64	51	62.8576
62	49	61.2642
61	48	60.4675
53	47	59.6708
52	38	52.5004
40	19	37.3630

TABLE 31.2

Dataset B

Y	X	pred_lgt Y	pred_prb Y
1	45	1.4163	0.8048
1	35	0.2811	0.5698
1	31	−0.1729	0.4569
1	32	−0.0594	0.4851
1	60	3.1191	0.9577
0	46	1.5298	0.8220
0	30	−0.2865	0.4289
0	23	−1.0811	0.2533
0	16	−1.8757	0.1329
0	12	−2.3298	0.0887

31.2.2 Illustration for the Simple Logistic Regression Model

Consider the simple logistic regression of response Y on X based on the 10 observations in dataset B in Table 31.2. Recall that the logistic regression model (LRM) predicts the logit Y and is *linear* in the same ways as the ordinary regression model. It is a linear model, as it is defined by the sum of a weighted predictor variable plus a constant, and has the linearity assumption that the underlying relationship between the *logit* Y and X is straight line. Accordingly, the definition of the simple logistic linear regression coefficient, *linear-RC(logit) for predictor variable X is the expected constant change in the logit Y associated with a unit change in X.*

The smooth plot of logit Y versus X is inconclusive regarding the linearity between the logit Y and X, undoubtedly due to only 10 observations. However, the plot of residual versus predicted logit Y suggests that the resultant regression model in Equation (31.3) is reliable. (The smooth plot is discussed in Chapter 8. Plots for these data are not shown.) Accordingly, the model provides a level of assurance that the estimated linear-RC(logit) value of 0.1135 is a reliable point estimator of the true linear logistic regression coefficient of X. Thus, for each unit change in X, between observed X values of 12 and 60, the expected constant change in the logit Y is 0.1135.

$$\text{Pred\_Logit } Y = -3.6920 + 0.1135 * X \qquad (31.3)$$

31.3 The Quasi-Regression Coefficient for Simple Regression Models

I present the quasi-regression coefficient, quasi-RC, for the simple regression model of Y on X. The *quasi-RC for predictor variable X is the expected change— not necessarily constant—in dependent variable Y per unit change in X*. The distinctness of the quasi-RC is that it offers a generalization of the linear-RC. It has a flexibility to measure nonlinear relationships between the dependent and predictor variables. I outline the method for calculating the quasi-RC and motivate its utility by applying the quasi-RC to the ordinary regression illustration. Then, I continue the use of the quasi-RC method for the logistic regression illustration, showing how it works for linear and nonlinear predictions.

31.3.1 Illustration of Quasi-RC for the Simple Ordinary Regression Model

Continuing with the simple ordinary regression illustration, I outline the steps for deriving the quasi-RC(ord) in Table 31.3 (columns are numbered from the left):

1. Score the data to obtain the predicted Y, pred_Y (column 3, Table 31.1).
2. Rank the data in ascending order by X and form the pair (X_r, X_{r+1}) (columns 1 and 2, Table 31.3).
3. Calculate the change in X: $X_{r+1} - X_r$ (Column 3 = Column 2 - Column 1, Table 31.3).
4. Calculate the change in predicted Y: $\text{pred\_Y}_{r+1} - \text{pred\_Y}_r$ (Column 6 = Column 5 - Column 4, Table 31.3).
5. Calculate the quasi-RC(ord) for X: change in predicted Y divided by change in X (Column 7 = Column 6/Column 3, Table 31.3).

The quasi-RC(ord) is constant across the nine (X_r, X_{r+1}) intervals and equals the estimated linear RC(ord) value of 0.7967. In superfluity, it is constant for

TABLE 31.3

Calculations for Quasi-RC(ord)

X_r	X_r+1	change_X	pred_Y_r	pred_Y_r+1	change_Y	quasi-RC(ord)
	19	.	.	37.3630	.	.
19	38	19	37.3630	52.5005	15.1374	0.7967
38	47	9	52.5005	59.6709	7.1704	0.7967
47	48	1	59.6709	60.4676	0.7967	0.7967
48	49	1	60.4676	61.2643	0.7967	0.7967
49	51	2	61.2643	62.8577	1.5934	0.7967
51	53	2	62.8577	64.4511	1.5934	0.7967
53	58	5	64.4511	68.4346	3.9835	0.7967
58	62	4	68.4346	71.6215	3.1868	0.7967
62	78	16	71.6215	84.3688	12.7473	0.7967

each and every unit change in X within each of the X intervals: between 19 and 38, 38 and 47, …, and 62 and 78. This is no surprise as the predictions are from a linear model. Also, the plot of pred_Y versus X (not shown) indicates a perfectly positively sloped straight line whose slope is 0.7967. (Why?)

31.3.2 Illustration of Quasi-RC for the Simple Logistic Regression Model

By way of further motivation for the quasi-RC methodology, I apply the five steps to the logistic regression illustration with the appropriate changes to accommodate working with logit units for deriving the quasi-RC(logit) in Table 31.4:

1. Score the data to obtain the predicted logit Y, pred_lgt Y (Column 3, Table 31.2).
2. Rank the data in ascending order by X and form the pair (X_r, X_{r+1}) (columns 1 and 2, Table 31.4).
3. Calculate the change in X: $X_{r+1} - X_r$ (Column 3 = Column 2 - Column 1, Table 31.4).
4. Calculate the change in predicted logit Y: pred_lgt Y_{r+1} - pred_lgt Y_r (Column 6 = Column 5 - Column 4, Table 31.4).
5. Calculate the quasi-RC for X (logit): change in predicted logit Y divided by change in X (Column 7 = Column 6/Column 3, Table 31.4).

The quasi-RC(logit) is constant across the nine (X_r, X_{r+1}) intervals and equals the estimated linear-RC(logit) value of 0.1135. Again, in superfluity, it is constant for each and every unit change within the X intervals: between 12 and 16, 16 and 23, …, and 46 and 60. Again, this is no surprise as the predictions

TABLE 31.4

Calculations for Quasi-RC(logit)

X_r	X_r+1	change_X	pred_lgt_r	pred_lgt_r+1	change_lgt	quasi-RC(logit)
	12	.	.	−2.3298	.	.
12	16	4	−2.3298	−1.8757	0.4541	0.1135
16	23	7	−1.8757	−1.0811	0.7946	0.1135
23	30	7	−1.0811	−0.2865	0.7946	0.1135
30	31	1	−0.2865	−0.1729	0.1135	0.1135
31	32	1	−0.1729	−0.0594	0.1135	0.1135
32	35	3	−0.0594	0.2811	0.3406	0.1135
35	45	10	0.2811	1.4163	1.1352	0.1135
45	46	1	1.4163	1.5298	0.1135	0.1135
46	60	14	1.5298	3.1191	1.5893	0.1135

are from a linear model. Also, the plot of predicted logit Y versus X (not shown) shows a perfectly positively sloped straight line with slope 0.1135. Thus far, the two illustrations show how the quasi-RC method works and holds up for simple linear predictions, that is, predictions produced by linear models with one predictor variable.

In the next section, I show how the quasi-RC method works with a simple nonlinear model in its efforts to provide an honest attempt at imparting regression coefficient-like information. A nonlinear model is defined, in earnest, as a model that is not a linear model, that is, it is not a sum of weighted predictor variables. The simplest nonlinear model—the probability of response—is a restatement of the simple logistic regression of response Y on X, defined in Equation (31.4).

$$\text{Probability of response } Y = \exp(\text{logit } Y)/(1 + \exp(\text{logit } Y)) \qquad (31.4)$$

Clearly, this model is nonlinear. It is said to be nonlinear in its predictor variable X, which means that the expected change in the probability of response varies as the unit change in X varies through the range of observed X values. Accordingly, the quasi-RC(prob) for predictor variable X is the expected change—not necessarily constant—in the probability Y per unit change in the X. In the next section, I conveniently use the logistic regression illustration to show how the quasi-RC method works with nonlinear predictions.

31.3.3 Illustration of Quasi-RC for Nonlinear Predictions

Continuing with the logistic regression illustration, I outline the steps for deriving the quasi-RC(prob) in Table 31.5 by straightforwardly modifying

TABLE 31.5

Calculations for Quasi-RC(prob)

X_r	X_{r+1}	change_X	prob_Y_r	prob_Y_ r+1	change_ prob	quasi- RC(prob)
	12	.	.	0.0887	.	.
12	16	4	0.0887	0.1329	0.0442	0.0110
16	23	7	0.1329	0.2533	0.1204	0.0172
23	30	7	0.2533	0.4289	0.1756	0.0251
30	31	1	0.4289	0.4569	0.0280	0.0280
31	32	1	0.4569	0.4851	0.0283	0.0283
32	35	3	0.4851	0.5698	0.0847	0.0282
35	45	10	0.5698	0.8048	0.2349	0.0235
45	46	1	0.8048	0.8220	0.0172	0.0172
46	60	14	0.8220	0.9577	0.1357	0.0097

the steps for the quasi-RC(logit) to account for working in probability units.

1. Score the data to obtain the predicted logit of Y, pred_lgt Y (column 3, Table 31.2).
2. Convert the pred_lgt Y to predicted probability Y, pred_prb Y (column 4, Table 31.2). The conversion formula is as follows: Probability Y equals exp(logit Y) divided by the sum of 1 plus exp(logit Y).
3. Rank the data in ascending order by X and form the pair (X_r, X_{r+1}) (columns 1 and 2, Table 31.5).
4. Calculate change in X: $X_{r+1} - X_r$ (Column 3 = Column 2 - Column 1, Table 31.5).
5. Calculate change in probability Y: pred_prb Y_{r+1} - pred_prb Y_r (Column 6 = Column 5 - Column 4, Table 31.5).
6. Calculate the quasi-RC(prob) for X: Change in probability Y divided by the change in X (Column 7 = Column 6/Column 3, Table 31.5).

Quasi-RC(prob) varies as X—in a nonlinear manner—goes through its range between 12 and 60. The quasi-RC values for the nine intervals are 0.0110, 0.0172, ... , 0.0097, respectively, in Table 31.5. This is no surprise as the general relationship between probability of response and a given predictor variable has a theoretical prescribed nonlinear S-shape (known as an ogive curve). The plot of probability of Y versus X in Figure 31.1 reveals this nonlinearity, although the limiting 10 points may make it too difficult to see.

Thus far, the three illustrations show how the quasi-RC method works and holds up for linear and nonlinear predictions based on the simple

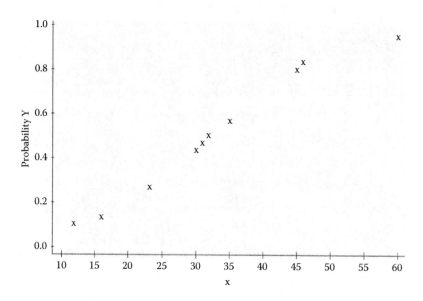

FIGURE 31.1
Plot of probability of Y versus X.

one-predictor variable regression model. In the next section, I extend the method beyond the simple one-predictor variable regression model to *the everymodel,* any multiple linear or nonlinear regression model, or any coefficient-free model.

31.4 Partial Quasi-RC for the Everymodel

The interpretation of the regression coefficient in the multiple (two or more predictor variables) regression model essentially remains the same as its meaning in the simple regression model. The regression coefficient is formally called the *partial* linear regression coefficient, partial linear-RC, which connotes that the model has other variables whose effects are partialled out of the relationship between the dependent variable and the predictor variable under consideration. The *partial linear-RC for predictor variable X is the expected constant change in the dependent variable Y associated with a unit change in the X when the other variables are held constant.* This is the well-accepted interpretation of the partial linear regression coefficient (as discussed in Chapter 12, Section 12.7).

The reading of the partial linear-RC for a given predictor variable is based on an implicit assumption that the statistical adjustment—which removes the effects of the other variables from the dependent variable and the predictor variable—produces a linear relationship between the dependent variable and the predictor variable. Although the workings of statistical adjustment are theoretically sound, it does not guarantee linearity between the adjusted-dependent and the adjusted-predictor variable. In general, an assumption based on the property of linearity is tenable. In the present case of statistical adjustment, the likelihood of the linearity assumption holding tends to decrease as the number of other variables increases. Interestingly, it is not a customary effort to check the validity of the linearity assumption, which could render the partial linear-RC questionable.

The quasi-RC method provides the partial quasi-RC as a trusty assumption-free alternative measure of the "expected change in the dependent variable" without reliance on statistical adjustment and restriction of a linear relationship between dependent variable and the predictor variable. Formally, the partial quasi-RC for predictor variable X is the expected change—not necessarily constant—in the dependent variable Y associated with a unit change in X when the other variables are held constant. The quasi-RC method provides a flexibility, which enables the data analyst to

1. Validate an overall linear trend in the dependent variable versus the predictor variable for given values within the other-variables region (i.e., given the other variables are held constant). For linear regression models, the method serves as a diagnostic to test the linearity assumption of the partial linear-RC. If the test result is positive (i.e., a nonlinear pattern emerges), which is a symptom of an incorrect structural form of the predictor variable, then a remedy can be inferred (i.e., a choice of reexpression of the predictor variable to induce linearity with the dependent variable).

2. Consider the liberal view of a nonlinear pattern in the dependent variable versus the predictor variable for given values within the other-variables region. For nonlinear regression models, the method provides an EDA (exploratory data analysis) procedure to uncover underlying structure of the expected change in the dependent variable.

3. Obtain coefficient-like information from coefficient-free models. This information encourages the use of "black box" machine-learning methods, which are characterized by the absence of regression-like coefficients.

In the next section, I outline the steps for calculating the partial quasi-RC for the everymodel. I provide an illustration using a multiple LRM to show how the method works and how to interpret the results. In the last section of this chapter, I apply the partial quasi-RC method to the coefficient-free GenIQ Model presented in Chapter 29.

31.4.1 Calculating the Partial Quasi-RC for the Everymodel

Consider the everymodel for predicting Y based on four predictor variables $X_1, X_2, X_3,$ and X_4. The calculations and guidelines for the partial quasi-RC for X_1 are as follows:

1. To affect the "holding constant" of the other variables $\{X_2, X_3, X_4\}$, consider the typical values of the *M-spread common region*. For example, the M20-spread common region consists of the individuals whose $\{X_2, X_3, X_4\}$ values are common to the individual M20-spreads (the middle 20% of the values) for each of the other variables; i.e., common to M20-spread for $X_2, X_3,$ and X_4. Similarly, the M25-spread common region consists of the individuals whose $\{X_2, X_3, X_4\}$ values are common to the individual M25-spreads (the middle 25% of the values) for each of the other variables.

2. The size of the common region is clearly based on the number and measurement of the other variables. A rule of thumb for sizing the region for reliable results is as follows: The initial M-spread common region is M20. If partial quasi-RC values seem suspect, then increase the common region by 5%, resulting in a M25-spread region. Increase the common region by 5% increments until the partial quasi-RC results are trustworthy. Note: A 5% increase is a nominal 5% because increasing each of the other-variables M-spread by 5% does not necessarily increase the common region by 5%.

3. For any of the other variables, whose measurement is coarse (includes a handful of distinct values), its individual M-spread may need to be decreased by 5% intervals until the partial quasi-RC values are trustworthy.

4. Score the data to obtain the predicted Y for all individuals in the common M-spread region.

5. Rank the scored data in ascending order by X_1.

6. Divide the data into equal-size slices by X_1. In general, if the expected relationship is linear, as when working with a linear model and testing the partial linear-RC linearity assumption, then start with five slices; increase the number of slices as required to obtain a trusty relationship. If the expected relationship is nonlinear, as when working with a nonlinear regression model, then start with 10 slices; increase the number of slices as required to obtain a trusty relationship.

7. The number of slices depends on two matters of consideration: the size of the M-spread common region and the measurement of the predictor variable for which partial quasi-RC is being derived. If the common region is small, then a large number of slices tends to produce unreliable quasi-RC values. If the region is large, then a

large number of slices, which otherwise does not pose a reliability concern, may produce untenable results, as offering too liberal a view of the overall pattern. If the measurement of the predictor variable is coarse, the number of slices equals the number of distinct values.

8. Calculate the minimum, maximum, and median of X_1 within each slice, form the pair (median $X_{slice\ i}$, median $X_{slice\ i+1}$).
9. Calculate the change in X_1: Median $X_{slice\ i+1}$ - Median $X_{slice\ i}$.
10. Calculate the median of the predicted Y within each slice and form the pair (median pred_$Y_{slice\ i}$, median $Y_{slice\ i+1}$).
11. Calculate the change in predicted Y: Median pred_$Y_{slice\ i+1}$ - Median pred_$Y_{slice\ i}$.
12. Calculate the partial quasi-RC for X_1: change in predicted Y divided by change in X_1.

31.4.2 Illustration for the Multiple Logistic Regression Model

Consider the illustration in Chapter 30 of cataloger ABC, who requires a response model built on a recent mail campaign. I build an LRM for predicting RESPONSE based on four predictor variables:

1. DOLLAR_2: dollars spent within last 2 years
2. LSTORD_M: number of months since last order
3. RFM_CELL: recency/frequency/money cells (1 = best to 5 = worst)
4. AGE_Y: knowledge of customer's age (1 = if known; 0 = if not known)

The RESPONSE model in Equation (31.5)

$$Pred\_lgt\ RESPONSE = -3.004 + 0.00210*DOLLAR\_2$$

$$- 0.1995*RFM\_CELL - 0.0798*LSTORD\_M + 0.5337*AGE\_Y \qquad (31.5)$$

I detail the calculation for deriving the LRM partial quasi-RC(logit) for DOLLAR_2 in Table 31.6.

1. Score the data to obtain Pred_lgt RESPONSE for all individuals in the M-spread common region.
2. Rank the scored data in ascending order by DOLLAR_2 and divide the data into five slices, in column 1, by DOLLAR_2.
3. Calculate the minimum, maximum, and median of DOLLAR_2, in columns 2, 3 and 4, respectively, for each slice and form the pair (median DOLLAR_2 $_{slice\ i}$, median DOLLAR_2 $_{slice\ i+1}$) in columns 4 and 5, respectively.

TABLE 31.6

Calculations for LRM Partial Quasi-RC(logit): DOLLAR_2

Slice	min_ DOLLAR_2	max_ DOLLAR_2	med_ DOLLAR_2_r	med_ DOLLAR_2_ r+1	change_ DOLLAR_2	med_ lgt_r	med_ lgt_r+1	change_ lgt	quasi-RC (logit)
1	0	43	.	40	.	.	-3.5396	.	.
2	43	66	40	50	10	-3.5396	-3.5276	0.0120	0.0012
3	66	99	50	80	30	-3.5276	-3.4810	0.0467	0.0016
4	99	165	80	126	46	-3.4810	-3.3960	0.0850	0.0018
5	165	1,293	126	242	116	-3.3960	-3.2219	0.1740	0.0015

4. Calculate the change in DOLLAR_2: Median DOLLAR_2$_{slice\ i+1}$ - median DOLLAR_2$_{slice\ i}$ (Column 6 = Column 5 - Column 4).

5. Calculate the median of the predicted logit RESPONSE within each slice and form the pair (median Pred_lgt RESPONSE$_{slice\ i}$, median Pred_lgt RESPONSE$_{slice\ i+1}$) in columns 7 and 8.

6. Calculate the change in Pred_lgt RESPONSE: Median Pred_lgt RESPONSE $_{slice\ i+1}$ - Median Pred_lgt RESPONSE $_{slice\ i}$ (Column 9 = Column 8 - Column 7).

7. Calculate the partial quasi-RC(logit) for DOLLAR_2: the change in the Pred_lgt RESPONSE divided by the change in DOLLAR_2 for each slice (Column 10 = Column 9/Column 6.)

The LRM partial quasi-RC(logit) for DOLLAR_2 is interpreted as follows:

1. For slice 2, which has minimum and maximum DOLLAR_2 values of 43 and 66, respectively, the partial quasi-RC(logit) is 0.0012. This means that for each unit change in DOLLAR_2 between 43 and 66, the expected constant change in the logit RESPONSE is 0.0012.

2. Similarly, for slices 3, 4, and 5, the expected constant changes in the logit RESPONSE within the corresponding intervals are 0.0016, 0.0018, and 0.0015, respectively. Note that for slice 5, the maximum DOLLAR_2 value, in column 3, is 1,293.

3. At this point, the pending implication is that there are four levels of expected change in the logit RESPONSE associated by DOLLAR_2 across its range from 43 to 1293.

4. However, the partial quasi-RC plot for DOLLAR_2 of the relationship of the smooth predicted logit RESPONSE (column 8) versus the smooth DOLLAR_2 (column 5), in Figure 31.2, indicates there is a single expected constant change across the DOLLAR_2 range, as the variation among slice-level changes is reasonably due to sample variation. This last examination supports the decided implication that the linearity assumption of the partial linear-RC for DOLLAR_2 is valid. Thus, I accept the expected constant change of the partial linear-RC for DOLLAR_2, 0.00210 (from Equation 31.5).

5. Alternatively, the quasi-RC method provides a trusty assumption-free estimate of the partial linear-RC for DOLLAR_2, *partial quasi-RC(linear)*, which is defined as the regression coefficient of the simple ordinary regression of the smooth logit predicted RESPONSE on the smooth DOLLAR_2, columns 8 and 5, respectively. The partial quasi-RC(linear) for DOLLAR_2 is 0.00159 (details not shown).

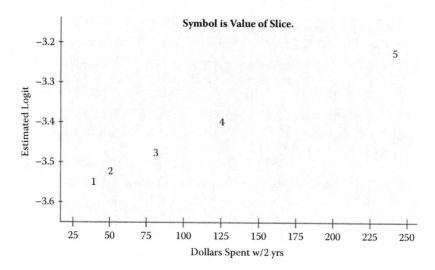

FIGURE 31.2
Visual display of LRM partial quasi-RC(logit) for DOLLAR_2.

In sum, the quasi-RC methodology provides alternatives that only the data analyst, who is intimate with the data, can decide on. They are (1) accept the partial quasi-RC after asserting the variation among slice-level changes in the partial quasi-RC plot as nonrandom; (2) accept the partial linear-RC (0.00210) after the partial quasi-RC plot validates the linearity assumption; and (3) accept the trusty partial quasi-RC(linear) estimate (0.00159) after the partial quasi-RC plot validates the linearity assumption. Of course, the default alternative is to accept outright the partial linear-RC without testing the linearity assumption. Note that the small difference in magnitude between the trusty and the "true" estimates of the partial linear-RC for DOLLAR_2 is not typical, as the next discussion shows.

I calculate the LRM partial quasi-RC(logit) for LSTORD_M, using six slices to correspond to the distinct values of LSTORD_M, in Table 31.7.

1. The partial quasi-RC plot for LSTORD_M of the relationship between the smooth predicted logit RESPONSE and the smooth LSTORD_M, in Figure 31.3, is clearly nonlinear with expected changes in the logit RESPONSE: -0.0032, -0.1618, -0.1067, -0.0678, and 0.0175.

2. The implication is that the linearity assumption for the LSTORD_M does not hold. There is not a constant expected change in the logit RESPONSE as implied by the prescribed interpretation of the partial linear-RC for LSTORD_M, -0.0798 (in Equation 31.5).

TABLE 31.7

Calculations for LRM Partial Quasi-RC(logit): LSTORD_M

Slice	min_LSTORD_M	m2ax_LSTORD_M	med_LSTORD_M_r	med_LSTORD_M_r+1	change_LSTORD_M	med_lgt_r	med_lgt_r+1	change_lgt	quasi-RC (logit)
1	1	1	.	1	.	.	-3.2332	.	.
2	1	3	1	2	1	-3.2332	-3.2364	-0.0032	-0.0032
3	3	3	2	3	1	-3.2364	-3.3982	-0.1618	-0.1618
4	3	4	3	4	1	-3.3982	-3.5049	-0.1067	-0.1067
5	4	5	4	5	1	-3.5049	-3.5727	-0.0678	-0.0678
6	5	12	5	6	1	-3.5727	-3.5552	0.0175	0.0175

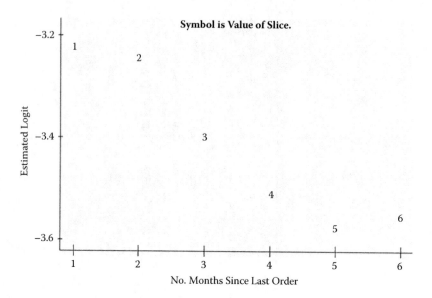

FIGURE 31.3
Visual display of LRM partial quasi-RC(logit) for LSTORD_M.

3. The secondary implication is that the structural form of LSTORD_M is not correct. The S-shaped nonlinear pattern suggests that quadratic or cubic reexpressions of LSTORD_M be tested for model inclusion.

4. Satisfyingly, the partial quasi-RC(linear) value of -0.0799 (from the simple ordinary regression of the smooth predicted logit RESPONSE on the smooth LSTORD_M) equals the partial linear-RC value of -0.0798. The implications are as follows: (1) The partial linear-RC provides the *average* constant change in the logit RESPONSE across the LSTORD_M range of values from 1 to 66; (2) the partial quasi-RC provides more accurate reading of the changes with respect to the six presliced intervals across the LSTORD_M range.

Forgoing the details, the LRM partial quasi-RC plots for both RFM_CELL and AGE_Y support the linearity assumption of the partial linear-RC. Thus, the partial quasi-RC(linear) and the partial linear-RC values should be equivalent. In fact, they are: For RFM_CELL, the partial quasi-RC(linear) and the partial-linear RC are -0.2007 and -0.1995, respectively; for AGE_Y, the partial quasi-RC(linear) and the partial-linear RC are 0.5409 and 0.5337, respectively.

In sum, this illustration shows that the workings of the quasi-RC methodology perform quite well on the linear predictions based on multiple predictor variables. Suffice it to say, by converting the logits into probabilities—as was done in the simple logistic regression illustration in Section 31.3.3—the quasi-RC approach performs equally well with nonlinear predictions based on multiple predictor variables.

31.5   Quasi-RC for a Coefficient-Free Model

The linear regression paradigm, with nearly two centuries of theoretical development and practical use, has made the equation form—the sum of weighted predictor variables ($Y = b_0 + b_1X_1 + b_2X_2 + \ldots + b_nX_n$)—the icon of predictive models. This is why the new machine-learning techniques of the last half-century are evaluated by the coefficients they produce. If the new coefficients impart comparable information to the regression coefficient, then the new technique passes the first line of acceptance. If not, the technique is all but summarily rejected. Ironically, some machine-learning methods offer better predictions without the use of coefficients. The burden of acceptance of the coefficient-free model lies with the extraction of something familiar and trusting. The quasi-RC procedure provides data analysts and marketers the comfort and security of coefficient-like information for evaluating and using the coefficient-free machine-learning models.

Machine-learning models without coefficients can assuredly enjoy the quasi-RC method. One of the most popular coefficient-free models is the regression tree (e.g., CHAID [chi-squared automatic interaction detection]. The regression tree has a unique equation form of "if … then" rules, which has rendered its interpretation virtually self-explanatory and has freed itself from a burden of acceptance. The need for coefficient-like information was never sought. In contrast, most machine-learning methods, like artificial neural networks, have not enjoyed an easy first line of acceptance. Even their proponents have called artificial neural networks black boxes. Ironically, artificial neural networks do have coefficients (actually, interconnection weights between input and output layers), but no formal effort has been made to translate them into coefficient-like information. The genetic GenIQ Model has no outright coefficients. Numerical values are sometimes part of the genetic model, but they are not coefficient-like in any way, just genetic material evolved as necessary for accurate prediction.

The quasi-RC method as discussed so far works nicely on the linear and nonlinear regression model. In the next section, I illustrate how the quasi-RC technique works and how to interpret its results for a quintessential everymodel, the nonregression-based, nonlinear, and coefficient-free GenIQ Model as presented in Chapter 29. As expected, the quasi-RC technique works with artificial neural network models and CHAID or CART regression tree models.

31.5.1 Illustration of Quasi-RC for a Coefficient-Free Model

Again, consider the illustration in Chapter 30 of cataloger ABC, who requires a response model to be built on a recent mail campaign. I select the best number 3 GenIQ Model (in Figure 30.3) for predicting RESPONSE based on four predictor variables:

1. DOLLAR_2: dollars spent within last 2 years
2. PROD_TYP: number of different products
3. RFM_CELL: recency/frequency/money cells (1 = best to 5 = worst)
4. AGE_Y: knowledge of customer's age (1 = if known; 0 = if not known)

The GenIQ partial quasi-RC(prob) table and plot for DOLLAR_2 are in Table 31.8 and Figure 31.4, respectively. The plot of the relationship between the smooth predicted probability RESPONSE (GenIQ-converted probability score) and the smooth DOLLAR_2 is clearly nonlinear, which is considered reasonable, due to the inherently nonlinear nature of the GenIQ Model. The implication is that partial quasi-RC(prob) for DOLLAR_2 reliably reflects the expected changes in probability RESPONSE. The interpretation of the partial quasi-RC(prob) for DOLLAR_2 is as follows: For slice 2, which has minimum and maximum DOLLAR_2 values of 50 and 59, respectively, the partial quasi-RC(prob) is 0.000000310. This means that for each unit change in DOLLAR_2 between 50 and 59, the expected constant change in the probability RESPONSE is 0.000000310. Similarly, for slices 3, 4, ¼, 10, the expected constant changes in the probability RESPONSE are 0.000001450, 0.000001034, ¼, 0.000006760, respectively.*

The GenIQ partial quasi-RC(prob) table and plot for PROD_TYP are presented in Table 31.9 and Figure 31.5, respectively. Because PROD_TYP assumes distinct values between 3 and 47, albeit more than a handful, I use 20 slices to take advantage of the granularity of the quasi-RC plotting. The interpretation of the partial quasi-RC(prob) for PROD_TYP can follow the literal rendition of "for each and every unit change" in PROD_TYP as done for DOLLAR_2. However, as the quasi-RC technique provides alternatives, the following interpretations are also available:

1. The partial quasi-RC plot of the relationship between the smooth predicted probability RESPONSE and the smooth PROD_TYP suggests two patterns. For pattern 1, for PROD_TYP values between 6 and 15, the unit changes in probability RESPONSE can be viewed as sample variation masking an expected constant change in

---

* Note that the maximum values for DOLLAR_2 in Tables 31.6 and 31.8 are not equal. This is because they are based on different M-spread common regions as the GenIQ Model and LRM use different variables.

TABLE 31.8

Calculations for GenIQ Partial Quasi-RC(prob): DOLLAR_2

Slice	min_DOLLAR_2	max_DOLLAR_2	med_DOLLAR_2_r	med_DOLLAR_2_r+1	change_DOLLAR_2	med_prb_r	med_prb_r+1	change_prb	quasi-RC (prob)
1	0	50	.	40	.	.	0.031114713	.	.
2	50	59	40	50	10	0.031114713	0.031117817	0.000003103	0.000000310
3	59	73	50	67	17	0.031117817	0.031142469	0.000024652	0.000001450
4	73	83	67	79	12	0.031142469	0.031154883	0.000012414	0.000001034
5	83	94	79	89	10	0.031154883	0.031187925	0.000033043	0.000003304
6	94	110	89	102	13	0.031187925	0.031219393	0.000031468	0.000002421
7	110	131	102	119	17	0.031219393	0.031286803	0.000067410	0.000003965
8	131	159	119	144	25	0.031286803	0.031383536	0.000096733	0.000003869
9	159	209	144	182	38	0.031383536	0.031605964	0.000222428	0.000005853
10	209	480	182	253	71	0.031605964	0.032085916	0.000479952	0.000006760

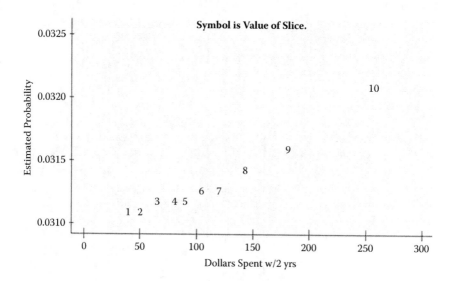

FIGURE 31.4
Visual display of GenIQ partial quasi-RC(prob) for DOLLAR_2.

probability RESPONSE. The "masked" expected constant change can be determined by the average of the unit changes in probability RESPONSE corresponding to PROD_TYP values between 6 and 15. For pattern 2, for PROD_TYP values greater than 15, the expected change in probability RESPONSE is increasing in a nonlinear manner, which follows the literal rendition for each and every unit change in PROD_TYP.

2. If the data analyst comes to judge the details in the partial quasi-RC(prob) table or plot for PROD_TYP as much ado about sample variation, then the partial quasi-RC(linear) estimate can be used. Its value of 0.00002495 is obtained from the regression coefficient from the simple ordinary regression of the smooth predicted RESPONSE on the smooth PROD_TYP (columns 8 and 5, respectively).

The GenIQ partial quasi-RC(prob) table and plot for RFM_CELL are presented in Table 31.10 and Figure 31.6, respectively. The partial quasi-RC of the relationship between the smooth predicted probability RESPONSE and the smooth RFM_CELL suggests an increasing expected change in probability. Recall that RFM_CELL is treated as an interval-level variable with a "reverse" scale: 1 = best to 5 = worst; thus, the RFM_CELL has clearly expected nonconstant change in probability. The plot has *double* smooth points at both RFM_CELL = 4 and RFM_CELL = 5, for which the *double-smoothed* predicted probability RESPONSE is taken as the average of the reported probabilities. For RFM_CELL = 4, the twin points are 0.031252

TABLE 31.9

Calculations for GenIQ Partial Quasi-RC(prob): PROD_TYP

Slice	min_PROD_TYP	max_PROD_TYP	med_PROD_TYP_r	med_PROD_TYP_r+1	change_PROD_TYP	med_prb_r	med_prb_r+1	change_prob	quasi_RC (prob)
1	3	6	·	6	·	·	0.031103	·	·
2	6	7	6	7	1	0.031103	0.031108	0.000004696	0.000004696
3	7	8	7	7	0	0.031108	0.031111	0.000003381	·
4	8	8	7	8	1	0.031111	0.031113	0.000001986	0.000001986
5	8	8	8	8	0	0.031113	0.031113	0.000000000	·
6	8	9	8	8	0	0.031113	0.031128	0.000014497	·
7	9	9	8	9	1	0.031128	0.031121	-0.000006585	-0.000006585
8	9	9	9	9	0	0.031121	0.031136	0.000014440	·
9	9	10	9	10	1	0.031136	0.031142	0.000006514	0.000006514
10	10	11	10	10	0	0.031142	0.031150	0.000007227	·
11	11	11	10	11	1	0.031150	0.031165	0.000015078	0.000015078
12	11	12	11	12	1	0.031165	0.031196	0.000031065	0.000031065
13	12	13	12	12	0	0.031196	0.031194	-0.000001614	·
14	13	14	12	13	1	0.031194	0.031221	0.000026683	0.000026683
15	14	15	13	14	1	0.031221	0.031226	0.000005420	0.000005420
16	15	16	14	15	1	0.031226	0.031246	0.000019601	0.000019601
17	16	19	15	17	2	0.031246	0.031305	0.000059454	0.000029727
18	19	22	17	20	3	0.031305	0.031341	0.000036032	0.000012011
19	22	26	20	24	4	0.031341	0.031486	0.000144726	0.000036181
20	26	47	24	30	6	0.031486	0.031749	0.000262804	0.000043801

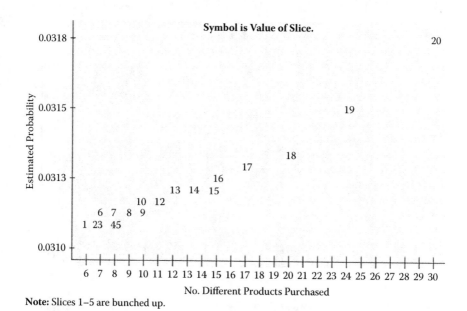

Note: Slices 1–5 are bunched up.

FIGURE 31.5
Visual display of GenIQ partial quasi-RC(prob) for PROD_TYP.

and 0.031137; thus, the double-smoothed predicted probability RESPONSE is 0.311945. Similarly, for RFM_CELL = 5, the double-smoothed predicted probability RESPONSE is 0.31204. The interpretation of the partial quasi-RC(prob) for RFM_CELL can follow the literal rendition for each and every unit change in RFM_CELL.

The GenIQ partial quasi-RC(prob) table and plot for AGE_Y are presented in Table 31.11 and Figure 31.7, respectively. The partial quasi-RC plot of the relationship between the smooth predicted probability RESPONSE and the smooth RFM_CELL is an uninteresting expected linear change in probability. The plot has double-smoothed points at both AGE_Y = 1, for which the double-smoothed predicted probability RESPONSE is taken as the average of the reported probabilities. For AGE_Y = 1, the twin points are 0.031234 and 0.031192; thus, the double-smoothed predicted probability RESPONSE is 0.31213. The interpretation of the partial quasi-RC(prob) for AGE_Y can follow the literal rendition for each and every unit change in AGE_Y.

In sum, this illustration shows how the quasi-RC methodology works on a nonregression-based, nonlinear, and coefficient-free model. The quasi-RC procedure provides data analysts and marketers with the sought-after comfort and security of coefficient-like information for evaluating and using coefficient-free machine-learning models like GenIQ.

TABLE 31.10

Calculations for GenIQ Partial Quasi-RC(prob): RFM_CELL

Slice	min_RFM_CELL	max_RFM_CELL	med_RFM_CELL_r	med_RFM_CELL_r+1	change_RFM_CELL	med_prb_r	med_prb_r+1	change_prb	quasi-RC (prob)
1	1	3	.	2	.	.	0.031773	.	.
2	3	4	2	3	1	0.031773	0.031252	-0.000521290	-0.000521290
3	4	4	3	4	1	0.031252	0.031137	-0.000114949	-0.000114949
4	4	4	4	4	0	0.031137	0.031270	0.000133176	.
5	4	5	4	5	1	0.031270	0.031138	-0.000131994	-0.000131994
6	5	5	5	5	0	0.031138	0.031278	0.000140346	.

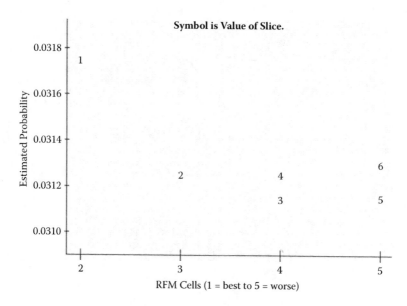

FIGURE 31.6
Visual display of GenIQ partial quasi-RC(prob) for RFM_CELL.

31.6 Summary

The redoubtable regression coefficient enjoys everyday use in marketing analysis and modeling. Model builders and marketers use the regression coefficient when interpreting the tried-and-true regression model. I restated that the reliability of the regression coefficient is based on the workings of the linear statistical adjustment. This removes the effects of the other variables from the dependent variable and the predictor variable, producing a linear relationship between the dependent variable and the predictor variable.

In the absence of another measure, model builders and marketers use the regression coefficient to evaluate new modeling methods. This leads to a quandary as some of the newer methods have no coefficients. As a counter-step, I presented the quasi-regression coefficient (quasi-RC), which provides information similar to the regression coefficient for evaluating and using coefficient-free models. Moreover, the quasi-RC serves as a trusty assumption-free alternative to the regression coefficient when the linearity assumption is not met.

I provided illustrations with the simple one-predictor variable linear regression models to highlight the importance of the satisfaction of linearity assumption for accurate reading of the regression coefficient itself, as well as its effect on the predictions of the model. With these illustrations, I outlined

TABLE 31.11

Calculations for GenIQ Partial Quasi-RC(prob): AGE_Y

Slice	min_ AGE_Y	max_ AGE_Y	med_ AGE_Y_r	med_ AGE_ Y_r+1	change_ AGE_Y	med_ prb_r	med_prb_ r+1	change_prb	quasi-RC (prob)
1	0	1	.	1	.	.	0.031177	.	.
2	1	1	1	1	0	0.031177	0.031192	0.000014687	.
3	1	1	1	1	0	0.031192	0.031234	0.000041677	.

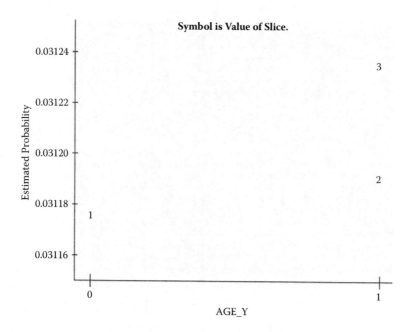

FIGURE 31.7
Visual Display of GenIQ Partial Quasi RC(prob) for AGE_Y

the method for calculating the quasi-RC. Comparison between the actual regression coefficient and the quasi-RC showed perfect agreement, which advances the trustworthiness of the new measure.

Then, I extended the quasi-RC for the everymodel, which is any linear or nonlinear regression or any coefficient-free model. Formally, the partial quasi-RC for predictor variable X is the expected change—not necessarily constant—in the dependent variable Y associated with a unit change in X when the other variables are held constant. With a multiple logistic regression illustration, I compared and contrasted the logistic partial linear-RC with the partial quasi-RC. The quasi-RC methodology provided alternatives that only the data analyst, who is intimate with the data, can decide on. They are (1) accept the partial quasi-RC if the partial quasi-RC plot produces a perceptible pattern; (2) accept the logistic partial linear-RC if the partial quasi-RC plot validates the linearity assumption; and (3) accept the trusty partial quasi-RC(linear) estimate if the partial quasi-RC plot validates the linearity assumption. Of course, the default alternative is to accept the logistic partial linear-RC outright without testing the linearity assumption.

Last, I illustrated the quasi-RC methodology for the coefficient-free GenIQ Model. The quasi-RC procedure provided me with the sought-after comfort and security of coefficient-like information for evaluating and using the coefficient-free GenIQ Model.

Index

A

Absolute deviations, mean of (MAD), 305
Absolute percentage errors, mean of (MAPE), 305
Accuracy, 227, *See also* Model assessment
    bootstrap method, 320
overfitting and, 415–417
    for profit model, 304–306
    for response model, 303–304
Adjusted correlation coefficient, 89, 95
Adjusted R-squared, 90, 165, 170, 181
Aggregated data, 108
All-subset selection, 179, 182
Alternative hypothesis, 26, 215
Analysis of variance (ANOVA), 189
Arithmetic functions, 185, 433, 434, 455
Arithmetic operators, 112, 393, 402, 459
Art, science, and poetry in regression modeling, 379
Artificial neural networks (ANNs), 431, 443, 444, 487
Automatic interaction detection (AID), 12, *See also* CHAID
Available-case analysis, 267–268
Average correlation, 225, 229–235
    assessing predictor variable importance, 225, 229, 232, 235
    desirable values, 232–233
    lifetime value model assessment, 229–235

B

Backward elimination (BE), 181, 452, 458, 464
Bar diagram, 340
Barricelli, Nils A., 392
Berkson, J., 381, 387
Best-of-generation model, 433

Beta regression coefficient, 221, *See also* Standardized regression coefficient
Bias, 7, 190, 370
    underfitted model and, 416
Biased model roulette wheel, 437
Big data, 8–10, 31, 178
    CHAID and, 34
    data mining definition, 13
    EDA and, 8
    filling in missing data and, 271
    fitting data to models, 178, 387, 388
    GenIQ Model and, 394
    good models and, 102
    p value and, 214, 216–217
    regression model suboptimality, 178, 384
    scatterplots and, 21–23
    smooth actual and predicted response points, 129
    spurious structures and, 11
    statistical regression models and, 178
    unnecessarily large sample sizes, 10
Binary logistic regression (BLR), 251–252
Binning, 48, 191
Black box, 383, 479
Blind data analysis, 180
Bonferroni method, 329
Boolean functions, 455, 459
Bootstrap, 317, 320–325
    Cum Lift estimates, 325–327
    decile analysis model validation, 325, 331–335
    model efficiency assessment, 331–334
    model implementation performance assessment, 327–331
    sample size and confidence interval determination, 326–331
Box, G. E. P., 7, 190
Box-and-whiskers plot (boxplot), 58–59, 74, 324, 340

497